Wallace, Darwin, and the Origin of Species

James T. Costa

Harvard University Press

Cambridge, Massachusetts
London, England
2014

Library of Congress Cataloging-in-Publication Data

Costa, James T., 1963- author.
    Wallace, Darwin, and the origin of species / James T. Costa.
      pages cm
    Includes bibliographical references and index.
    ISBN 978-0-674-72969-8 (alk. paper)
    1. Wallace, Alfred Russel, 1823–1913.   2. Darwin, Charles, 1809–1882. On
the origin of species.   3. Evolution (Biology)   4. Natural selection.   I. Title.
    QH26.C67 2014
    576.8'2—dc23

                                                              2013040287

This one is for my guys, Addison and Eli—
Indefatigable in all they do.

Contents

Alfred Russel Wallace: A Short Biography
by Andrew Berry
ix

Introduction
1

1
Granted the Law
Alfred Russel Wallace's Evolutionary Travels
15

2
The Consilient Mr. Wallace
Transmutation and Related Themes of Wallace's Species Notebook
65

3
Wallace and Darwin
Parallels, Intersections, and Departures on the Evolutionary Road
104

4

Two Indefatigable Naturalists

Wallace and Darwin's Watershed Papers

143

5

A Striking Coincidence

The Wallace and Darwin Papers of 1858 Compared

214

6

True with a Vengeance

From Delicate Arrangement to Conspiracy: A Guide

232

Coda

The Force of Admiration

265

Appendixes

283

Bibliography

301

Acknowledgments

317

Notes on the Text and Illustrations

319

Index

321

Alfred Russel Wallace: A Short Biography

by Andrew Berry

It is ironic that Alfred Russel Wallace's finest moment—his 1858 discovery of natural selection—has in many ways compromised his place in history, condemning him forever to footnotes in textbooks. It has resulted in his always being coupled with Darwin, but as a junior partner: he is destined always to play Watson to Darwin's Holmes. One of the many interesting features of the Wallace story, then, is how and why, relative at least to Darwin's, his star has dimmed so precipitously. After all, he was one of the true superstars of Victorian science; he was the codiscoverer of natural selection; he was the discoverer of what would become known as Wallace's Line, the biological discontinuity between Australasia and Asia; he was the father of a whole new science, biogeography; and he was arguably the leading tropical biologist of his day. Although Wallace missed out by a century or so on our era of papers with author lists that read like phone books, his bibliography runs to more than 700 publications, including some twenty books, several of them hefty two-volume tomes.

Wallace was born into genteel poverty in Usk, in what is now Wales, in 1823. Wallace's father, a qualified lawyer who never practiced and had a facility for losing inherited money through inept investing, had no particular Welsh connections. He had moved his large family—Alfred was eighth of nine—to Usk simply because, in his son's words, it was "where the living was as cheap as possible." The family returned to England, where Wallace received the rudiments of an education in Hertford before, at age fourteen, becoming apprenticed to his brother William in his land-surveying business. It was over the years that followed, as he trekked through the British countryside with his surveying pole, that Wallace became interested in the plants he encountered. His interest was amateurish and completely untutored, as he recalled: "I knew nothing of whatever as to genera and species" (Wallace 1905, 194). However, a lull in the surveying business gave him an opportunity to take his scientific education to a new level. In 1844 he taught for a year at a school in Leicester, where, in the public library, he met another young self-taught naturalist, Henry Walter Bates (of Batesian mimicry fame). Bates's beetle enthusiasm was much further developed than Wallace's botany: though only nineteen, Bates had actually published a scientific paper (charmingly entitled "Note on Coleopterous Insects Frequenting Damp Places"). Wallace was impressed by both Bates and his beetles. He too became a passionate coleopterist, and their correspondence becomes filled with breathless exclamations over species lists. In 1847 Wallace published his first scientific note—a trivial natural history observation, but a huge step forward for Wallace—and Wallace and Bates were beginning to think big. Wallace in particular had been impressed by a book on transmutation (as evolution was then called), *The Vestiges of the Natural History of Creation,* published anonymously in 1844. Poorly articulated and scientifically incoherent, the *Vestiges* nevertheless brought evolution, a scurrilous, godless idea, into the public eye. Writing to Bates, Wallace called the *Vestiges'* evolutionary ideas an "ingenious hypothesis." Inspired, Wallace and Bates decided to become professional biologists. In 1848 they traveled to Brazil to collect material for their own assault on the questions surrounding the generation of biological diversity. They would sell duplicate specimens to underwrite the whole enterprise.

It is remarkable how quickly the two neophytes adapted to the business of collecting and cataloging biodiversity in one of its great, hyperdiverse citadels. Curiously the commercial aspects of what he was doing may have been key in the development of Wallace's scientific interests. Geography, Wallace discovered, mattered. He complained about previous naturalists who had failed to record on which side of the river they had made their observations because he was acutely attuned to small but significant regional differences. After all, he could get a better price for a new form than for yet another specimen of an old one. It was this sensitivity to geographic differentiation that both underpinned the development of his evolutionary ideas and eventually matured into his work on biogeography. Wallace and Bates split up, dividing the territory between them. Wallace headed up the Rio Negro, where, often entirely dependent upon the hospitality and assistance of local people, he struggled to collect, preserve, and record. Disease nearly killed him on a couple of occasions; he faced extraordinary logistical challenges as a collector in unexplored regions; his younger brother Herbert came out to assist him but died of yellow fever in Bélem on his way home. In 1852 it was time to head back to England. Wallace had assembled a remarkable collection, including, for example, some 10,000 bird skins. Most exciting, though, were some thirty live animals he brought downriver with him: these, surely, would be his passport to the scientific big time in London. Just imagine the impact of a live toucan at a London scientific salon! The first hiccough occurred when he reached Manaus to find that the specimens he had been sending downriver to his agent in London had been held up in customs. No matter; Wallace would accompany his biological treasure trove all the way home. But Wallace's visions of success in scientific London were not to be realized.

Several weeks into the voyage, poor stowage of flammable materials in the hold caused the ship to catch fire in the middle of the Atlantic. There then followed what is surely one of the most poignant episodes in all of the history of science. Wallace barely had to time to grab a small case of notes and drawings; he received bad rope burns on his hands in his haste to clamber into the lifeboat as the ship, her timbers dried out by the tropical sun, went up like a pyre. The two lifeboats circled around the burning wreck in the forlorn hope that the smoke would attract rescue. Wallace therefore watched as his dreams of scientific success in London went up in flames and, worst of all, as his living animals—pets, really, as he had cared for them all the way down the river—fled the flames to the bowsprit, the last part of the ship not on fire, only ultimately to be engulfed by the inexorable flames. To make matters even worse, it was ten days before Wallace and the crew were rescued. Wallace nevertheless managed to put a positive retrospective spin on the experience: "During the night I saw several meteors, and in fact could not be in a better position for observing them, than lying on my back in a small boat in the middle of the Atlantic."

Back in England Wallace scrambled to salvage something from the disaster. He published a short book on the *Palm Trees of the Amazon* and, despite the loss of his journals, a travelogue. Neither was successful. If Wallace were to realize his ambition to become a player in scientific London, he would have to do it all over again. This time he headed in the other direction, to Southeast Asia. The eight years (1854–1862) of Wallace's travels in what are today Malaysia, Indonesia, and New Guinea rank among the greatest scientific journeys, and the book that resulted, *The Malay Archipelago*, is also justly celebrated as a classic of travel writing. It was also during these years that Wallace came of age as a biologist: in just a few years, he produced three major insights. Hitherto he had published natural history and taxonomic notes, but in 1855 he suddenly emerged in print as a fully fledged evolutionary theorist. The paper, known as the Sarawak Law paper because Wallace wrote it while in Sarawak, North Borneo, unveiled what is essentially half the theory of evolution. Wallace observed that "allied forms" occur in contiguous geographic regions—kangaroos all occur in and around Australia, not elsewhere—and that "allied forms" in the fossil record similarly occur in contiguous strata. This neat congruence between the space and time determinants of biological diversity clearly argues for a genealogical process: there are kangaroos in Australia and not elsewhere because they are all ultimately derived from an Australian common ancestor, the first kangaroo. What, of course, was missing from this vision of an evolutionary process was a mechanism to entrain genealogical differentiation along adaptive paths. Wallace knew that his paper was something of a scientific bombshell—this was part of his long-term strategy to become a player—and he anxiously and excitedly awaited the reaction of the scientific community. There was none. His bombshell had apparently been roundly ignored. His agent in London told him to stop theorizing and get back to collecting—there was no money in ideas.

Next came a big-picture empirical moment. Stymied by the lack of available transport to Sulawesi, Wallace found himself island-hopping between Bali and Lombok, where he noticed that Lombok's birds were of Australasian stock whereas Bali's were Asian. Wallace had identified the boundary between two of the major biogeographic regions on the planet, a boundary that has subsequently come to be called "Wallace's Line." Wallace was also still struggling with the evolutionary questions that had stimulated his earlier journey with Bates. In February 1858 while collecting on the island of Halmahera in the Moluccas, Wallace was stricken with a high fever, probably malaria. Through the flickering delirium he continued to grapple with the problem. How could species become so perfectly adapted to their environments? What ensured that the bill of a hummingbird is long and narrow to enable it to probe the depths of a tubular flower? Wallace was also struck by the anthropological gradient he was traveling along: the shift from the Asian peoples of most of Southeast Asia to the distinctive Melanesians of New Guinea. He recalled reading Thomas Malthus, the political economist, whose vision was of permanent competitive struggle as human populations outstripped available resources. Suddenly it all came together: Wallace glimpsed natural selection. As soon as the fever had passed, he wrote out a brief summary of his ideas, giving his address as the neighboring island of Ternate, the local center of civilization. That brief summary is his most famous paper, "On the Tendency of Varieties to Depart Indefinitely from the Original Type," often referred to as the Ternate essay. Wallace's next problem was where to place the paper. He had been disappointed by the lack of response to his Sarawak Law paper; how best to ensure that this one was properly appreciated? He was in correspondence with Charles Darwin, who had read the Sarawak Law paper but failed to recognize in Wallace a competitor. Knowing Darwin to be interested in the species problem, Wallace sent the manuscript to Darwin in the hope that he would deem it worth passing on to the geologist Charles Lyell. Lyell, as one Britain's most distinguished scientists, was ideally placed to make sure that the paper was published in an appropriately prominent place. He would also be sure to be interested in the topic, having laid out careful arguments against transmutationism in his Principles of Geology.

It is not clear exactly when Darwin received Wallace's letter, but he first acknowledged it on 18 June 1858. Darwin, who had been quietly developing and consolidating his evolutionary ideas over the previous twenty years, was mortified. Science, even for Victorian gentlemen, is about being first, and Darwin saw his precedence usurped by Wallace. In fact the most remarkable—and serendipitous—aspect of this episode, one of the most celebrated in the history of science, is that Darwin had a chance to respond. Wallace had sent his every other manuscript directly to a journal editor; had he done this with the Ternate essay, Darwin would have discovered, several months later when the paper was published, that he had been scooped. What happened next is both well known and controversial. Darwin's friends, Lyell and the botanist Joseph Hooker, contrived an arrangement that they hoped would preserve Darwin's claim to precedence and yet not do Wallace an injustice. They presented a joint paper, Wallace's and some unpublished material from Darwin, at the Linnean Society meeting of 1 July 1858. Evolution by natural selection was published. Darwin knuckled down to produce On the Origin of Species, which appeared in November 1859. Wallace, thousands of miles away, was not consulted and did not know what had transpired until several months after the events. Did he feel cheated by finding that his paper had been tacked onto Darwin's? Not in the least. He was still desperate to make a name for himself in scientific circles, and here, in the form of copublication with Darwin, who was older and well established, was his passport to the scientific big time. In October 1858 he wrote to his mother: "I have received letters from Mr. Darwin and Dr. Hooker, two of the most eminent naturalists in England, which has highly gratified me. I sent Mr. Darwin an essay on a subject on which he is now writing a great work. He showed it to Dr. Hooker and Sir C. Lyell, who thought so highly of it that they immediately read it before the Linnean Society. This assures me the acquaintance and assistance of these eminent men on my return home." Wallace had made it.

Wallace returned from Southeast Asia in 1862. This time his collections, including several thousand species new to science and two living birds of paradise, made it to London without incident. Courtesy astute marketing and investment by his agent, Wallace was now well off, and, better, he was a full-fledged member of the scientific élite. He plunged into his new life with vigor, publishing a series of remarkable papers rich in insight. In one, for example, on a group of butterflies in Southeast Asia, he gave a definition of species that is strikingly similar to today's "Biological Species Concept," an idea typically attributed to Ernst Mayr in the 1940s: "Species are merely those strongly marked races or local forms which when in contact do not intermix, and when inhabiting distinct areas, are generally believed to have had a separate origin, and to be incapable of producing a fertile hybrid offspring" (1864). He started his great work on biogeography that would culminate in his two-volume masterwork, *The Geographical Distribution of Animals* (1876).

Wallace once described himself as "more Darwinian than Darwin" because of his rigid insistence on the primacy of natural selection in evolution, but there were nevertheless several disagreements between Wallace and Darwin. The most prominent of these was over the evolution of our own species. Darwin had been famously cagey on the subject in *On the Origin of Species,* but his readership was more than capable of reading between the lines: humans, Mr. Darwin was implicitly asserting, were not after all made in the image of God, but were merely modified great apes. Wallace was less inhibited and published a provocative paper in 1864 on human evolution. By the end of the decade, however, he had come out in print denying that natural selection was sufficient for the evolution of the human brain. It remains controversial whether or not he had changed his mind on the subject or merely refined ideas that he had held for some time. In the 1864 paper he advanced the idea that bodily evolution has been suspended in humans in light of the flexible power of what Wallace calls "mind." Essentially he is highlighting the preeminent role of cultural—as opposed to biological—evolution in the history of our species: if we desire

to move into an inhospitable polar environment, we use appropriate cloth-
ing rather than waiting for the required mutations that would endow us
with insulating body hair:

> At length, however, there came into existence a being in whom that subtle
> force we term mind, became of greater importance than his mere bodily
> structure. Though with a naked and unprotected body, this gave him cloth-
> ing against the varying inclemencies of the seasons. Though unable to com-
> pete with the deer in swiftness, or with the wild bull in strength, this gave
> him weapons with which to capture or overcome both. Though less capable
> than most other animals of living on the herbs and the fruits that unaided
> nature supplies, this wonderful faculty taught him to govern and direct na-
> ture to his own benefit, and make her produce food for him when and
> where he pleased. From the moment when the first skin was used as a cov-
> ering, when the first rude spear was formed to assist in the chase, the first
> seed sown or shoot planted, a grand revolution was effected in nature, a
> revolution which in all the previous ages of the earth's history had had no
> parallel, for a being had arisen who was no longer necessarily subject to
> change with the changing universe—a being who was in some degree supe-
> rior to nature, inasmuch, as he knew how to control and regulate her action,
> and could keep himself in harmony with her, not by a change in body, but
> by an advance of mind. (Wallace 1864, 167–168)

His view that natural selection could not account for our own species (first
enunciated in 1869 in a book review of the latest edition of Lyell's *Principles of
Geology*) related to "mind." If "mind" has essentially taken over the direction
of human evolution from natural selection, what accounts for the evolution of
"mind" itself? Two strands of thinking underpinned Wallace's conclusion.

Beginning in 1865 he attended séances and was soon a convinced spiritualist. In hindsight—now we know that most spiritualist mediums were
typically merely petty fraudsters—this seems a bizarre step for a scientist
to take. In fact, Wallace was certainly not alone within the scientific community at the time in this enthusiasm. For scientists who had rejected standard religion but who were living in a highly religious society, perhaps
spiritualism represented an intellectually respectable—Wallace for one
insisted that the claims of spiritualism were empirically verifiable—way to
fill the void left in their lives by the absence of church attendance. As a
spiritualist Wallace had to postulate nonmaterial processes in human evolution: there exists no material mechanism for the generation of souls.

Wallace's other reason for doubting the sufficiency of natural selection for human evolution is more scientific. He was impressed by the mental capabilities of many of the people he had lived and worked with during his twelve years of travel. These people were deemed "savages" by the Victorians and were accordingly considered inferior. Wallace, because he had been dependent on "savages" and established close relationships with them in both Brazil and Southeast Asia, had an enlightened, nonracist perspective. But this created an evolutionary problem for him. He appreciated that the man with whom he had shared a hut on an island off New Guinea had the potential to match the most urbane and educated Victorian gentleman in his intellectual and aesthetic pursuits, but he recognized too that that barely clad islander would never have the opportunity to take advantage of this potential. He would spend his entire life in his native forests, using his brain for problems no more exalted than the day-to-day issue of how and where to find food. Wallace, with a good understanding of natural selection, knew that natural selection endows organisms with only what they *need.* How then could humans in "uncivilized" places have been endowed with such extraordinary abilities that they would never have the opportunity to use? Today we recognize that human mental abilities are largely a by-product of natural selection in favor of a large brain with remarkable learning capabilities. Wallace chose instead to invoke some kind of teleological process—an implicitly supernatural one—to account for this apparent mismatch between the abilities of "savages" and the opportunities they have to implement them.

Darwin, needless to say, was not happy about Wallace's defection on this, the key issue, writing to Wallace, "I hope you have not murdered too completely your own and my child."

Wallace used his scientific prominence as a springboard to become engaged in the social issues of the day. Always sympathetic to the underdog, he was an early socialist. For many years he was the president of the Land Nationalization Society, which traced all economic and social iniquity to the private ownership of land. "To allow one child to be born a millionaire and another a pauper is a crime against humanity, and, for those who believe in a deity, a crime against God." He wrote passionately on what today we call conservation. He championed votes for women: "As long as I have thought or written at all on politics, I have been in favour of woman suffrage. None of the arguments for or against have any weight with me, except the broad one, which may be thus stated:—'All the human inhabitants of any one country should have equal rights and liberties before the law; women are human beings; therefore they should have votes as well as men.'" In many ways Wallace was the prototypical socially engaged scientist. Not for him the monasticism of the ivory tower.

Many of these post-*Origin* developments have contributed to the dimming of Wallace's star. Contrast his publication strategy with Darwin's. Wallace published plenty of science for sure, but he also published extensively on politics, spiritualism, and much else besides; Darwin, in contrast, kept steadily building on the foundation he had laid in *On the Origin of Species*. Posterity, it seems, prefers a scientist who sticks to science. Critical too is Wallace's failure to be consistent in his application of natural selection. His disavowal of natural selection in human evolution makes him seem, in retrospect, halfhearted—not a real evolutionist. Other factors have contributed as well. Wallace was almost pathologically modest and always deferred to Darwin as the senior member of the pairing, even titling his major work on evolution *Darwinism* (1889). For most people autobiography is an opportunity to airbrush; for Wallace it was an exercise in searing self-criticism. What other autobiography includes a section on "certain marked deficiencies in my mental equipment"?

Despite all his successes and his extraordinary abilities, Wallace never did quite make it. His personal story after his return from Southeast Asia is telling. He came back well off but hemorrhaged money as he made ever-more-hopeless investments. Socially awkward, he was never able to land a regular job and made ends meet by doing what is surely the lowliest of all academic tasks, grading exams. His financial situation became so precarious that Darwin kindly intervened, successfully petitioning for a state pension for Wallace. Even Darwin had some difficulty recruiting his friends and colleagues to Wallace's cause. Joseph Hooker, irked at Wallace's insistence on treating spiritualism as a legitimate field of scientific inquiry, was unwilling to support the petition, writing to Darwin, "Wallace has lost caste considerably." That says it all: Wallace, despite his best efforts, was destined always to be an outsider. History famously is written by insiders, so perhaps we should not be surprised at posterity's neglect of Wallace.

Wallace, Darwin, and the Origin of Species

.

Introduction

THIS BOOK IS OFFERED as one biologist's apologia for Alfred Russel Wallace (1823–1913), autodidact explorer-naturalist whose epic journeys, geographical and intellectual, contributed profoundly to revolutionary new understandings of earth history and of the life upon earth in the mid- to late nineteenth century. It is occasioned by the Wallace Centennial and an opportunity to pore over Wallace's never-before-published "Species Notebook," the most important field notebook from his Southeast Asian explorations. I had the privilege to transcribe, annotate, and analyze this notebook in a work published by Harvard University Press under the title *On the Organic Law of Change: A Facsimile Edition and Annotated Transcription of Alfred Russel Wallace's Species Notebook of 1855–1859*. In the process I gained a new appreciation for Wallace's accomplishments (Costa 2013a). I had long admired Wallace, of course, knowing him as codiscoverer of the principle of natural selection, founder of evolutionary biogeography, and author of the classic travel memoir *The Malay Archipelago*. I also knew of Wallace's later turn toward spiritualism and his reservations over the applicability of natural selection to the evolution of human cognition. Both of these coexisted uncomfortably in my mind (though clearly they coexisted comfortably in his) with his stature as a fierce defender of natural selection and the reality of evolution generally, his watershed works in zoological geography, and his contributions to the understanding of animal coloration and years-long, friendly, and insightful debate with Darwin over sexual selection.

1

What I did not realize was how divided the scientific community seemed to be over Wallace's legacy. To some, his convictions about spiritualism and human evolution damned him—an apostate Darwinian expelled from the community of serious evolutionary scientists. Some in this camp have disparaged even his central scientific accomplishment, the discovery of natural selection, as a chance or incomplete discovery rather than the fruits of determined labors, and so of unequal merit with Darwin's accomplishments. To those of another camp, all of the real evolutionary insights are actually Wallace's, ignominiously stolen by Darwin and the machinations of his privileged circle. In the view of these partisans all the glory for the momentous discovery of evolution by natural selection should by rights go to Wallace, with Darwin cast at best as an evolutionary convert with unfocused ideas on the subject until Wallace provided direction and at worst as a charlatan guilty of intellectual theft. I could not help but think views at each end of this spectrum did a disservice to both Wallace and Darwin.

Undertaking to read all I could by and about Wallace, I sought to better understand for myself this giant of seeming contradictions. My first guide was *Infinite Tropics*, the Wallace anthology edited by my friend Andrew Berry (2002). Delighted to discover a fellow enthusiast for the history of evolutionary thinking in Andrew, I benefited from many an enlightening conversation about Wallace and Darwin over the years. It seemed eerily providential, despite my deep skepticism of such things, that I became aware of the notebook's existence at the Linnean Society in a paper I was reading while Andrew and I happened to be en route to that very institution. I soon proposed to undertake analysis and publication of the Species Notebook. Immersing myself in it, I came to see that this notebook opens a window into Wallace's pre–*On the Origin of Species* evolutionary thinking like no other document. One thing led to another, ultimately coming to fruition in *On the Organic Law of Change*—my effort to present and explain the Species Notebook in a way that I can only hope Wallace would have approved of.

In the process I came to appreciate the remarkable parallels in the thinking of Wallace and Darwin as they clarified their pro-transmutation ideas and formulated plans for pitching the argument. I came, too, to appreciate that history has not been as kind to Wallace as it should have been—one whose star shone brightly by the time of his death in 1913 and yet dimmed considerably in the ensuing century. Yet restoring Wallace to his proper place in the sun need not and should not come at the expense of Darwin. As I will show in this book, both of these remarkable individuals were our first guides to the evolutionary process; both went up that mountain and were electrified by what they saw. Both set out to collect evidence to bolster their vision and planned and strategized over how best to present a convincing pro-transmutation argument. Thus when I say this book is an apologia for Wallace, it is not an apologia built on the alleged wrongs or failings of Darwin, whose insights and accomplishments are clear. Rather, I aim to show Wallace in a new light, unfamiliar to most readers: a Wallace whose own pre-*Origin* insights and accomplishments are equally clear and whose laurels were perhaps even harder won considering how this explorer, self-taught and without the benefits of personal wealth, social standing, or connections, was audacious enough to conceive a plan to launch himself from the British Isles to the tropics of the West and East in pursuit of one of the burning questions of the day: the origin of species. Such big questions were the purview of the gentleman-naturalists, pursued in the halls and salons of science, not working specimen-collectors scribbling in jungle huts. Think, too, of Wallace laboring through years of trial and tribulation, triumph and tragedy, and against all odds succeeding in this very quest just thirteen years after setting out to do so.

With the notebook itself now transcribed and annotated, this book offers a detailed analysis. Wallace's Species Notebook makes for fascinating reading, not least for its many "transmutational" speculations, questions, comments, and arguments. Wallace reveals his plans for a pro-transmutation book, unrealized owing to the events of 1858–1859, but in his sketch we can get a sense of how he would have approached his own version of *On the Origin of Species*. Here I focus on Wallace's transmutational thinking and highlight the similarities and differences with that of Darwin, an analysis that underscores Wallace's standing as a deeply creative thinker. Indeed, in those pre-*Origin* years Wallace is the only thinker besides Darwin who conceived of a branching model of gradual transmutation in concert with gradual Lyellian change in the earth and climate, and who saw the evidence for this inherent in many lines of empirical observation: morphology, embryology, habit, fossils, geographical distribution, variation, domestic varieties, and more.

In the chapters that follow, I first step back and trace the development of Wallace's interest in the species question and his pursuit of this in Amazonia and the Malay Archipelago. The Species Notebook itself is then explored,

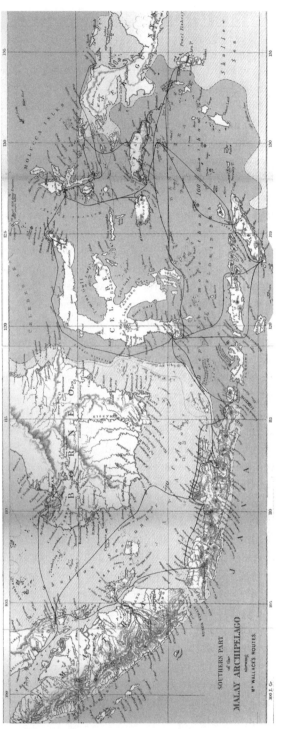

A map showing Wallace's travels in Southeast Asia, 1854 to 1862, from *The Malay Archipelago* (Wallace 1869). During his eight years in the archipelago Wallace traveled some 14,000 miles in sixty to seventy separate journeys, often traveling by prau. Courtesy of the Staatsbibliothek zu Berlin.

first in terms of its many transmutational themes and then, in greater detail, in reference to Darwin's own treatments of the same subjects. Here informative intersections, parallels, and departures between the two are brought to light, as well as Wallace's plan to engage directly with the anti-transmutation arguments of Charles Lyell in the landmark *Principles of Geology* and his plan for a pro-transmutation book. The ideas of these naturalists are then considered through an analysis of the key papers—Wallace's "Sarawak Law" paper and his and Darwin's 1858 papers presented to the Linnean Society of London on 1 July 1858—with an eye to the similarities and differences in their approach to and arguments for transmutation and natural selection.

This inevitably raises the question of the "delicate arrangement" of the Linnean Society readings and the "conspiracy theory." I offer a guide to charges of dishonesty leveled against Darwin, from the timing of the receipt of Wallace's Ternate essay (could it have been received weeks earlier than Darwin claimed?) to the accusation that Darwin appropriated key elements of his theory from Wallace (the verdict is . . .). I conclude that discussion with musings on what might have been had Wallace sent his Ternate essay directly to a journal rather than to Darwin. What emerges from the whole is a renewed appreciation for the sweeping scope of thought of both Wallace *and* Darwin as each labored, in ignorance of the extent of one another's ideas, on a common transmutational vision that each knew had revolutionary implications. I begin here, though, with the story of a notebook, Wallace's Species Notebook, which prompted this book. It is a notebook that has itself traveled far and that records ideas farther-ranging still.

. . .

There sits demurely on a shelf in central London a modest and worn notebook clad in marbled boards. Outwardly the notebook is rather ordinary; its contents are anything but. In fact this notebook's ordinariness belies the fact that it is an evolutionary manifesto: intended to revolutionize our understanding of life on earth, it was a scientific bombshell that was never deployed. The notebook records the earliest insights and speculations into transmutation, as evolution was called then, by the self-taught naturalist Alfred Russel Wallace, who would just about scoop Darwin and become one of the most famous scientists of his day. Wallace's unique blend of pluck, perseverance, creativity, and genius makes him the quintessential nineteenth-century naturalist-explorer and this notebook the quintessential field notebook: part journal, part daybook, part record and memorandum book, part travelogue, it is chock full of ideas, critiques, speculations, schemes, narrative, and natural history observations. Designated manuscript no. 180 of the library of the Linnean Society of London, the notebook's quiet existence in Burlington House is a far cry from the huts, houses, steamships, and praus of the archipelago where it journeyed with Wallace, ranging from Singapore and Sarawak in the west to New Guinea and the Aru Islands in the east from 1854 through his triumphant return home in 1862. In the interval, the notebook hopscotched some 14,000 miles across the vast Malay Archipelago with the tides and currents and monsoon winds. One can almost fancy on this notebook the lingering scent of . . . what? Perhaps orangutan, durian, arrack, the spice islands, sago cakes, gunpowder, camphor, the spray of the Coral Sea? It seems appropriate that so well-traveled a notebook should record thoughts just as far-ranging—Wallace's insights into the "mystery of mysteries," as it was memorably described: the origin of species.

In the years leading up to Darwin's *On the Origin of Species,* Wallace was sketching in splendid isolation ideas for his own book on the subject. The *Origin,* in fact, is why Wallace's planned manifesto, which the late historian of science Lewis McKinney suggested may have been entitled *On the Organic Law of Change,* never came to pass. More precisely it was the revelation that he had hit upon the very same idea as Darwin after fatefully sending to the elder naturalist his manuscript from Ternate in early 1858, and news that Darwin was already working toward a book on the subject, that led Wallace to shelve his plan. His own book-length treatment was not to appear for some thirty years, until 1889, and even then the title he chose, *Darwinism,* reflected his long-standing preference for the back seat. But Wallace was perhaps too quick to defer to Darwin; there is no question of Darwin's own genius and long labor over the species question, certainly, and the two became friends, but Wallace's tendency to politely skirt the evolutionary limelight has led to an underappreciation of his own genius. Some have even dismissed Wallace as someone who was simply lucky, stumbling by chance upon a great idea and unworthy of laurels in comparison with Darwin's patient labors. This view of Wallace is patently false, doing a disservice to him and to the history of evolutionary thought. Wallace's Species Notebook reveals the depth and breadth of his insight like no other document in those pre-*Origin* years.

Wallace Voyaging

Alfred Russel Wallace was no mere traveler-collector; he did not travel to collect so much as he collected to travel, proceeds from his specimen sales back in London funding his continued explorations. His object? The species question. Ever since reading the anonymously published *Vestiges of the Natural History of Creation* in 1845, at the age of twenty-two, he was convinced that the central transmutational argument of the book was correct. Species must change, but how to prove it? And by what mechanism? That he actually succeeded in his quest to solve the mystery of species origins little more than a decade into this quest is nothing short of astonishing. The Species Notebook shows him in the years leading up to his discovery, made in 1858, steadily accumulating evidence and crafting arguments in defense of the transmutation thesis.

Though a self-taught, working naturalist with little by way of social standing or connections, Wallace was kindred spirit with the "philosophical naturalists" of his day, interested in the big questions. He wrote paper after paper during his travels, a compact library in tow wherever he went. He published no fewer than sixty papers and letter extracts during his eight years in the East, including the two now seen as landmark works in the history of evolutionary thought: his 1855 Sarawak Law paper (*"Every species has come into existence coincident both in space and time with a pre-existing closely allied species"*) and the 1858 Ternate essay announcing his discovery of the process of species change, natural selection. Where did he find the time and energy between his endless travel arrangements and stiff schedule of collecting, not to mention documenting, preparing, and shipping his prodigious collections? Tending to the specimens alone, his bread and butter, must have been daunting; his insects numbered in the tens of thousands, plus the skins and skeletons of a veritable museum of birds, mammals, reptiles, amphibians, and the odd fish. Then there was the incessant battle with insect pests and other scavengers seeking to make a meal of his specimens.

When it came to writing, Wallace was at his most prolific when he was laid up (fevers, sore ankles, and infected and festering insect bites being a way of life) or his pace was slowed by rainy seasons. The Species Notebook was his constant companion and would be remarkable even if its contents were limited to his collecting adventures and natural history notes: accounts of hunting orangutans, the orphaned infant orang he tried to raise, lovely birds of paradise, bats that filled the twilight sky, and striking insect finds recorded in this notebook all enlivened his published works, especially his now-classic travel memoir *The Malay Archipelago* of 1869. In the notebook we also see Wallace the reformer, offering practical and prescriptive remedies for problems faced by working naturalists, for example, a proposal for halting the proliferation of taxonomic "synonyms" (the confusing duplication of scientific names, the bane of every naturalist) through the creation of an international body to regulate the naming of new species and adjudicate taxonomic squabbles, or a scheme for a shared multi-institutional library of natural history, freely accessible to naturalists. Wallace was far ahead of his time with such proposals; today they are realized in the form of the International Commission for Zoological Nomenclature, and a multitude of libraries electronically pool their resources through such organizations as the online Biodiversity Heritage Library.

Besides Wallace the observer and problem solver there are the "evolutionary" entries, in which we see Wallace the synthetic thinker, philosopher, and discoverer. Critiques of the prevailing arguments of the day for benevolent design and harmony in nature are interspersed with observations of island species, anatomical structure, domestic varieties, relationships of fossil species, embryology, instinct, and more. These wide-ranging notebook entries together constitute an extended argument in favor of transmutation—the centerpiece of which is Wallace's twenty-four-page attack on Charles Lyell's anti-transmutation arguments in the landmark *Principles of Geology*.

Mr. Lyell Says . . .

Famed geologist Charles Lyell was the preeminent naturalist of Britain, and his long attack on transmutation in the wildly successful *Principles of Geology* was widely taken as the final word on the matter. To Wallace he was just the person to engage with: to undermine what many considered the definitive arguments against species change was critical to paving the way to acceptance of the revolutionary idea. In the Species Notebook Wallace engages in a long "dialog" with Lyell, aiming to refute his arguments point by point. Wallace's general approach was to copy out offending passages from the *Principles* and follow each with his own rhetorical replies. Lewis McKinney first pointed out Wallace's intentions: in more than one place Wallace takes Lyell to task for not taking his "uniformitarian" vision of changes in the earth to its logical conclusion. Under the heading "Note for Organic law of change," Wallace wrote that "the inorganic world is the result of a series of changes from the earliest periods produced by causes still acting," and so "it would be most unphilosophical to conclude . . . that the organic world was subjected to other laws." After a related comment, Wallace noted to himself that he should "introduce this and disprove all Lyells [sic] arguments first at the commencement of my last chapter." That mention of a "last chapter" implies, of course, multiple chapters, that is, a book. This is resonant with a comment Wallace made in a letter to his friend and traveling companion on his earlier South American expedition, Henry Walter Bates. Bates had written from Amazonia to compliment Wallace on his 1855 paper. Finally getting around to replying in December 1857 from the island of Ambon, Wallace wrote that he had "prepared the plan & written portions of an extensive work embracing the subject in all its bearings & endeavouring to provide what in the paper I have only indicated."

In "embracing the subject in all its bearings" Wallace would have taken the bull by the horns in his planned book, first rebutting Lyell and then elaborating on the central argument of the Sarawak Law paper, namely, that species show striking patterns of relationship both in space (in terms of geographical distribution) and in time (in terms of their distribution in the fossil record). Closely related species are by and large proximate to one another in space and time; to Wallace this strongly indicated that new species are somehow derived directly from preexisting species. Wallace knew just how to skewer Lyell's anti-transmutationism. Let us consider just a few of his points, a sampler relating to the fossil record, island life, and domestication and the supposed limits of variability of species and varieties.

In his Species Notebook Wallace not only argues for the "progressive change" of fossils over time, he also has a remarkably modern grasp of the idea of lineages branching from lineages. Lyell asserted that a fossil mammal found among the reptiles of the Mesozoic dealt a fatal blow to the idea of a progressive succession of groups in the fossil record. Not so, says Wallace: "all that is required for the progression is that some reptiles should appear before Mammalia & birds or even that they should appear together. In the same manner reptiles should not appear before fishes but it matters not how soon after them" (Wallace's emphases). "Not one fact contradicts the progression," Wallace declares; "each group goes on progressing after other groups have branched from it. They then go on in parallel or diverging series." In modern terms this is evolutionary-tree thinking.

The significance of the unique species characteristic of remote islands is equally clear to Wallace. Having read about the species of the Galápagos in Darwin's *Journal of Researches* (later known as *Voyage of the Beagle*), Wallace notes that on such islands we find species that are found nowhere else, yet they resemble those from the nearest mainland. He asks, "If they are special creations why should they resemble those of the nearest land? Does not that fact point to an origin from that land?" He realizes that the unique species likely descended from ancient colonists that made it (by chance) to the islands from the nearest mainland. He notes that this is why older islands have more such species: more time for colonization and subsequent slow modification. He charges Lyell with inconsistency in arguing on the one hand that natural processes are responsible for shaping the earth's features and causing the extinction of existing species, and on the other hand that supernatural processes are responsible for the origin of new species: "it would be an extraordinary thing if while the modification of the surface . . . by natural causes now in operation & the extinction of species was the natural result of the same causes, yet the reproduction & introduction of new species required special acts of creation, or some process which does not present itself in the ordinary course of nature." Wallace's comment here deliberately invokes the subtitle of the *Principles of Geology:* "Being an attempt to explain the former changes of the Earth's surface, by reference to causes now in operation."

It is interesting, finally, that Wallace uses domesticated breeds in his pro-transmutation arguments. Lyell famously based a series of anti-transmutation arguments on the supposed limited nature of variability and change as exemplified by domestic breeds. Species only vary within limits, he maintained, as no domestic variety had been transmutated into a new species. The limited nature of the capacity for change was reinforced, in Lyell's view, by their apparent tendency to revert to a more generalized "parental" type when they run wild—think of the generalized appearance of mutts that might be produced by the crossing of several different dog breeds. Wallace sidestepped these issues in his 1858 Ternate essay, but in the Species Notebook he rejects Lyell's arguments. First he points to strikingly different dog varieties as *themselves* evidence for great capacity for change: "is not the change of one original animal to two such different animals as the Greyhound & the bulldog a transmutation?" Wallace points out that these breeds differ more from each other than do, say, donkeys and zebras. According to Lyell varieties of some species may differ from each other even more than some species do "without shaking our confidence in the reality of species," by which he meant the permanent and unchanging nature of species. But isn't this confidence misplaced, Wallace asks. "Is it not a mere . . . prejudice like that in favor of the stability of the earth which [Lyell] has so ably argued against?"

"In fact," he demands, "what positive evidence have we that species only vary within certain limits?" In a long passage Wallace does a thought experiment: imagine all dog breeds but one become extinct, and that remaining one is spread far and wide around the world and used as stock to develop new breeds. Then suppose that all breeds but one of *those* (the one "farthest removed from the original") in turn become extinct, and the process begins again. "Does it not seem probable that again new varieties would be produced," asks Wallace, "and have we any evidence to show that at length a check would be placed on any further change & ever after the species remain perfectly invariable?" He concluded with another Lyellian statement of the change in species and varieties that can be realized over long periods of time: "changes which we bring about artificially in short periods may have a tendency to revert to the parent stock. This is considered a grand test of a variety. But when the Change has been produced by nature during a long series of generations, as gradual as the changes of Geology, it by no means follows that it may not be permanent & thus true species be produced."

There are many other pro-transmutation arguments in the Species Notebook critiquing Lyell and other authors. Frustratingly there are also gaps: no entries bearing on the struggle for existence, for example, or on the mechanism of natural selection or its discovery by Wallace in February of 1858 while he was collecting on the island of Gilolo (now Halmahera). He later wrote that natural selection came to him in a flash of insight in a fevered state and that over the course of the next couple of evenings he fleshed out the idea in the form of the essay he fatefully sent to Darwin from Ternate the following March. He was then off to New Guinea and did not return to Ternate until the following August. Soon after his return Wallace received word from Darwin and Hooker about the dramatic effect of his essay, and the reading hastily arranged by Lyell and Hooker along with unpublished writings on the subject by Darwin at the Linnean Society on 1 July 1858. On that date Wallace was still in New Guinea, making the best of a bad situation with illness and terrible collecting conditions (one of his assistants, Jumaat, succumbed to disease and died in late June, and halfway around the world in England on the very day of the Linnean Society readings a grief-stricken Darwin was burying his infant son, also felled by disease).

The Force of Admiration

With the eventual revelation of Darwin's earlier discovery of the mechanism of species change and progress toward a treatise on the subject, Wallace abandoned his own planned book. Today we tend to see Wallace and Darwin as the protagonists in this drama, with Lyell in a supporting role. Yet it is clear that as far as Wallace was concerned, Lyell was his "evolutionary foil," and there is good evidence that both the Sarawak Law and Ternate papers were aimed at Lyell, as his book would have been. *On the Organic Law of Change* would have likely been an elaboration of the Sarawak Law paper, along with the mechanism behind the slow transmutation of species. Wallace did not name this mechanism, which has become known as natural selection—the name that Darwin gave it. In his "book that should have been," Wallace was pursuing what nineteenth-century philosopher William Whewell termed a "consilience" argument, tying together many seemingly unrelated strands of evidence in support of transmutation, each successive strand interwoven with the others to produce an evidentiary fabric of great strength. It is a powerful way to argue; significantly, *On the Origin of Species* was constructed along similar consilience lines. Indeed, Wallace's and Darwin's paths of discovery and explication of their ideas were more similar than has been realized.

Upon reading the *Origin*, which he received with Darwin's compliments while still in remote Southeast Asia, Wallace lavished praise on the elder naturalist's efforts in letters to friends and family, lauding the *Origin*'s "vast accumulation of evidence, its overwhelming argument, and its admirable tone and spirit." He enthusiastically concluded one letter declaring that "the force of admiration can no further go!!!" The Species Notebook reveals a Wallace about whom the same can be said—the force of *our* admiration can no further go, seeing, through the lens of this much-traveled notebook, Wallace's tenacity, creativity, and impressively deep insight into the then-revolutionary idea of species change.

The Species Notebook thus underscores Wallace's great stature as cofounder, with Darwin, of modern evolutionary biology. Wallace did not always see eye-to-eye with Darwin on evolutionary matters over the years, and in the modern view both of them had their share of blunders as well as keen insights. By the end of his long life, Wallace had authored more than a thousand articles and some twenty-two books on a wide variety of subjects, the best known of which include the two acclaimed books already mentioned (his travel memoir *The Malay Archipelago* and *Darwinism*, his spirited defense of evolution by natural selection), as well as his landmark

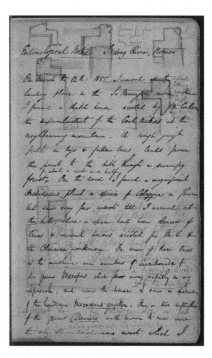

Cover and first page of Wallace's "Species Notebook" of 1855–1859. Wallace often used his notebooks, including this one, as a double notebook by making entries from both ends simultaneously, inverted with respect to one another (designated "recto" and "verso" notebooks in Costa 2013a). He labeled this recto side of the notebook "Notes Vertebrata," though in fact many entomological observations are given as well. The entry on the first page records Wallace's arrival at the "landing place" on the Sadong River in Sarawak on 12 March 1855, commencing a period of fruitful insect and orangutan collecting in the region around the Simunjon coal mines. This entry was made shortly after the Sarawak Law paper was written and mailed off to the *Annals and Magazine of Natural History*. Courtesy of the Linnean Society of London and the A. R. Wallace Literary Estate.

The Geographical Distribution of Animals (1876), a founding text of modern biogeography. Wallace had also been showered with honors and awards, including the Order of Merit, the greatest honor that can be bestowed upon a civilian by the British crown.

Wallace's Species Notebook passed to his son William, who in turn presented it to the Linnean Society in 1936. And there it sits on a paneled shelf, a portal to a lost time and place and a most remarkable record of the thinking of a most remarkable naturalist.

ONE

Granted the Law

Alfred Russel Wallace's Evolutionary Travels

ALFRED RUSSEL WALLACE'S road to accepting the idea of transmutation of species was short: one reading of *Vestiges of the Natural History of Creation* in 1845 was sufficient to convince him of the essential correctness of the idea. His journey to deep insight into the phenomenon, however—his intellectual journey adducing the evidence in support of the idea and figuring out the mechanism behind it—was rather longer, but not exceedingly so: his Sarawak Law paper was published just ten years later, and the Ternate essay announcing his discovery of the mechanism of species change three years after that. When Wallace wrote to his friend Henry Walter Bates in December 1845 that the "development hypothesis" was "an incitement to the collection of facts, and an object to which they can be applied when collected," what he had primarily in mind was facts pertaining to his "favourite subject," namely, "the variations, arrangements, distribution, etc., of species" (Wallace Correspondence Project [WCP] 346)—in essence, facts of geographical distribution of species. In this chapter I trace Wallace's "evolutionary" thinking, to use a modern term, from his initial conviction of the reality of transmutation to the triumphant announcement of its mechanism in his Ternate essay. His evolutionary travels began in earnest in Leicester, but the road took him to Amazonia and then Southeast Asia—an intellectual journey that can be traced through Wallace's papers, journals, and notebooks in this period, with special emphasis on the "Species Notebook" (Linnean Society MS 180, Linnean Society of London). In this most important of his field notebooks, kept between 1855 and 1860 or so, Wallace reveals his far-ranging and creative insights into the species question (Costa 2013a).

Inspiration and Context

Beyond mere species variation, the geographical distribution of species and varieties lay at the heart, Wallace felt sure, of the species problem. This had long been recognized in a botanical context through the efforts of the polymath Prussian explorer and naturalist Alexander von Humboldt (1769–1859), whose writings Wallace eagerly consumed in his teens. Humboldt had devised the technique of "botanical arithmetic" to sleuth geographical patterns in species richness relative to genera and levels of endemism (Browne 1980). He urged comparative analysis of biota, such as new-world versus old-world species, to gain insight into the big philosophical question of the "creative power" behind species. The Swiss botanist Augustin Pyramus de Candolle (1778–1841), an admirer of Humboldt, famously declared in his *Essai élémentaire de géographie botanique* of 1820 that "all of the theory of geographical botany rests on the particular idea one holds about the origin of living things and the permanence of species." Wallace had a taste for botany since about 1838, when, at the age of fifteen, he purchased his first botanical manual: the fourth (1841) edition of John Lindley's *Elements of Botany.* The book proved to be disappointing since it was more a treatise on principles of plant anatomy and classification than identification manual, but Wallace annotated it heavily with, among other things, notes from John Claudius Loudon's *Encyclopedia of Plants* and Humboldt's *Personal Narrative of Travels in South America;* he even copied out passages from Charles Darwin's *Journal of Researches*—passages that celebrated the "gorgeous beauty" and "luxuriant verdure" of tropical forests (McKinney 1972, 3–5). What better place to hunt for clues than in what seemed the very manufactory of species: the teeming tropics, with species richness dizzyingly far beyond anything that a denizen of the north-temperate zone could imagine?

Two other authors read by Wallace by the mid-1840s left a powerful impression on him: "Mr. Vestiges," the then-anonymous author of the above-mentioned sensational *Vestiges of the Natural History of Creation* (first published in 1844), and geologist Charles Lyell (1797–1875), author of the landmark *Principles of Geology* (first published 1830–1833). In some ways antithetical to one another, these two works nonetheless presented to Wallace new and exciting insights into the nature of the earth and its inhabitants. *Vestiges* presented a sweeping view of cosmological, planetary, and organic transmutation that extended even to people and social institutions—more than scandalous, some saw it as seditious. But the book was nothing if not a sensation, as meticulously documented in James Secord's aptly titled history of the book's reception, *Victorian Sensation* (Secord 2000). *Vestiges* also had the beneficial effect of airing the tainted subject, bringing discussions of transmutation into drawing rooms and parlors across a broad social cross section of the country (Secord 2000, 1–6). Edinburgh writer and publisher Robert Chambers, the anonymous author, was wise to keep his identity a secret, judging from the vehement condemnation of the book from pulpits to Parliament and the scientific salons of England.

Why such vehemence? In the first half of the nineteenth century, the doctrine of transmutation had become repugnant to the scientific and clerical establishment (often one and the same) owing to its connection with the radical French "transformists" of the late eighteenth and early nineteenth centuries, with atheistic implications. Zoologist Jean-Baptiste de Lamarck (1744–1829) was the best-known exponent of French transformism, and his ideas found currency in some thinkers across the channel—two prominent examples being Erasmus Darwin (1731–1802) and Scottish zoologist Robert Edmond Grant (1793–1874) (Corsi 1978, 2–5; Sloan 1985, 73–80). Erasmus, grandfather of Charles Darwin, was a famous physician and poet who put his ideas about the origin of life and organic transmutation into verse; Grant was Charles's first scientific mentor when Charles was a medical student at Edinburgh (and surely provided the young Darwin's second exposure to transmutationist thinking after his grandfather's writings). Charles Darwin was no transmutationist at Edinburgh or Cambridge in the years that followed (1828–1831), however; the respectable naturalist-clerics who taught Darwin at Cambridge were strictly orthodox on this count, and some vehemently denounced *Vestiges*. But by the time *Vestiges* came out in 1844, Darwin had long since changed his mind and secretly embraced a view of species change. Like Wallace he agreed with the basic transmutational premise of *Vestiges*, though he deplored its lack of philosophical or scientific rigor.

Not so the twenty-two-year-old Wallace, who was more broadly enthusiastic. Victorian England was abuzz with the discreditable *Vestiges*, and Wallace eagerly devoured the book. In a November 1845 letter to Bates, he wrote, "Have you read 'Vestiges of the Natural History of Creation,' or is it out of your line?" Bates evidently was not as impressed as Wallace was, provoking Wallace to write a month later:

> I have rather a more favorable opinion of the "Vestiges" than you appear to have—I do not consider it as a hasty generalisation, but rather as an ingenious hypothesis strongly supported by some striking facts and analogies but which remains to be proved by more facts & the additional light which future researches may throw upon the subject.—it at all events furnishes a subject for every observer of nature to turn his attention to; every fact he observes must make either for or against it, and it thus furnishes both an incitement to the collection of facts & an object to which to apply them when collected—I would observe that many eminent writers gave great support to the theory of the progressive development of species in Animals & plants. (WCP346)

The genealogical model of descent that Wallace came to embrace was likely inspired by the treelike diagram of relationships in *Vestiges*, a diagram that represented embryological development but with transmutational implications (Figure 1.1). The author was careful to point out that the diagram "shews only the main ramifications; but the reader must suppose minor ones, representing the subordinate differences of orders, tribes, families, genera, &c." (Chambers 1844, 212).

This process of differentiation set up an inquiry into how changes in embryological development can lead to transmutation: "it is apparent that the only thing required for an advance from one type to another in the generative process is that, for example, the fish embryo should not diverge at A, but go on to C before it diverges, in which case the progeny will be, not a fish, but a reptile" (Chambers 1844, 213). Geographical distribution also figured prominently in the *Vestiges*, as holding essential clues to the process of progressive development. Mr. Vestiges provided a set of "general conclusions regarding the geography of organic nature" in his chapter discussing the "Progress of Organic Creation":

> (1.) There are numerous distinct foci of organic production throughout the earth. (2.) These have everywhere advanced in accordance with the local conditions of climate &c., as far as at least the class and order are concerned, a diversity taking place in the lower gradations. No physical or geographical reason appearing for this diversity, we are led to infer that, (3.) it is the result of minute and inappreciable causes giving the law of organic

development a particular direction in the lower subdivisions of the two kingdoms. (4.) Development has not gone on to equal results in the various continents, being most advanced in the eastern continent, next in the western, and least in Australia, this inequality being perhaps the result of the comparative antiquity of the various regions, geologically and geographically. (Chambers 1844, 190–191)

About the same time as reading *Vestiges*, Wallace read Lyell's *Principles of Geology*. *Principles* was already in its sixth edition when *Vestiges* came out, so the transmutationist foil for Lyell was Lamarck. Still, the near-contemporaneous reading of *Vestiges* and *Principles* by Wallace is important. Mr. Vestiges and Lyell disagreed on many things, but they seemed to agree on the critical importance of geographical distribution for understanding

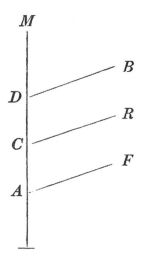

Figure 1.1. Tree of embryological differentiation of four vertebrate classes (F = fish, R = reptiles, B = birds, and M = mammals) from *Vestiges of the Natural History of Creation*, inspired by William Carpenter's concept of successive differentiation of vertebrate embryos (Carpenter 1841, 196–197). "The foetus of all the four classes may be supposed to advance in an identical condition to point A. The fish there diverges and passes along a line apart, and peculiar to itself, to its mature state at F. The reptile, bird, and mammal, go on together to C, where the reptile diverges in like manner, and advances by itself to R. The bird diverges at D, and goes on to B. The mammal then goes forward in a straight line to the highest point of organization at M" (Chambers 1844, 212). Courtesy of the Staatsbibliothek zu Berlin.

the nature of species. Lyell opened his chapter on "Laws which Regulate the Geographical Distribution of Species" in volume 2 of *Principles* declaring:

> Next to determining the question whether species have real existence, the consideration of the laws which regulate their geographical distribution is a subject of primary importance to the geologist. It is only by studying these laws with attention, by observing the position which groups of species oc- cupy at present, and inquiring how these may be varied in the course of time by migrations, by changes in physical geography, and other causes, that we can hope to learn whether the duration of species be limited, or in what manner the state of the animate world is affected by the endless vicis- situdes of the inanimate. (Lyell 1832, 66)

Lyell's influence on Wallace went far beyond an incitement to study the laws of distribution, however. The *Principles* also articulated powerful statements on the history of the earth and its inhabitants. Earth history was seen in terms of slow, steady change, a product of the long-continued action of natural forces still seen in action today—a principle called "actu- alism" in Wallace's day, and "uniformity" today. Wallace wholly embraced this exciting new view of the planet's transformations, but the other great statement articulated by the *Principles* had the opposite effect: Lyell's strong anti-transmutationism. Lyell went to great lengths to argue against the doctrine point by point, and Wallace's answers to many of these points are first articulated in the Species Notebook.

Collectors in Paradise

Such reading clearly had an effect on Wallace, and he began to think big— geographically and philosophically speaking. In the fall of 1847 he visited the collections of the Muséum National d'Histoire Naturelle in Paris in the company of his francophone sister Fanny, and soon afterward the insect room of the British Museum. Not long after his visit to the "fair city of Paris" he wrote to Bates: "I begin to feel rather dissatisfied with a mere lo- cal collection; little is to be learnt by it. I [should] like to take some one family, to study thoroughly—principally with a view to the theory of the origin of species. By that means I am strongly of [the] opinion that some definite results might be arrived at." Wallace then asked Bates to help "in choosing one that it will be not difficult to obtain the greater number of the known species" (WCP348). Wallace might have become a "museum-man," as they were called in those days, poring over pinned specimens from far-flung regions housed in row upon row of orderly cabinets in the great museums. But he no doubt realized early on that the breathtaking riches of the museum were in large measure unsuitable for the kind of analysis he had in mind. They boasted rich diversity, yes, and in some taxonomic groups perhaps even complete sets of species, but knowledge of many groups and regions was spotty. Then, too, the specimens often bore imprecise information on the locality in which they had been collected. He was to later exhort his fellow naturalists to pay closer attention to precisely where specimens came from, as he realized that only with such information could the long-sought patterns underlying geographical distribution of species and varieties be divined.

No, mere museum work would not do, and it was around this time Wallace determined to gather such information himself, hatching with kindred spirit Bates an audacious plan (Bates 1863, iii; Wallace 1905, 1:254). Over the subsequent autumn and winter of 1847–48 their plans to become collector-naturalists in some exotic locale took shape. Where to go? W. H. Edwards's effusively romantic book *A Voyage Up the River Amazon* (1847) was the deciding factor—by chance they met Edwards, who was encouraging and even wrote them letters of recommendation. Despite what they must have soon realized were the many inaccuracies of Edwards's book once they arrived in Amazonia, the choice was a fruitful one in many respects. Wallace and Bates were fortunate to be taken on by the able agent Samuel Stevens of Bloomsbury Street, London (Stevenson 2009), who helped them get outfitted and provided all manner of advice. The two met in London that March, booked passage on a trans-Atlantic ship, and departed from Liverpool in April. They soon arrived in the New World, literally and figuratively, landing at Pará, Brazil, on 26 May 1848, eager to explore "some of the vast and unexamined regions of the province . . . said to be so rich and varied in its productions of natural history" (Wallace and Bates 1849, 74).

Of the duo Wallace at that time was perhaps the more philosophically inclined, and he seems to have been especially keen on the question of species origins. He took to heart the pronouncements of Humboldt, Lyell, and others about the importance of geographical distribution in his pursuit of the species question—indeed, there is perhaps no clearer statement of this than Wallace's declaration in his 1853 travel memoir *A*

Narrative of Travels on the Amazon and Rio Negro: "There is no part of natural history more interesting than the study of the geographical distribution of animals" (Wallace 1853, 469). But it is important to bear in mind that Wallace's starting point was a conviction that the idea of species change was fundamentally correct—that is, he started with the working assumption of transmutation, and proceeded to collect data and observations in the area he thought most likely to yield insights. Lewis McKinney put it well when he commented that "zoogeography did not lead Wallace to evolution; on the contrary, evolution led him to the phenomena of the geographical distribution of organisms" (McKinney 1966, 357). And observe he did. In Wallace's most important writings from his four years in Amazonia we see him circling around the species question, putting his finger on fundamental problems of variation and distribution in his attempt to ferret out the underlying patterns he thought must be there, and would once elucidated throw open a window on the mystery of species origins.

Though by necessity a working collector, Wallace sold only his duplicate specimens to finance his expeditions, saving most of the fruits of his labors for his personal scientific studies. His collecting efforts over four years in Amazonia were directed at problems of species distribution in one form or another (George 1964; Brooks 1984). He was fascinated by species that appeared to be restricted to particular locales, as seen in his pursuit of the Guianan cock-of-the-rock *Rupicola rupicola* (Cotingidae), which is restricted to a small area of granitic highlands occurring in the mountains of Guiana (Figure 1.2). Yet it was groups of species that held the greatest potential for shedding light on the species problem, by potentially providing a series that might map onto geography in illuminating ways. His working hypothesis while collecting in South America had an important geological component: geologically younger, newer, dynamically changing areas may, he thought, be inhabited by more recently derived species transmutated from related older species that occupy older geological formations. The great Amazon basin was a testing-ground for this idea, since the vast valley largely consisted of young alluvial sediments derived from the much older granitic uplands found to its west, south, and north.

Thus he sought, for example, an elusive white umbrella-bird *(Cephalopterus)* he had heard occurred in the older uplands, way up the River Uaupés, a species that if found would have fetched him a considerable sum for its beauty and rarity but which could also be compared with the more common black umbrella-bird of the younger lowlands, and furnish evidence bearing on his idea of species relationships between lowland and upland (geologically younger and older areas, respectively). After his herculean efforts ascending the river in early 1852, it appeared that the chimeric white bird, if it existed at all, was but an odd and occasional variant.

Figure 1.2. Cock-of-the-rock *Rupicola rupicola* (Cotingidae). In 1850 Wallace collected these brilliant orange birds in the isolated and rugged granitic uplands on the upper reaches of the Rio Negro River. He commented in *Narrative of Travels on the Amazon and Rio Negro* that this species is "an example of a bird having its range defined by a geological formation" (Wallace 1853, 474), as compared to others with ranges defined by barriers such as the great rivers of Amazonia. (From Darwin 1871, 2:88, courtesy of Kathy Horton, Museum of Comparative Zoology, Harvard University.)

Wallace was similarly frustrated by his efforts to map the distribution of *Mauritia* palms in upland versus lowland areas. He observed five *Mauritia* species, all of which occur in the uplands, though three are narrow endemics. One of the species also occurred in the lowlands, an overlapping and not contiguous distribution that muddied the waters of interpretation for Wallace.

He did, however, believe he had found one clear example that supported his idea of young species in geologically recent lowlands derived from older species found in the geologically more ancient uplands. A year after his return to England, in December of 1853, he read a paper entitled "On the Habits of the Butterflies of the Amazon Valley" to the Entomological Society of London. In this paper Wallace alluded to his lowland/upland species hypothesis, describing how in "the space between [the uplands] the Amazon rolls its mighty flood through a vast alluvial valley, which is everywhere clothed with dense forests of lofty timber trees. The whole of this valley lies in the very centre of the tropics, and enjoys a climate in which a high and uniform temperature is combined with a superabundance of moisture. . . . These seem to be the conditions most favourable to the development and increase of Lepidopterous insects" (Wallace 1854). But a most remarkable statement is buried within this paper, given almost in passing. Writing of highly variable groups of *Heliconia* butterfly species, Wallace ventured:

> All these groups are exceedingly productive in closely allied species and varieties of the most interesting description, and often having a very limited range; and as there is every reason to believe that the banks of the lower Amazon are among the most recently formed parts of South America, we may fairly regard those insects, which are peculiar to that district, as among the youngest of species, the latest in the long series of modifications which the forms of animal life have undergone. (p. 258)

Note Wallace's use of the term "modifications" and the phrase "youngest of species." Although his observation can also be interpreted in the context of successive creations, he had a transmutational view. This was an astonishing thing to suggest publicly. Brooks (1984, 50–52) suggested that Wallace may have been emboldened by an anonymously published essay by Herbert Spencer, published in the *Leader* for 30 March 1852 and which Wallace almost certainly had read. This essay was a spirited defense of transmutation, challenging readers to consider "which, then, is the most rational hypothesis; that of special creations which has neither a fact to support it nor is even definitely conceivable; or that of modification, which is not only definitely conceivable; but is countenanced by the habitudes of every existing organism" (Spencer 1852; reprinted in Spencer 1891, 1–7).

Wallace may as well have read "habitats" for "habitudes," underscoring how for him the evidentiary basis for transmutation was to be found in geographical distribution. Whether Wallace's audience at the Entomological Society noticed his embrace of "modifications"—the development hypothesis—is not known, as there does not appear to have been any discussion of the matter. Some in the audience were likely unhappy with Wallace, judging by the tenor of Edward Newman's remarks that very month in his closing presidential address to the society. Newman, who was to become Wallace's friend and supporter (though ultimately a friendly critic, never accepting Wallace's evolutionary ideas), lamented how some fellow naturalists deprecated and dismissed the work of those deemed "speciesmen" or "mere collectors," and their tendency to draw comparisons "disadvantageous" to their fellows. Later in this address he specifically referenced Wallace, wishing him "God speed" on the eve of his departure for points east (Brooks 1984, 53–54). The criticisms of Wallace likely stemmed as much from his low social standing as from his embrace of heterodox ideas. This was not to be the last time such a criticism was leveled against him.

Although Wallace was apparently convinced that geographical distribution somehow held the key to the species question, it is fair to say that he did not embark upon his Amazonian adventure single-minded about mapping, in detail, the distribution of species and their variants, something that is clear from his own lament that he regretted not having been more careful in recording such information in his South American collecting (Wallace 1905, 1:377). This lament was manifested in an exhortation to his fellow naturalists in a paper read before the Zoological Society in December of 1852, shortly after his return home. In "On the Monkeys of the Amazon" (Wallace 1852) he puts his finger on the relationship between the ranges of several monkey species through the vast Amazon basin and the great river systems coursing through it: the Amazon, the Rio Negro, and the Madeira. The evolutionary implications of the delineation of species distributions by such impassable geographical barriers, dubbed the Riverine Barrier hypothesis today, are clear. In pointing this out, however, Wallace underscored the importance of geographical knowledge, without which such patterns would remain unknown:

In the various works on natural history and in our museums, we have generally but the vaguest statements of locality. S. America, Brazil, Guiana, Peru, are among the most common; and if we have "River Amazon" or "Quito" attached to a specimen, we may think ourselves fortunate to get anything so definite: though both are on the boundary of two distinct zoological districts, and we have nothing to tell us whether the one came from the north or south of the Amazon, or the other from the east or the west of the Andes. Owing to this uncertainty of locality, and the additional confusion created by mistaking allied species from distant countries, there is scarcely an animal whose exact geographical limits we can mark out on the map.

On this accurate determination of an animal's range many interesting questions depend. Are very closely allied species ever separated by a wide interval of country? What physical features determine the boundaries of species and of genera? Do the isothermal lines ever accurately bound the range of species, or are they altogether independent of them? What are the circumstances which render certain rivers and certain mountain ranges the limits of numerous species, while others are not? None of these questions can be satisfactorily answered till we have the range of numerous species accurately determined. (Wallace 1852, 109–110)

He continued (not altogether accurately): "During my residence in the Amazon district I took every opportunity of determining the limits of species." Certainly this was true of the monkeys, leading him to the realization that "the Amazon, the Rio Negro and the Madeira formed the limits beyond which certain species never passed." But, as we have just seen in his own words, for at least his first two years in Amazonia Wallace neglected detailed locality information. Despite those early lapses in record-keeping, however, and despite the disastrous loss of most of his notes and specimens "from the last two and most interesting" years of his travels when, as he was homeward bound, his ship burned, it is clear that Wallace emerged from the Amazon more convinced than ever of the reality of transmutation, producing several penetrating papers looking at the issue through the lens of geographical distribution. He was doggedly on the trail of solving the mystery of species, and had the good fortune to select as a new collecting locale perhaps the single best theater on earth in which to gain insight into the intimate relationship between species and biogeography as played out over geological time: the vast archipelago consisting of peninsular Malaysia, Singapore, Indonesia, and western New Guinea, spanning some forty degrees of longitude.

Sarawak and the Law

Wallace arrived in Singapore on 20 April 1854, his journey facilitated by the Royal Geographical Society. A comment made years later in his autobiography reveals that the pressing species question was not far from his mind: "From my first arrival in the East I had determined to keep a complete set of certain groups from every island or distinct locality which I visited for my own study on my return home, as I felt sure they would afford me very valuable materials for working out the geographical distribution of animals in the archipelago, and also throw light on various other problems" (Wallace 1905, 1:385).

For his first few months he collected in and around Singapore as he made plans for travels further afield, the first of which he embarked upon in late July 1854. Heading to Malacca on the west coast of peninsular Malaysia, Wallace collected there for nine weeks before returning to Singapore to plan for his next, longer, sojourn. He again contacted Sir James Brooke, the so-called White Rajah of Sarawak, who had earlier "kindly promised him every assistance," as Wallace put it in his application for aid for his journey from the Royal Geographical Society (reprinted in McKinney 1972, appendix 1). Brooke hospitably invited Wallace to collect in Sarawak, a small province in northern Borneo. Wasting no time, Wallace landed in Kuching on 1 November 1854, collecting for some months along the Sarawak and Santubong Rivers and upriver to Bau. He was invited to spend Christmas as Brooke's guest at his home overlooking the Sarawak River, where he enjoyed an extensive library and spirited philosophical discussions. It seems likely that the respite with Brooke and his circle gave Wallace an opportunity to further read and reflect. Spenser St. John, Brooke's private secretary and later biographer, recalled how Brooke greatly enjoyed philosophical discussions with Wallace (St. John 1879, 274). Brooke kindly offered Wallace and his assistant, Charles Allen, the use of a small bungalow on stilts, located at the mouth of the Sarawak River at the foot of imposing Santubong Mountain.

Here, as Wallace waited out the rainy season, he composed his latest reflections on the species question, goaded into action by reading a paper by Edward Forbes in the October 1854 *Proceedings of the Royal Institution*. Forbes's paper propounded his so-called Polarity Theory, a quasi-mystical theory postulating that species diversity ebbs and flows over earth history according to a preordained and idealistic cycle. Wallace said later that he was "annoyed to see such an ideal absurdity put forth when such a

simple hypothesis will explain all the facts." Wallace's response, "On the
Law Which Has Regulated the Introduction of New Species," was written
in February 1855 and published in the *Annals and Magazine of Natural
History* in September of that year. (See Chapter 4 for a detailed discussion
of this paper.) Now known as the Sarawak Law paper, it was far and away the
most cogently argued evolutionary paper published up to that time that
never explicitly mentions evolution (or transmutation, development, modification, etc.), though the "gnarled oak" analogy Wallace gave surely made its
evolutionary thesis clear. Wallace described species relationships in time and
space being "as intricate as the twigs of a gnarled oak or the vascular system
of the human body," with "the stem and main branches being represented by
extinct species" and the "vast mass of limbs and boughs and minute twigs
and scattered leaves" representing living species. Wallace's paper is a marvel
of clear and careful inductive reasoning, as we shall discuss presently, putting his finger on the profound correspondence of species relationships in
space and time, nicely encapsulated by the "law" itself: *"Every species has
come into existence coincident both in space and time with a pre-existing
closely allied species,"* in Wallace's own italics. Just what was Wallace poring
over during that wet season in Sarawak to produce such a paper? Wallace
connected the biogeographical and paleontological dots in a way no one had
thought to do before; what was the source of this synthesis so brilliant that it
provoked Charles Lyell into opening the first of what would become seven
notebooks on the "species question" (Wilson 1970)? Nowhere does Wallace
state what his inspiration was in those dreary rain-soaked weeks, beyond
Forbes's paper. His journals are unhelpfully blank for that time, and his later
account of writing the paper, in his autobiography, is rather general:

> Having always been interested in the geographical distribution of animals
> and plants, having studied Swainson and Humboldt, and having now myself
> a vivid impression of the fundamental differences between the Eastern and
> Western tropics; and having also read through such books as Bonaparte's
> "Conspectus" . . . and several catalogues of insects and reptiles in the Brit-
> ish Museum (which I almost knew by heart), giving a mass of facts as to the
> distribution of animals over the whole world, it occurred to me that these
> facts had never been properly utilized as indications of the way in which
> species had come into existence. The great work of Lyell had furnished me
> with the main features of the succession of species in time, and by combin-
> ing the two I thought that some valuable conclusions might be reached.
> (Wallace 1905, 1:354–355)

There have been efforts to elucidate the immediate sources and inspirations for the Sarawak Law paper, in particular by analyzing Wallace's reading and collecting activity in the months leading up to the writing of the paper. His geological influences are clear: Lyell, certainly, but also F. J. Pictet's *Traité de Paléontologie* (on which Wallace took extensive notes while in Singapore; LINSOC MS 179, 17; McKinney 1972, 47–48). While his insect- and bird-collecting was of interest, including the discovery of a new species of birdwing butterfly (which he named in honor of James Brooke: *Ornithoptera* [now *Trogonoptera*] *brookiana*) and collections of trogons which he compared with similar birds he had collected in Amazonia, the Sarawak Law paper—both argument and observations adduced in support of the argument—may be best seen as something that had been gestating for years by that time, rather than something suddenly precipitated by an utterly new insight in Sarawak. In the paper itself Wallace comments that the idea for such a "law" (which we can take as transmutation) first occurred to him ten years previously, which would have been 1844–1845, the very period he first read *Vestiges*. The Sarawak Law paper is more the result of ten years of contemplation of the issue than any unique discovery or insight in his first months in Borneo.

The very first sentence of the Sarawak Law paper frames the problem biogeographically. Its opening paragraph is a positive endorsement of the uniformitarian vision of earth history articulated by Lyell's *Principles,* and indeed the Lyellian view is found throughout the paper (Costa 2013b). Wallace, like Darwin, saw that Lyell provided evidence for, not against, transmutation, however much he might rail against Lamarck—particularly in the sequence of species in the fossil record, a pattern referred to at the time as the "law of geological succession." Wallace combined geological with geographical data and put the two sets of observations together in the form of a set of "propositions in Organic Geography and Geology," which, he said, "give the main facts on which the hypothesis is founded." Wallace thus brilliantly articulated a joint biogeography of space and time, precisely the same intersection that had occurred to Darwin in connection with South American biogeography and paleontology. In 1837–1838 the leading British comparative anatomist, Richard Owen (1804–1892), analyzed Darwin's South American fossils and documented how Darwin had found gigantic, extinct representatives of groups still living in South America, while noted ornithologist John Gould's analysis of Darwin's birds revealed that the unique "productions" of the isolated Galápagos Islands had a clear affinity with mainland South American species. Proximity, or regionality: this was the best predictor of species relationship, both "laterally" in a biogeographical sense and "vertically" in terms of the fossil record. An appreciation of the joint pattern of relationship in space and time was thus crucial to both Wallace and Darwin.

Wallace did not stop at his statement of the law; he turned to evaluate observed patterns in diverse areas of investigation in a "consilience" argument (though he did not use that term). "This law agrees with, explains and illustrates all the facts connected with the following branches of the subject," he wrote. "1st. The system of natural affinities. 2nd. The distribution of animals and plants in space. 3rd. The same in time, including all the phaenomena of representative groups, and those which Professor Forbes supposed to manifest polarity. 4th. The phaenomena of rudimentary organs" (Wallace 1855, 186). After showing how observations in each of these areas support his "law," Wallace concludes the paper with a remarkable sentence: "Granted the law," he declared, "and many of the most important facts in Nature could not have been otherwise, but are almost as necessary deductions from it, as are the elliptic orbits of the planets from the law of gravitation." Putting his law on a level with the law of universal gravitation was bold.

McKinney (1972) suggested that the extensive notes on Lyell's *Principles* found in the Species Notebook were made prior to the Sarawak Law paper. This seems unlikely: not only do these entries come well after the first date given in the notebook (12 March 1855), which itself comes after the paper was written, but there is only one notebook entry regarding Lyell that can also be found in the Sarawak Law paper (and even there Lyell's name is not used explicitly). It is the very general reference to Lyell's uniformitarian argument previously mentioned from the opening paragraph of the paper: "the present state of the earth, and the organisms now inhabiting it, are but the last stage of a long and uninterrupted series of changes which it has undergone that the inorganic world is the result of a series of changes still acting." The same point was made in the Species Notebook, page 35: "we must in the first place assume that the regular course of nature from early Geological Epochs to the present time has produced the present state of things & still continues to act."

Wallace may have felt he had all but thrown down the gauntlet with this paper, and so was disappointed that the response seemed to be a resounding silence. Some readers probably did miss his underlying point. In view of Wallace's earlier direct reference to modification in his "Butterflies of the Amazon Valley" paper of 1853–1854, he managed in the Sarawak Law paper to construct the strongest case yet for species' transitioning one into another without ever explicitly mentioning transmutation, modification, or development. It is possible to talk about a subject while avoiding specific words that may be otherwise commonly used by others in regard to that subject, but could he have deliberately avoided using them because of their negative reception through works such as *Vestiges?* In any case, his use of the word "creation" in the Sarawak Law paper no fewer than eight times might have led some readers to assume he thought that the new species that arise coincident in time and space with other, closely related species did so by special creation, the prevailing model among naturalists of the day being that earth history is a record of successive creations and extinctions.

Wallace's coining of the term "antitypes" may have similarly contributed to a misreading of his paper: to Wallace "antitypes" were, in modern terms, common ancestors (he even referred to "common antitypes" in several places), but the very word "type" stems from the then-prevalent context of typological species thinking, as in archetypes. Writing about new species "created on the type" of existing species was precisely the way many naturalists of the time imagined new species to come into existence. Putting Wallace's language into this context may help explain why his paper seemed to make barely a stir and why Darwin, making notes on the paper, was able to dismiss it as "nothing very new," and "it seems all creation with him" (Beddall 1988a, 13). Some, however, were stirred indeed by the paper—his agent, Stevens, reported grumbling in the learned societies, whose fellows wished Wallace would stop theorizing and get on with collecting more facts. But others could appreciate, if not altogether agree with, its contents; Lyell was moved to open a notebook on the species question as a result of reading it, and notes from the Sarawak Law paper filled the very first page (Wilson 1970)—Lyell clearly recognized the paper's evolutionary implications. Darwin later commented in a letter to Wallace that "two very good men," Charles Lyell and Edward Blyth, "specially called my attention to it" (Darwin Correspondence Project, letter 2192). But Wallace was not to receive this encouraging news until early 1858.

More Insights from Borneo

In the meantime, shortly after sending off his Sarawak Law paper Wallace arrived at a new collecting locality on Sarawak's Simunjon River, near a new coal-mining operation. Moving into a small dwelling with two rooms and a veranda, Wallace and his assistant Charles Allen settled in for an extended period of collecting. It is here that the Species Notebook begins, and the first several pages report progress getting settled and the initial insect finds amid the downed timber around the coal-works. They stayed for nine months, a period that was to furnish Wallace with a bounty of provocative observations, observations that inevitably led him to continue his bad habit of "theorizing."

The first opportunity for continued theorizing was provided by Wallace's encounter with orangutans while at Simunjon. One of the main reasons Wallace had traveled to Borneo was to observe and collect orangutans. The report of his first encounter with orangs on page 5 of his Species Notebook opens, "This was a white day for me." The date was Monday, 19 March 1855. The entries for the orangutans, or the "mias," as the Dyaks called them, go on for some twenty-five pages; in the end Wallace collected fifteen orang specimens and published five papers relating to them, correcting a good deal of misinformation such as the idea that there were three species. From his comparative analysis Wallace concluded there was but one species, albeit one with considerable variation that included the male-specific secondary sexual facial ridges; today two species are recognized. Three of his orang papers were published in successive months in the *Annals and Magazine of Natural History* (May, June, and July 1856; Wallace 1856a,c,d), and were likely written at the end of the year when Wallace spent several weeks as a guest at James Brooke's hilltop retreat twenty miles from Kuching, or early in 1856 from Singapore, where he arrived on 25 January 1856 in preparation for a journey to Macassar.

His "theorizing" comes in his first orang paper for the *Annals*, "On the Habits of the Orang-Utan of Borneo," published in July 1856. The paper opens with geographical distribution, and then turns to their habits and habitat. Commenting on the sizable canine teeth of the males, Wallace observes that these are not used in defense, nor are they used to tear at prey, as these great apes are strictly vegetarian. "Here then we have an animal which lives solely and exclusively on fruits or other soft vegetable food, and yet has huge canine teeth. It never attacks other animals, and is rarely attacked itself, but when it is, it uses, not these powerful teeth, but its arms and legs to defend itself," he wrote (notwithstanding that his *Malay Archipelago* includes an illustration of an orang biting an attacking Dyak, suggesting the teeth can be used defensively). He added that the females lack these prominent teeth, yet it is the females, encumbered by carrying their young, who would most benefit from having such teeth for defensive use. "Do you mean to assert, then, some of my readers will indignantly ask, that this animal, or any animal, is provided with organs which are of no use to it? Yes, we reply, we do mean to assert that many animals are provided with organs and appendages which serve no material or physical purpose." Here, then, we see that Wallace has used the canines as an opportunity for waxing theoretical. To quote further,

> We conceive it to be a most erroneous, a most contracted view of the organic world, to believe that every part of an animal or of a plant exists solely for some material and physical use to the individual,—to believe that all the beauty, all the infinite combinations and changes of form and structure should have the sole purpose and end of enabling each animal to support its existence,—to believe, in fact, that we know the one sole end and purpose of every modification that exists in organic beings, and to refuse to recognize the possibility of there being any other. Naturalists are too apt to *imagine,* when they cannot *discover,* a use for everything in nature. . . . The separate species of which the organic world consists being parts of a whole, we must suppose some dependence of each upon all; some general design which has determined the details, quite independently of individual necessities. (Wallace 1856a, 30–31)

Naturalists assume there is a use for everything in nature and imagine those uses when they cannot discover them, Wallace says. Suggesting "some general design" determines the details would be acceptable to his readers, and Wallace even mentions a "Supreme Creator" and makes statements such as the following, suggestive of the spiritualism that he so publicly embraced later: "The talented author of the 'Plurality of Worlds' [Philosopher William Whewell's essay *Of the Plurality of Worlds* (1853) argued against the likelihood that life existed on other planets, something that Wallace was to weigh in on later in life] has some admirable remarks on this subject. He says, 'In the structure of animals, especially that large class best known to us, vertebrate animals, there is a general plan, which, so far as we can see, goes beyond the circuit of the special adaptation of each animal to its mode of living; and is a rule of creative action.'" What might have been considered more incendiary in his paper came next:

> It is a remarkable circumstance, that an animal so large, so peculiar, and of such a high type of form as the Orang-Utan, should yet be confined to such

a limited district,—to two islands, and those almost at the limits of the range of the higher mammalia. . . . One cannot help speculating on a former condition of this part of the world which should give a wider range to these strange creatures, which at once resemble and mock the "human form divine,"—which so closely approach us in structure, and yet differ so widely from us in many points of their external form. And when we consider that almost all other animals have in previous ages been represented by allied, yet distinct forms,—that the bears and tigers, the deer, the horses, and cattle of the tertiary period were distinct from those which now exist, with what intense interest, with what anxious expectation must we look forward to the time when the progress of civilization in those hitherto wild countries may lay open the monuments of a former world, and enable us to ascertain approximately the period when the present species of Orangs first made their appearance, and perhaps prove the former existence of allied species still more gigantic in their dimensions, and more or less human in their form and structure! (Wallace 1856a, 31)

Here Wallace brought his Sarawak Law to bear on humans by suggesting that like all other animal groups the orangs—which mock the "human form divine" and "so closely approach us in structure"—must have ancestral "allied, yet distinct forms." If one reads between the lines, humans, too, must have ancestral allied, yet distinct forms. Note the final quoted sentence: could the (relatively) diminutive orangs be descended from "allied species still more gigantic in their dimensions, and more or less human in their form and structure," and if so do humans, too, derive from this ancestor? Incendiary, one would think, yet regardless of whether Wallace was being mischievous or provocative (despite his talk of the "Supreme Creator"), the result was once again a deafening silence. The only response, if it can be viewed as such, was an editing out of these speculations when the *Zoologist* reprinted parts of the *Annals* orang papers.

Crossing Another Line

Arriving in Singapore on 25 January 1856, Wallace unfortunately just missed his boat to the eastern archipelago and was stranded in Singapore for four months. He made the most of his time, writing among other things. Two papers of interest are his "Observations on the Zoology of Borneo" and "Attempts at a Natural Arrangement of Birds." The first paper is significant in that it gave the first suggestion of Wallace's theorizing about the former state of the Malay Archipelago: "the districts nearest to Sumatra and to the peninsula of Malacca possess an ornithological fauna so little peculiar as to furnish strong presumptive evidence of a closer connexion between these countries having existed at no very distant geological epoch. What is known of the whole island, indeed, favours the same view" (Wallace 1856d, 5113). Wallace has a sense that a former connection between these islands and peninsular Malaysia must be responsible for the similarity in the fauna.

The "Birds" paper (Wallace 1856e) was more overtly evolutionary. Brooks (1984, 112) suggested that this paper was written while Wallace was in Sarawak, and the date of publication (September 1856) is consistent with this, following on the heels of the orang papers published in the *Annals* over the previous few months and which were written in Sarawak. In it we see Wallace bringing his Sarawak Law model of species change to bear on classification. What is noteworthy from the point of view of this chapter is Wallace's clear understanding of how transmutation and extinction over time jointly produce empirical patterns of species relationships— their phylogenetic relationships, in modern terms—which can be represented as a treelike pattern of linkages and branches. Wallace was building on naturalist Hugh Strickland's (1841) method of using such treelike diagrams to express relationships, albeit in Strickland's case the supposition was that the species being linked are unchanging. It follows naturally from the Sarawak Law that all species and species groups that have ever lived must link one to another in time and space, but extinction results in gappy patterns of affinity among living groups. "It is an article of our zoological faith," Wallace wrote, "that all gaps between species, genera, or larger groups are the result of the extinction of species during former epochs of the world's history, and we believe this view will enable us more justly to appreciate the correctness of our arrangement" (Wallace 1856e, 206). Wallace's diagram of bird affinities (Figure 1.3) for the "Fissirostres" (cleft-billed birds that use their feet solely for rest, like swifts and kingfishers) and "Scansores" (climbing birds like woodpeckers and relatives) can be seen as a modern unrooted phylogenetic tree, where branch lengths vary according to closeness of relationship. "It is intended," he wrote, "that the distances between the several names should show to some extent the relative amount of affinity existing between them; and the connecting lines show in what direction the affinities are supposed to lie" (Wallace 1856e, 206).

The Species Notebook does not at first appear to contain much of direct relevance to this paper, except for some notes on pages 58–59 on walking and hopping ability of several bird groups. On page 59 Wallace asks, "<u>Crows</u>? ?<u>Colius</u> walk like parrots" (emphasis his) and then "On what muscles &c. does this difference of action depend," ending with "Good character for a primary division"—he has in mind divisions like Fissirostres, Scansores, and so on. However, on closer inspection the paper echoes overarching questions of adaptation and affinity: the relationship between habit and structure. In that light the paper is clearly linked to Wallace's earlier musings on this subject, for example, the insightful discussion in his book *A Narrative of Travels on the Amazon and Rio Negro,* which begins: "In all works on

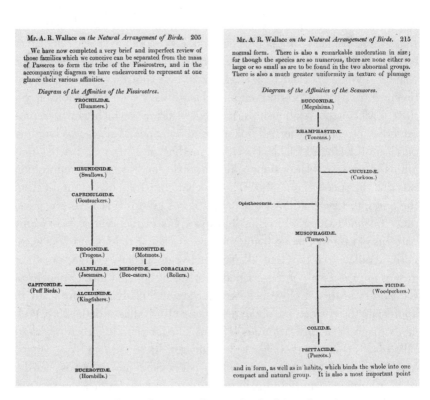

Figure 1.3. Unrooted tree diagrams showing bird relationships from "Attempts at a Natural Arrangement of Birds" (Wallace 1856e). Courtesy of the Natural History Museum, London, and Biodiversity Heritage Library: www .biodiversitylibrary.org.

Natural History, we constantly find details of the marvellous adaptation of animals to their food, their habits, and the localities in which they are found. But naturalists are now beginning to look beyond this, and to see that there must be some other principle regulating the infinitely varied forms of life" (Wallace 1853, 83–84). Wallace's observation in the Species Notebook (p. 53) on the diet of African and Indian hornbills—identical in structure, yet differing in diet—relates directly to this issue, providing him with a piece of evidence that structure and habit are not always correlated and calling into question the prevailing natural theology view of perfect adaptation.

It would have been illuminating had he elaborated on his ideas about the diagrams, his "arrangements" and "agglutination or juxtaposition" of bird taxa in a treelike fashion. Comparing the diagrams in Figure 1.3, note that Scansores has fewer representative groups than Fissirostres, and the branch lengths are longer. The implication is that Scansores is an older group than Fissirostres, with more time for divergence to lengthen branches and extinction to cause gaps, increasingly obscuring the evolutionary relationships among the member families. The families are more closely related to one another than any are to other bird groups, "yet they present so many important points of difference, as to show that they are in reality very distant from each other, and that an immense variety of forms must have intervened to have filled up the chasms, and formed a complete series presenting a gradual transition from one to the other." Wallace declared, "We should be inclined to consider therefore that they [the divergent Scansores] form widely distant portions of a vast group, once perhaps as extensive and varied as the whole of the existing Passeres" (Wallace 1856e, 208–209). To Wallace the true Passeres (songbirds and relatives) seem to represent the youngest group, "extensive and varied," the Fissirostres older, and the Scansores the oldest group. With the benefit of hindsight we can see the evolutionary interpretation of these patterns that must have been in Wallace's mind, but nowhere in this paper is development or modification explicitly advanced. For example, Wallace notes that in arranging bird taxa using the treelike approach he advocates, obtaining the correct arrangement "will not always be so easy a matter as it appears at first sight, for you will most likely find that you have set down some conflicting affinities, or that you have mistaken some mere analogies for affinities" (Wallace 1856e, 207).

Wallace almost certainly held an evolutionary interpretation of "affinity" and "analogy," but these terms were coined without reference to transmutation—naturalists wrote of affinity and analogy in the context of "relationship" on a plan of creation. This was the default view of many naturalists of the time, including Richard Owen, then at the Royal College of Surgeons, who coined the terms "homology" and "analogy" (Wood 1995). Botanist John Lindley (1799–1865), whose book *Elements of Botany* first excited Wallace's natural history interests more than a decade before and whom Wallace credits in the bird paper with inspiring his tree device (see Lindley 1841, 85–86), is typical among naturalists in using these terms without the evolutionary meaning they bear today, even while recognizing their utility for accurate classification. Lindley's paper "Remarks upon the Botanical Affinities of Orobanche" (1837) opened with the author declaring that he doubts if there is any branch of natural history more difficult than systematic botany when it comes to "distinguish[ing] between the relations of affinity and analogy" (Lindley 1837, 409). Wallace surely thought of affinity and analogy in evolutionary terms, but deliberately or not, Wallace was holding back on articulating an explicit evolutionary interpretation of avian relationships. Indeed, the *Zoologist* printed a summary of Wallace's birds paper directly following one of the extracts of his orang papers, declaring Wallace's views on the subject "clear, masterly, and correct." The author would not have been so glowing had Wallace taken the paper to its logical conclusion and pointed out that his Sarawak Law and a process of species change over time explained the patterns so described and diagrammed in the paper.

The next stops on Wallace's evolutionary travels were Bali and Lombock, two islands just to the east of Java that Wallace had not initially intended to visit but now, after four months of waiting in Singapore, would serve as stepping-stones to Aru via Macassar, Celebes (now Ujung Pandang, Sulawesi). It was to prove a most fortuitous unplanned visit. Wallace left Singapore for Bali on 25 May 1856, arriving at Buleleng, on the north-central coast of Bali a few weeks later on 13 June. After two days he hopped over to neighboring Lombock, crossing the fifteen-mile-wide Lombock Strait and landing at Ampanam. It did not take Wallace long to notice that the birds he found there were allied to Australian forms. This realization was driven home as he collected in different areas, encountering enormous cockatoo flocks and a striking mound-building *Megapodius*, both Australian groups. The pattern was further seen in the mammalian fauna. That August he wrote to Stevens: "The islands of Baly and Lombock . . . though of nearly the same size, of the same soil aspect, elevation and climate and within sight of each other, yet differ considerably in their productions, and in fact belong to two quite distinct Zoological provinces, of which they form extreme limits" (Wallace 1857a, 5414). He had accidentally stumbled upon perhaps the only spot in the world where two great zoogeographic provinces approach one another so closely and yet remain distinct. Wallace's explanation for this is not the accepted one today, but he nonetheless had the correct basic idea. With a Lyellian geological model of elevating and subsiding land levels in mind, he believed that the depth of the sea between the two areas was correlated with the time since the areas had been connected by dry land. Areas of uplift were compensated by areas of subsidence: "according to the system of alternate bands of elevation and depression that seems very generally to prevail, the last great rising movement of the volcanic range of Java and Sumatra was accompanied by the depression that now separates them from Borneo and from the continent," as he put it in a later paper (Wallace 1860a, 179). The modern view also recognizes uplift and subsidence, but over much longer time periods. Rising and falling sea levels are more important at a more recent time scale. Thus islands on either side of the divide in the Malay Archipelago reside on shallow continental shelves and become upland provinces of a large and continuous land mass when sea level drops enough to expose the intervening low-lying areas. Bali becomes, in effect, the southeasternmost province of Asia, while Lombock becomes part of the western frontier of a greater Australian landmass. The difference is subtle but important: in the view Wallace then held, only land rises and falls over time, as opposed to *both* land and sea levels rising and falling.

Stevens had part of Wallace's August 1856 letter on Bali and Lombock printed in the *Zoologist,* including the brief account of his zoogeographic discovery. The birds, he said, "throw great light on the laws of geographical distribution of animals in the East" (Wallace 1857a, 5414). Although more deeply impressed by this discovery than the letter suggests, Wallace was not to publish a detailed treatment on the subject for quite a while, eventually communicating a paper to the Linnean Society via Darwin (Wallace 1860a). It is safe to say that for the time being the discovery reinforced his conviction that geographical distribution is a window into earth history, jointly shaped by geological changes in the earth and the process of transmutation ever acting upon it.

Nine days after writing to Stevens, Wallace boarded a ship for Macassar, located at the southern end of Celebes in the center of the archipelago. The passage took only a few days, landing on 2 September 1856, but what turned

out to be disappointing collecting led Wallace to seize an opportunity to sail
to the Aru Islands a few months later, a distant locale he had long wanted to
visit in hopes of obtaining birds of paradise. Trading praus made the
thousand-mile journey to Aru once annually, late in the year, riding the
western or winter monsoon winds. The trading lasted six months, returning
the traders on the eastern or summer monsoon in July–August. En route he
visited the Ké (Kai) Islands for nearly a week, arriving in Aru on 8 January
1857. Wallace set up house, so to speak, in the trading village of Dobbo for
his first few months, then moved to Wokan followed by the inland settle-
ment of Wanumbai. It was there, during another period of forced inactivity
(this time from ulceration and infection of insect bites on his legs) that Wallace paused to write a paper on one of his prized finds: the Greater bird of
paradise, *Paradisea apoda*. Observations of this magnificent bird are found
in the Species Notebook (pp. 71–72), and entries on other birds follow.

Geographical distribution looms large in the paper published the following December in the *Annals and Magazine of Natural History*. Wallace
had learned from the traders that the Greater bird of paradise is limited
to the southern peninsula of New Guinea and the Aru Islands, while a
smaller, related species *(P. papuana)* occurs in the northern peninsula of
New Guinea. "It is interesting to observe," Wallace wrote in his *Annals*
paper, "that though the Ké Islands and Goram approach nearer to New
Guinea than Aru, no species of the Paradise birds are found upon them,—
pretty clearly showing that these birds have not migrated to the islands
beyond New Guinea in which they are now found. I have, in fact, strong
reasons for believing, from geographical, geological, and zoological evidence,
that Aru is but an outlying portion of New Guinea, from which it has been
separated at no very distant epoch" (Wallace 1857b, 416).

Wallace perceived that just as Bali was once a province of Asia, so too
was Aru a sometime province of Australia–New Guinea, an assertion he
reinforced with evidence laid out in a later paper for the Royal Geographical Society. "On the Arru Islands" was communicated on 22 February
1858. In this paper Wallace made much of the peculiar winding saltwater
channels that dissect the islands. Wallace correctly deduced that these
channels, based on their winding nature and uniformity of width and
depth, not only resemble river channels, they *are* river channels. Now
flooded by the sea so that salt water flows through them, these are essentially fossil rivers, a clue, Wallace realized, to the former continuity
between Aru and mainland New Guinea: "The physical features here de-

scribed are of the greatest interest, and probably altogether unique, for I have been unable to call to mind any other islands in the world which are completely divided by salt-water channels, having the dimensions and every other character of true rivers. What is the real nature of these, and how they have originated, are questions which have occupied much of my attention, and which I have at length succeeded in answering, to my own satisfaction at least" (Wallace 1858a, 165). He adduced evidence from species distribution as well, drawing a parallel with Britain and the continent in "On the Zoological Geography of the Malay Archipelago" (Wallace 1860a). A relatively close-lying island such as Britain separated by shallow seas has fewer bird species than the continent, but all of them are also found on the continent. So, too, he argued, did Aru have fewer bird species than New Guinea, but few distinct species; nearly all in Aru were also to be found in New Guinea.

On 2 July 1857 Wallace departed Dobbo with a flotilla of native praus, riding the eastern monsoon at such great speed that they landed at Macassar, a thousand miles to the west, only nine days later. There he found some seven months' worth of mail awaiting him, including a letter from his old friend Bates congratulating him on the Sarawak Law paper, published nearly two years previously, and one from Darwin dated 1 May 1857. It was in Macassar that his geological and zoological musings on Aru would take full form in a series of papers later fired off one after another to the *Annals*, the *Zoologist*, and the Royal Geographical Society, but not before sorting things out: "Arriving safe at Macassar, and taking up my old quarters, I had a most fatiguing task,—to open out, clean and pack my collections (more than seven thousand specimens), which occupied my whole time for three weeks," as he described later (Wallace 1858b, 5891). In his earlier *Annals* paper he had commented that "a host of new species burst upon me, revealing the richness of the country, and its intimate connexion with New Guinea" (Wallace 1857c, 476), reiterating this in the one for the *Zoologist,* in which he noted, "The Entomology, the Ornithology, and certain peculiarities in the physical geography of these islands, prove to me that at no distant period (geologically) they formed a portion of the southern peninsula of [New Guinea]" (Wallace 1858b, 5889). Wallace was riding high: the letter from Bates heaped praise on his Sarawak Law paper, and there was the letter from Darwin informing him that from Wallace's previous letter and Annals paper, he could "plainly see that we have thought much alike," and agreed "to the truth of almost every word" of the paper (Darwin

Correspondence Project, letter 2086). This news, combined with his growing knowledge of the profound geological changes that had taken place in
the Malay Archipelago, shaping the ebb and flow—and origin and extinction—of species, must have been terribly exciting.

In his *Annals* paper he took aim at his fellow naturalists, especially Lyell:

> Let us now examine if the theories of modern naturalists will explain the
> phaenomena of the Aru and New Guinea fauna. We know (with a degree of
> knowledge approaching to certainty) that at a comparatively recent geological
> period, not one single species of the present organic world was in existence;
> while all the *Vertebrata* now existing have had their origin still more recently.
> How do we account for the places where they came into existence? Why are
> not the same species found in the same climates all over the world? The gen-
> eral explanation given is, that as the ancient species became extinct, new ones
> were created in each country or district, adapted to the physical conditions of
> that district. Sir C. Lyell, who has written more fully, and with more ability,
> on this subject than most naturalists, adopts this view. (Wallace 1857c, 481)

Wallace then pointed out that the prevailing theory implies that similar
organisms should be found in similar habitats or environments: we should
"find a general similarity in the productions of countries which resemble
each other in climate and general aspect, while there shall be a complete
dissimilarity between those which are totally opposed in these respects."
He went on to demolish the argument by pointing out that New Guinea
and Borneo are virtually identical—large, topographically varied, heavily
forested tropical islands straddling the equator and about a thousand miles
apart—and yet their flora and fauna are radically different. New Guinea
and Australia, on the other hand, differ in size, topography, climate, and
other characteristics and yet exhibit striking similarities in flora and fauna.
"We can hardly help concluding, therefore," Wallace maintained, "that
some other law has regulated the distribution of existing species than the
physical conditions of the countries in which they are found, or we should
not see countries the most opposite in character with similar productions,
while others almost exactly alike as respects climate and general aspect,
yet differ totally in their forms of organic life" (Wallace 1857c, 481). He
then drew the reader's attention to his Sarawak Law and how this explains
the distribution of Aru and New Guinean species: "in a former Number
of this periodical we endeavoured to show that the simple law, of every
new creation being closely allied to some species already existing in the

same country, would explain all these anomalies, if taken in conjunction with the changes of surface and the gradual extinction and introduction of species, which are facts proved by geology" (p. 482); indeed, "On the Natural History of the Aru Islands" is an extension of the Sarawak Law paper.

Wallace stayed in Macassar about four months, a tremendously creative period during which he kept up his paper salvos bearing evolutionary arguments. Two more are worth notice here: the brief "Note on the Theory of Permanent and Geographical Varieties" (ultimately published in the *Zoologist* the following January) and the slightly longer "On the Habits and Transformations of a Species of *Ornithoptera*," read 7 December 1857 and published the following April (Wallace 1858c,d). Although the birdwing butterfly Wallace found in Aru is explicitly named only in the latter paper, it may be the inspiration for the former as well. Recall Wallace's previous birdwing discovery in Borneo, which he named for James Brooke. In the eastern part of the archipelago there were two described species: *O. priamus* from Amboyna (modern Ambon), described in 1758 by Linnaeus, and *O. poseidon* from New Guinea, described just in 1845 by Doubleday. Wallace's new birdwing was precisely intermediate in coloration between *O. priamus* and *O. poseidon*. In the "Note," Wallace put his finger on the dilemma of distinguishing species and varieties, and more specifically the untenable view held by many naturalists that species are divinely created whereas varieties arise through secondary laws. "If an amount of permanent difference, represented by any number up to 10, may be produced by the ordinary course of nature, it is surely most illogical to suppose, and very hard to believe, that an amount of difference represented by 11 required a special act of creation to call it into existence" (Wallace 1858b, 5888). Wallace led the reader through an informative exercise:

> Let A and B be two species having the smallest amount of difference a species can have. These you say are certainly distinct; where a smaller amount of difference exists we will call it a variety. You afterwards discover a group of individuals C, which differ from A less than B does, but in an opposite direction; the amount of difference between A and C is only half that between A and B: you therefore say C is a variety of A. Again you discover another group D, exactly intermediate between A and B. If you keep to your rule you are now forced to make B a variety, or if you are positive B is a species, then C and D must also become species, as well as all other per-

manent varieties which differ as much as these do: yet you say some of these groups are special creations, others not. Strange that such widely different origins should produce such identical results. (Wallace 1858b, 5888)

We might read *O. priamus* for "A" and *O. poseidon* for "B"—two species with a very small difference between them. The Aru species is "exactly intermediate," "D," as Brooks (1984, 91) pointed out. The precise definition of species and varieties, which speaks to their origin, is at the heart of these remarks. Wallace was stirring the pot, and stirred more vigorously with his birdwing paper (1858d) written about the same time: while the adults of the Aru *Ornithoptera* Wallace found are intermediate in coloration between two other species, the larvae and pupae appear to be very similar with those of butterflies of another genus altogether, the swallowtail genus *Papilio*. (Observations of this caterpillar are found in the verso section of the Species Notebook, p. 35, followed by observations of *Papilio* caterpillars on verso pp. 36–37.) "It would thus appear that there are no characters in the larva or pupa to separate *Ornithoptera* from *Papilio;* but the large size of the perfect insects, their long and powerful legs, the large anal valve of the males, their uniform and characteristic form, their striking colours and their limited geographical range, are, I think, sufficient reasons why the genus should be kept distinct" (Wallace 1858d, 273). It is difficult enough drawing the line between species and varieties, but what to do when even different genera have profound similarities? Clearly, the difference between variety, species, and genus is one of degree.

It is important to mention one other striking observation Wallace made on his Aru journey, though it did not immediately result in a paper. Wallace had long been fascinated by ethnography, predating his time in Amazonia (see Wallace's discussion of human racial variation in his 28 December 1845 letter to Bates; WCP346; McKinney 1972, 172). He became very familiar with the Malay groups found in the western archipelago, a familiarity that put the denizens of the eastern archipelago into striking contrast. He recorded physical, behavioral, and "moral" features of the peoples he encountered throughout his travels, but his interest was all the keener once he had traversed the entire region. The Species Notebook records many such ethnological observations, and his 1869 memoir *The Malay Archipelago* includes a chapter on "The races of man in the Malay Archipelago" and an extensive (twenty-five-page) comparative lexicon for a diversity of native groups.

Ethnologically speaking, in the eastern archipelago Wallace found a human racial divide as striking to him as the faunal divide of Bali and Lombock. In the Ké Islands, where Wallace spent a week before moving on to Aru, he encountered for the first time people of Papuan ethnicity. It was from Aru that Wallace commented in his journal (LINSOC MS 178b, entry 71) that "the human inhabitants of these forests are not less interesting to me than the feathered tribes." With shock reminiscent of Darwin's upon first encountering the Fuegians in their native environment, Wallace seemed repulsed by the appearance and means of existence of the Aru natives, declaring them "on the whole a miserable set of savages. They live much as all people in the lowest state of human existence & it seems to me now a more miserable life than ever I have thought it before." Humans were a unitary species, Wallace believed, and clearly consisted of several well-marked "varieties" (as well as several intermediate and intermixed forms). What was their origin? What could account for such radical differences? His view was influenced by Sir William Lawrence's *Lectures on Physiology, Zoology and the Natural History of Man* (1819; abbreviated to *Lectures on Man*) and James Cowles Prichard's *Researches into the Physical History of Mankind*, first published in 1813. These are works he discussed and recommended to Bates in 1845. Remarkably for their time, these works contained transmutational speculations about the origin of human races, for example, positing that races stemmed from an accumulation of small variations (Brooks 1984, 12). Prichard even identified Africa as the locus of human origin: "On the whole there are many reasons which lead us to the conclusion that the primitive stock of men were probably Negroes, and I know of no argument to be set on the other side" (Prichard 1851, 5:238–239). Lawrence was forced to retract his 1819 book for its provocative speculations about human origins, but it was reprinted in 1822 and became widely available.

Wallace's friend the ethnologist Robert Gordon Latham (1812–1888), who contributed to Wallace's summary of Amazonian tribal languages in the *Narrative of Travels on the Amazon and Rio Negro* (Wallace 1853), was also a disciple of Lawrence and Prichard. Latham expanded on their views in his *Natural History of the Varieties of Man* (1850), a work Wallace was well familiar with. Races were, to these men, "permanent varieties" of a single species, and Latham added further nuance to this in his discussion of transitional forms between races. Aru and the Moluccas were thought to be a region with just such transitional forms, and Latham pointed to Gilolo (modern Halmahera) and environs as holding special promise: "The probable source . . . of the Papuan population must be sought for in the parts about Gilolo" (Latham 1850, 212). There is no evidence that Wallace's keen interest in these islands is attributable to Latham per se, but Wallace was certainly aware of Latham's ideas about their significance (see also Brooks 1984, 165–168, on Wallace's reading of Latham). "The Malay & the Papuan appear to be as widely separated as any two human races can be. . . . It is a most interesting question & one to which I shall direct my attention in all the islands of the Archipelago I may be enabled to visit," Wallace recorded (first *Malay Journal,* LINSOC MS 178a, entry 63). He was to be disappointed: at Aru and Ké the Papuan and Malay races were clearly distinct, with no evidence of "transitional" forms. He looked ahead to Ternate and Gilolo but there too could not find convincing evidence of transitions.

His first detailed commentary on the subject of racial origins was not to appear for nearly seven years, in his paper "On the Varieties of Man in the Malay Archipelago" in the *Transactions of the Ethnological Society of London* (Wallace 1865a). His object, he said, was "to give some account, from personal observation, of the inhabitants of the chief islands of the Malay Archipelago, and by a comparison with the published descriptions of the inhabitants of the surrounding countries, to endeavour to arrive at some definite conclusion as to their mutual relations or their common origin." He never did solve that problem, of course; the Malays were assumed to have an Asian origin, and the origin of the Papuans was merely pushed back by linking them with a supposed former Polynesian race that dwelt on a now-sunken Pacific continent (later construed as the mythical land of Lemuria or Mu, though not by Wallace). "My solution of the difficulty [of the origin of this ancient Polynesian race] depends chiefly upon the evidence for the existence at a comparatively recent period (geologically speaking) of a Pacific continent" (Wallace 1865a, 212).

McKinney stressed that the question of human racial origins was increasingly on Wallace's mind in late 1857, and points out that the very last entry in his journal before his key insight into natural selection (probably made within weeks of his "eureka moment") contains ethnological observations. As we shall see, Wallace's musings about human varieties in particular may have helped catalyze his insight. McKinney perhaps takes this a bit far, however, in arguing that Wallace's interest in human origins generally and racial groups in particular was a prime motivating factor in his interest in the species question to begin with. He cited as evidence Wallace's many ethnographic entries in the Species Notebook (McKinney 1972, 173–175). McKinney's appendix of "Excerpts on Ethnology from Wallace's Species Notebook" amounts to little more than two pages of brief entries, however, from a notebook that is nearly 250 pages long recto and verso. Wallace was certainly struck by the diversity of racial "types" he encountered as he traversed the Malay Archipelago, pondered the respective origin of these groups, certainly appreciated the close kinship of humans and other primates, and recorded a great many cultural and physical features of the peoples he encountered. The puzzle of the origins of human racial diversity was important, to be sure, but for this to be seen as the primum mobile for his travels in both South America and Southeast Asia, a far more sustained program of analysis is to be expected. His writings in this regard are paltry in comparison with the number that bear on the species question more generally from a biogeographical perspective. While he may have declared that the human inhabitants were not less interesting to him than the "feathered tribes," he certainly wrote far and away more on lessons gleaned from the feathered and six-legged tribes in his journals, Species Notebook, and papers.

The Centrifugal Governor of Species

Catching up on his correspondence while in Macassar, in September of 1857 Wallace wrote to Darwin. No doubt in an elated mood because he had received kudos on his 1855 Sarawak Law paper from Darwin and Bates, and with the myriad provocative observations from his Aru travels, Wallace evidently felt encouraged by Darwin's letter and wrote him a lengthy reply. Only a fragment of this letter is extant, but Darwin's subsequent reply in turn indicates that Wallace discussed his insights into the historical connections of landmasses in the eastern archipelago and shared to some extent his views on species and varieties. He also evidently asked Darwin if he planned on treating humans in his forthcoming book. Wallace was fresh from the Aru Islands, and human racial diversity and origins were on his mind.

In November 1857 Wallace was off again, this time to Ambon, in the Moluccas. He stayed the month of December before heading to Ternate, but not before writing Bates a revealing letter, replying to Bates's letter of 1 May 1857. It is here that Wallace mentions his intention to write a book on the species question. Bates, recall, had congratulated him on his masterful Sarawak Law paper. Wallace now wrote, "To persons who have not thought much on the subject, I fear my paper on the succession of species will not appear so clear as it does to you. . . . That paper is, of course, only the announcement of the theory, not its development. I have prepared the plan & written portions of an extensive work embracing the subject in all its bearings & endeavouring to provide what in the paper I have only indicated." He also mentioned to Bates that he had been "much gratified" by a letter from Darwin, "in which [Darwin] says that he agrees with 'almost every word' of my paper. He is now preparing for publication his great work on Species & Varieties, for which he has been collecting information 20 years. He may save me the trouble of writing the 2nd part of my hypothesis, by proving that there is no difference in nature between the origin of species & varieties, or he may give me trouble by arriving at another conclusion, but at all events his facts will be given for me to work upon. Your collections and my own will furnish most valuable material to illustrate & prove the universal applicability of the hypothesis" (letter dated 4 January 1858; WCP366). *On the Organic Law of Change* (see Costa 2013a) is the work that Wallace was planning, but as yet he was still lacking a mechanism for deriving species from varieties. He continued on to Ternate, arriving on his thirty-fifth birthday, just four days after writing the letter to Bates.

The little island had outsized significance in the commerce of the day as the center of the spice trade, and Wallace soon secured the kind assistance of the local magnate named Duivenboden, a Dutch trader who befriended Wallace and helped him secure a "fixer-upper" of a house. "A few repairs were soon made, some bamboo furniture & other necessaries obtained, and after a visit to the Resident & Police Magistrate, I found myself an inhabitant of the earthquake tortured island of Ternate & able to commence operations & prepare the plan of my campaign for the ensuing year," he recorded (second *Malay Journal*, LINSOC MS 178b, entry 123). He was impressed with the furniture, commenting in the Species Notebook, "Bamboo—In Ternate excellent chairs, very strong are made at 1/4 guilder each. Arm chairs 1/2 guilder & bamboo sofas 1g[uilder]" (p. 64; emphasis Wallace's). His "ensuing year" was to become two and a half years with Ternate as the base of operations for explorations of the Moluccas and western New Guinea. But it was near the outset of his stay on this "earthquake tortured island" that he at last chanced upon the mechanism by which varieties become species. The resulting paper was a seismic event of its own.

Wallace planned on taking a long journey to New Guinea seeking birds of paradise, but had to await passage on one of Duivenboden's schooners, due to depart in a month's time or so. Killing time, in late January Wallace headed off to the neighboring and much larger island of Gilolo (Halmahera), planning on collecting for about a month in what was entomological terra incognita. The collecting went well enough—in his first walk he noted in his journal that he "obtained a few insects quite new" and "was very well pleased with my prospects of making a fine collection"—although we also learn in other sources (e.g., his autobiography) that he was ill much of his time on Gilolo. His interests in ethnology still piqued from Aru, in his journal Wallace also commented on examining the natives with "much interest, as they would help to determine whether, independent of mixed races, there is any transition from the Malay to the papuan type" (second *Malay Journal*, entry 127). His journal next simply records his return to Ternate on 1 March. Frustratingly, there is no indication in any of his notebooks or journals of a momentous discovery or illness. It is clear, however, that while on Gilolo Wallace gained the insight he had sought so long: he cracked the species question in a flash of insight while laid low by fever.

Equally nonchalantly on his Ternate return Wallace recorded his disappointment in not receiving an expected shipment from England when the mail steamer appeared on March 9, and he had to scramble to pull together the supplies and other items he would need for the long New Guinean trip planned for later that month. What he did receive in the mail, however, is well characterized as momentous: a letter from Darwin, replying to his letter of September 1857 from Macassar. Darwin's letter, dated 22 December 1857, held news that must have been exciting for Wallace given his recent cracking of the species mystery: this was the letter in which Darwin consolingly informed Wallace that "two very good men," Lyell and Blyth, had called his attention to Wallace's Sarawak Law paper. Could the timing have been better for receiving such news, with his freshly written manuscript reporting a mechanism that tidily explained the Sarawak Law? Is it any surprise that he decided to post his essay to Darwin straightaway? Wallace's essay, "On the Tendency of Varieties to Depart Indefinitely from the Original Type," was dated "Ternate, February, 1858." He was in fact on Gilolo, as first shown by McKinney (1972); the erroneous locality has been variously attributed to oversight, practicality (the trade center Ternate was his base and the main mail hub in that part of the archipelago), and deception (wishing to associate his grand theory with famous Ternate and not anonymous Gilolo for posterity). A simple explanation that seems to have been overlooked is that Wallace temporally compressed the chain of events from the fever fit in which the idea was conceived, his initial recording of the idea, and making out a fair copy for mailing. In some accounts of the event Wallace separates writing out the paper from subsequently copying it, such as in this letter to the German naturalist A. B. Meyer: "As soon as my ague fit was over I sat down, wrote out the article, copied it, and sent it off by the next post to Mr. Darwin" (Meyer 1895). Wallace returned to Ternate on 1 March, and the manuscript was posted later that month or in early April. Is it not possible that Wallace wrote out the paper on Gilolo but subsequently produced a fair copy for mailing once he was back on Ternate? Then as a matter of course he would have signed it "Ternate," as both his base, the locale from which it was to be mailed, and literally the site of producing the fair copy (which was very likely to have had some changes from the original, if only minor editorial ones), but leaving the original month of February as the timing of his great insight.

The structure and contents of the Ternate essay are discussed in detail in Chapter 4 (the precise timing of the mailing of the essay has been the subject of much speculation, incidentally, and is discussed in Chapter 5). Here its key elements are summarized to show how it fits, chronologically and conceptually, into Wallace's continuing investigations. Running to just over 4,000 words, Wallace's essay is a lucid statement of the causal mechanism behind his Sarawak Law. He does not give the mechanism a name, but later, in 1868, declared his preference for "survival of the fittest"—far superior, he argued with Darwin, than "natural selection" with its inevitable personification of the process by implying a "selector" (Beddall 1988b; Paul 1988). But Wallace did not jump right into a triumphant announcement of this mechanism. He built his case in an interesting manner. Just as the Sarawak Law opened with Lyell, implicitly, so too does the Ternate essay—and also implicitly. It opens with domestication, but only to raise and then toss aside the standard criticism of the time that domestic varieties are "unstable," reverting, once feral, to "type" and thereby demonstrating that species can only vary within certain limits and no further. This was powerfully argued by Lyell in the *Principles*, but here Wallace counters that this criticism rests on the assumption that domestic varieties are just like naturally occurring varieties, an assumption he rejects. The tendency for domestic varieties to "revert" teaches us nothing about varieties in a state of nature, because they are altogether contrived, artificial, unnatural. This nicely sets up the key thesis, as indicated in the essay's title: domestic varieties may "depart" (diverge, differ) only so far from the parental type, but natural varieties can and do depart from the parental type *indefinitely*.

To explain how, Wallace brings in a populational, ecological, and geological vision: in areas that remain physically unchanged over time, populations cannot change or, if something results in the population of some species increasing, others will respond by decreasing. This is a zero-sum model of population sustainability understood in terms of available resources. Given the powerful inherent tendency for populations to increase, maintenance of the population at what in modern terms we call carrying capacity means that annual mortality must be immense. Which population members survive? Wallace points out that varieties (here meaning individual variants), which are constantly occurring, will always yield some that are better or worse able to survive than others—it is the strongest, healthiest, best suited to procure the sustenance necessary in that environment that flourish. The weaker, by the same token, succumb.

In an unchanging environment populations would be maintained in a steady state, with nothing appearing to change. In reality there is immense mortality, but to outward appearances the population remains more or less constant. He then turns to the effect of altered physical conditions: this sets the stage for selection of a superior variety which, given enough time, will eventually completely supersede the parental form because it is better suited to the new conditions. Should the environment change further still, a still different variety will be favored, and so on. Under each set of new conditions, eventually "the superior variety alone remains," Wallace wrote. Moreover, he made the important point that under such conditions the newly favored varieties could not return to the original form, precisely because conditions suited to that form no longer exist. The progenitors of each such new variety are completely replaced, or driven to extinction, but note that each new variety arises at a locus immediate to its parental variety: the Sarawak Law, *"Every species has come into existence coincident both in space and time with a pre-existing closely allied species"* was now neatly explained.

Where did Wallace's insight come from? He later cited the influential 1798 *Essay on Population* by Thomas Robert Malthus (1766–1834), which he had read some dozen years before, as a catalyst. Malthus largely addressed human struggles—the internecine strife of tribe versus tribe and nation versus nation: population pressure and the conflict this can engender. Not long before Wallace's insight he had been in Aru and environs, where his ethnological interests led him to study the local Malay and Papuan peoples as discussed earlier in this chapter. Several scholars have highlighted Wallace's ruminations on human struggle shortly before seeing how this applied to natural populations (McKinney 1972, 82–83; Brooks 1984, 184–186; Moore 1997, 303–306)—in fact, as McKinney pointed out, in all four later-published accounts of his momentous discovery Wallace related how he had been thinking about population checks in relation to human populations before transferring the struggle concept to animal populations. Malthus's ideas were likely reinforced in Wallace's mind by his reading of Lyell's *Principles of Geology* and Darwin's *Voyage of the Beagle,* which provided lucid explications of Malthusian population struggle (Lyell 1835, 3:108–109; Darwin 1845, 174–176).

Having described the mechanism we now know as natural selection, Wallace then returned in his essay to domestication as if anticipating the degree to which his readers would be hung up on the idea that reversion of domestic varieties constitutes evidence refuting his claims. He argued once again that there is no parallel between domestic and natural varieties, but in so doing he seems to throw the baby out with the bathwater in asserting that "we see, then, that no inferences as to varieties in a state of nature can be deduced from the observation of those occurring among domestic animals." "No inferences" is going rather far, belied by Darwin's inference that the process of human-mediated creation of these varieties is analogous to the creation of new varieties and species in nature. The two would continue to disagree about the lessons that may or may not be gleaned from domestication; in a later letter Darwin explained that when his book comes out, "then you will see what I mean about the part which I believe selection played with domestic productions" (Darwin Correspondence Project, letter 2405).

Wallace next commented that the transmutation principle posited by Lamarck (and incidentally Mr. Vestiges) was now explained as a naturalistic process, obviating Lamarck's ill-defined inherent "tendency" to progressive change. The dynamic, Wallace pointed out, "is exactly like that of the centrifugal governor of the steam engine, which checks and corrects any irregularities almost before they become evident." The centrifugal governor, brilliant in its simplicity, regulates engine speed by linking the throttle valve to a spinning rod fitted with balls that pivot freely in accordance with centrifugal force as the rod rotates. When the engine speed exceeds a preset level, the whirling balls are centrifugally upraised and tip down the throttle, slowing the engine. As speed drops below the desired level, the balls trace a smaller circle closer to the central rod, their lower position tipping the throttle open and increasing speed.

Wallace envisioned this mechanism similarly maintaining varieties; in both the analogy and the natural process he envisioned, a simple checks-and-balances system maintains the status quo (referred to today as stabilizing selection). It is this very status quo of form, despite the regular occurrence of variation, that provides evidence for the action of selection, Wallace later pointed out: "The proof that there is a selective agency at work is, I think, to be found in the general stability of species during the period of human observation, notwithstanding the large amount of variability that has been proved to exist. If there were no selection constantly going on, why should it happen that the kind of variations that occur so frequently under domestication never maintain themselves in a state of nature? . . . There seems no reason for this but that [many variations] are quickly eliminated through the struggle for existence—that is, by natural selection" (Wallace 1891a, 518).

Changing physical conditions (presumably over geological time) alter the "rules" for which variants are to be favored, just as an engineer can change the rules for the centrifugal governor and increase or decrease speed. It should be noted, however, that Wallace's centrifugal governor analogy was presented in the context of correcting any "unbalanced deficiency" in a variety that chances to appear. This explains, Wallace says, "that balance so often observed in nature,—a deficiency in one set of organs always being compensated by an increased development in others—powerful wings accompanying weak feet, or great velocity making up for the absence of defensive weapons." His point is well taken, but also a truism that amounts to a statement on the adaptedness of organisms. More instructive (and insightful) would have been recognition that this "balance" is achieved in different ways in different species or lineages, reflecting the unpredictable direction of change or available variation. Thus "lack of defensive weapons" might be compensated by great velocity or other adaptive solutions, such as crypsis or mimicry. Also, note that Wallace's emphasis is on elimination: a winnowing process that corrects any unbalanced deficiencies. Nonetheless, Wallace clearly articulates a dynamic we can recognize as natural selection.

One final aspect of the Ternate essay that is of interest here is Wallace's sense of the treelike pattern of relationships that his process yields over time, a pattern likely inspired by his reading of *Vestiges* (and perhaps Lindley). This is better expressed in the Sarawak Law and birds papers, but in the Ternate essay he uses the word *diverge* or *divergence* three times, in all cases referring to a variety becoming increasingly dissimilar from its parental or ancestral form: "But this new, improved, and populous race might itself, in course of time, give rise to new varieties, exhibiting several diverging modifications of form, any of which, tending to increase the facilities for preserving existence, must, by the same general law, in their turn become predominant. Here, then, we have *progression and continued divergence*" and "An origin such as is here advocated will also agree with the peculiar character of the modifications of form and structure which obtain in organized beings—the many lines of divergence from a central type." The latter quote in particular evokes an image of an unrooted tree much like those given in "Attempts at a Natural Arrangement of Birds" (Wallace 1856e), one of which shows a branch with a branch (Figure 1.3). His description in the Ternate essay might also be read, however, as a means of representing diverse forms descended from a common ancestor (the "central type"), and therefore sharing a fundamental similarity by virtue of this.

It should be pointed out that Wallace's idea of divergence is not identical with Darwin's, which has built into it a mechanism that would tend to force lineages apart and create an arborescent pattern of relationship: Darwin's "ecological division of labor" is presented as a principle of divergence, responsible for generating diversity rather than simply having species transmutate into new species (Darwin 1859, 111–126; reviewed by Ospovat 1981, 170–209). This is discussed in detail in Chapter 4; here suffice it to say that Wallace's divergence includes some lineage splitting, but this seems to be an occasional occurrence, with most of the action lying in within-lineage (what we would call anagenetic) change rather than constantly splitting lineages (cladogenetic change in modern terms). This may be why the trees drawn by Wallace in the birds paper (the only place he drew trees) bear few branches, though they correspond to specific taxonomic groups and are not abstract depictions of a general evolutionary process.

Winding down his Ternate essay, Wallace declared, "We believe we have now shown that there is a tendency in nature to the continued progression of certain classes of varieties further and further from the original type." He concludes by pointing out the explanatory power of his mechanism of continued divergence from the parental type, expressed in lucid style:

This progression, by minute steps, in various directions, but always checked and balanced by the necessary conditions, subject to which alone existence can be preserved, may, it is believed, be followed out so as to agree with all the phenomena presented by organized beings, their extinction and succession in past ages, and all the extraordinary modifications of form, instinct, and habits which they exhibit. (Wallace 1858b, 62)

This is a grand vision, explaining "all the phenomena of organized beings"—their history in past ages, and their present adaptations. Wallace was triumphant, having solved the mystery of the origin of species little more than a decade after he first announced to Bates his plan to do so. Recall his words from 1847: "I [should] like to take some one family to study thoroughly," he wrote, "principally with a view to the theory of the origin of species" (WCP348).

But the reception of his essay was some months away, and there was work to be done. On 25 March Wallace was off again, this time to New Guinea via Gilolo aboard one of Mr. Duivenboden's praus, in pursuit of birds of paradise. He arrived in Dorey, in western New Guinea, in short order, but it was to be a disappointing if not disastrous trip. Birds of paradise were a hot commodity, and the locals preferred to sell specimens (and just about all available food) to the Prince of Tidore, who had appeared in early May and commenced to snap them up. Worse, Wallace was incapacitated by a severely infected ankle which he had injured while collecting, on top of which he and his group were beset with dysentery and other maladies, sadly resulting in the death of one of the young native collectors who accompanied him from Ternate. Bitterly disappointed, Wallace was back at Ternate by mid-August with little to show for his three months in New Guinea.

He next made a second short-lived visit to Gilolo, cut short because he was stymied by "interminable tracts of reedy grass ten feet high"—the existence of which he conjectured about in the Species Notebook earlier that year, in an entry on "Plains in the Tropics" dated 20 January 1858 (Species Notebook, pp. 108–109). The collecting may have been disappointing, but his visit to the village of Sahoe while he awaited passage back to Ternate proved to be very interesting, for here he believed he discovered "the exact boundary between the Malay and Papuan races, and at a spot where no other writer had expected it" (Wallace 1869, 323). He entered observations of these indigenes of northern Gilolo (Alfures or Alfuros) in the Species

Notebook (p. 134) and his third *Malay Journal* (LINSOC MS 178c, entry 154), where he concluded:

> I hear much of the excessively light colour & beauty of the indiges of Northern Celebes, & think an examination of them will throw much light on the origin of the Gilolo population; & this idea is strengthened by the singular fact that it is only those parts opposite Menado [northern Gilolo is separated from Menado, in northern Celebes, by the Molucca Passage] which possess an indigenous population,—the southern peninsula of Gilolo with the islands of Batchian & Obi being either uninhabited or occupied only by settlers traders or fishermen from Ternate Tidore & northern Gilolo. This is a most interesting point for future enquiry & investigation.

Back on Ternate, Wallace was preparing for a longer trip to Batchian (modern Bacan) when news came at last regarding his Ternate essay. To his delight he received not one letter but two: a letter from Darwin with an enclosed letter from Hooker. His subsequent letter to his mother (dated 6 October 1858) says it all: "I have received letters from Mr. Darwin and Dr. Hooker, two of the most eminent naturalists in England, which has highly gratified me. I sent Mr. Darwin an essay on a subject on which he is now writing a great work. He showed it to Dr. Hooker and Sir C. Lyell, who thought so highly of it that they immediately read it before the Linnean Society. This assures me the acquaintance and assistance of these eminent men on my return home" (WCP369).

The Travels Continue

Wallace's "evolutionary travels" did not end with the welcome news of his Ternate essay's reception, his great insight into the species question. Over three years of travels lay before him—hopscotching among the Moluccas, Timor, and Celebes, and even returning briefly to western New Guinea. He produced many more papers while in the east, some brief letter extracts and others more detailed scientific analyses; for our purposes, two papers are noteworthy. The last dated entry in the Species Notebook appears to come from Bacan (Wallace's Batchian), just south of Gilolo: observations on black ants tending aphids (Species Notebook, pp. 137–140), but mention of species, locales, and so on makes it possible to at least approximately date many subsequent entries. Mid-May 1859 found Wallace on the island of Timor, awaiting passage to Menado on the northern peninsula of

Celebes (Sulawesi). Here, prompted by ornithologist Philip Sclater's essay "On the Geographical Distribution of Birds," Wallace presented a fuller analysis of the remarkable biogeographical patterns manifest across the Malay Archipelago. He sent his paper to Charles Darwin—after his Ternate essay he was indeed assured the "acquaintance and assistance" of such eminent men as Darwin, who, with his priority over the discovery of natural selection assured, felt quite magnanimous toward his younger colleague. Darwin duly had Wallace's latest paper read to the Linnean Society on November 3, 1859.

It was a characteristically masterful analysis that put Wallace's massive knowledge base of faunal distribution into the context of geological time, arguing for a geologically recent connection between the various islands in the eastern portion of the archipelago and a likewise connection between those in the western portion. The islands on each side of the abyssal divide were once sizable (and separate) land areas, each set of islands connected much more recently than an inferred very ancient connection *across* the divide. The clue to this history is provided by the flora and fauna of the region, Wallace noted: "Geology can detect but a portion of the changes the surface of the earth has undergone. It can reveal the past history and mutations of what is now dry land; but the ocean tells nothing of her bygone history. Zoology and Botany here come to the aid of their sister science, and by means of the humble weeds and despised insects inhabiting its now distant shores, can discover some of those past changes which the ocean it-self refuses to reveal" (1860a, 181–182). Wallace embraced the geological model of Lyell and Darwin that the earth's surface is constantly in a state of flux, with some regions subsiding and others elevating—the dominant model before plate tectonics. (Wallace initially took this to an extreme, as did bota-nist Joseph Hooker and others, in applying the idea of land bridges nearly everywhere, not to mention invoking entire lost continents; for this reason he initially rejected Darwin's steadfast belief that chance transoceanic dis-persal explained much about the biogeography of island systems.) Darwin must have been pleased that Wallace pointed to his own work on coral atolls in support of this conception even though he disapproved of applying it to the scale of land-bridges and continents: "By the invaluable indications which Mr. Darwin has deduced from the structure of coral reefs," Wallace wrote, "by the surveys of the ocean-bed now in progress, and by a more extensive and detailed knowledge of the geographical distribution of ani-mals and plants, the naturalist may soon hope to obtain some idea of the continents which have now disappeared beneath the ocean, and of the general distribution of land and sea at former geological epochs" (1860a, 182). Wallace conceived of a dynamic earth with dynamically changing species upon it, so also noteworthy in this paper is his ready application of the evolutionary insights he had recently gained:

> We really require no speculative hypothesis, no new theory, to explain these phenomena; they are the logical results of well-known laws of nature. The regular and unceasing extinction of species, and their replacement by allied forms, is now no hypothesis, but an established fact; and it necessarily produces such peculiar faunae and florae in all but recently formed or newly disrupted islands, subject of course to more or less modification according to the facilities for the transmission of fresh species from adjacent continents. (Wallace 1860a, 182)

Wallace was to formalize this analysis of the faunal discontinuity of the region in an equally masterful paper read to the Royal Geographical Society in June 1863 (a year after his return to England) and published in the society's *Journal* later that year (Wallace 1863a). It was there that he produced a map of the archipelago and illustrated the discontinuity with a red line that Thomas Henry Huxley would dub "Wallace's Line" five years later, in 1868.

In late May 1859 Wallace departed Timor for northern Celebes via Banda, a tiny island south of Ceram, Ambon, and Ternate. He arrived at Menado on 10 June 1859 and made this his base for the next three months. Collecting was fairly good, despite another bout of fever, and he succeeded in his main quest: collecting the maleo bird *(Macrocephalon maleo),* a megapodiid. Maleo birds lay their eggs in gigantic mounds of their own creation, in this case of black lava sand, which in the equatorial sun becomes a natural incubator. The maleo bird was largely the subject of the next paper on Wallace's evolutionary journey: "The Ornithology of Northern Celebes," penned from "Amboyna, Oct. 1859" and sent to Philip Sclater for publication in the *Ibis* (Wallace 1860b). This species was an unusual member of the megapodiid family in using sand for its nesting substrate. They descend, he reports, "by scores and hundreds" on certain beaches in August and September of each year. Wallace was fascinated by the "peculiar habits" of megapodiids as a group, "departing widely from those of all other birds"—note the significant use of the word "depart" in the context of the Ternate essay. Puzzling over the "peculiar habits" of maleo birds led him to mull over the nature of instinct and habit:

It has been generally the custom of writers on natural history to take the habits and instincts of animals as the fixed point, and to consider their structure and organization as specially adapted to be in accordance with them. But this seems quite an arbitrary assumption, and has the bad effect of stifling inquiry into those peculiarities which are generally classed as "instincts" and considered as incomprehensible, but which a little consideration of the *structure* of the species in question, and the peculiar physical *conditions* by which it is surrounded, would show to be the inevitable and logical result of such structure and conditions. (Wallace 1860b, 145–146; emphasis in original)

Far from considering odd and divergent "instincts" to be unexplainable, Wallace essentially argued that structure and function can be understood in the context of adaptation to physical conditions. Given the constraints of food and predators, "the *Megapodiidae must* behave as they do. They must quit their eggs to obtain their own subsistence,—they must bury them to preserve them from wild animals,—and each species does this in the manner which slighter modifications of structure render most convenient" (emphasis in original). What is more, such widely diverging habits are not isolated, but connected by steps to less divergent form: "I am decidedly of opinion," he continued, "that in very many instances we can trace such a necessary connexion, especially among birds, and often with more complete success than in the case which I have here attempted to explain." He next makes reference to the explanatory power of his and Darwin's recently announced mechanism of change and adaptation, but curiously he refers to it as *Darwin's* theory: "For a perfect solution of the problem we must, however, have recourse to Mr. Darwin's principle of 'natural selection,' and need not then despair of arriving at a complete and true 'theory of instinct.'" This appears to be the first instance of many to come where Wallace refers to the theory as Darwin's, reflecting his long-continued deference to Darwin's priority (e.g., Gardiner et al. 2008, 45–47). Darwin tended to refer to "my theory," and on occasion to it jointly belonging to both of them, as in his letter of 18 May 1860, mentioning one Patrick Matthew's claim to have discovered "*our* view of natural selection" (Darwin Correspondence Project, letter 2807; emphasis added).

"Ornithology of Northern Celebes" appears to be the last of Wallace's papers bearing evolutionary musings written from the archipelago. Extensive travels lay ahead, of course, including some triumphs (procuring with much difficulty several Red birds of paradise, *Paradisea rubra,* on the island of Waigiou off the far west coast of New Guinea) and even more setbacks: at various times his crew abandoned him, and others of his men were marooned and then saved a month later owing to Wallace's efforts in sending a rescue party. The warlike and cannibalistic New Guineans struck fear into the hearts of the Moluccans, and it was enormously difficult for Wallace to assemble a crew to travel to Waigiou with him. In November 1860 Wallace departed Waigiou for Ternate. Over the next year he collected on several islands in the eastern archipelago, including Timor, Banda, Bouru, and northern Celebes. He departed this region in late June 1861, heading to Jakarta, in Java, where he was to spend the next several months preparing, at last, to return home. He landed at the town of Bangka, Sumatra, on 3 November 1861, and then traveled on to Singapore in late January 1862. A month later, 20 February, Wallace departed Singapore for London with a first-class steamer ticket, courtesy of the Zoological Society of London, with live birds of paradise in his possession. He landed in England on 1 April; it had been an astonishingly eventful and productive eight years to the month since his arrival in the Malay Archipelago in 1854.

Wallace had received a copy of the Darwin's *On the Origin of Species* in early 1860 in Amboyna, while recuperating from his latest collecting fiasco on Ceram. He read and reread the book some five or six times, he wrote to his friend George Silk in September of that year. It did not take long to convince Wallace that the book was a watershed event in the history of ideas, as expressed by Wallace in a letter to Bates a few months later: "I know not how or to whom to express fully my admiration of Darwin's book. . . . Mr. Darwin has created a new science and a new philosophy, and I believe that never has such a complete illustration of a new branch of human knowledge been due to the labour and researches of a single man. Never have such vast masses of facts been combined into a system, and brought to bear upon the establishment of such a grand and new and simple philosophy!" (WCP374). In a subsequent letter to his brother-in-law Thomas Sims he referred to Darwin as the "Newton of Natural History" (WCP3351; emphasis in original), and declared to his friend George Silk that the *Origin* "will live as long as the 'Principia' of Newton" and that Darwin's name "should, in my opinion, stand above that of every philosopher of ancient or modern times. The force of admiration can no further go!!!" (WCP373).

Return of the Prodigious Son

And so I close this overview of the first great chapter in the development of Wallace's evolutionary ideas. And just the first chapter it is: in an important sense, this conclusion to Wallace's long sojourn in the Malay Archipelago—physical and intellectual—is more beginning than end. Wallace rightly regarded his epic journey as "the central controlling incident" of his life, and he was received with laurels aplenty on his return home. The riches of his collections, notebook, and journals would serve him well for years to come as he published a steady stream of papers and book after book beginning with *The Malay Archipelago* in 1869, his best-selling travel memoir that went to ten editions and has never been out of print. His leading scientific books include *Contributions to the Theory of Natural Selection* (1870), the two-volume *Geographical Distribution of Animals* (1876), *Tropical Nature, and Other Essays* (1878), *Island Life: Or, The Phenomena and Causes of Insular Faunas and Floras* (1880), *Darwinism: An Exposition of the Theory of Natural Selection with Some of Its Applications* (1889), and *Natural Selection and Tropical Nature: Essays on Descriptive and Theoretical Biology* (1891b). There were many other books and papers, on a diversity of scientific, nonscientific, and even pseudoscientific topics, some of which had Wallace swimming against prevailing social and scientific currents (diminishing his stature in the eyes of the scientific establishment). He was perhaps at his zenith in scientific standing in 1868–1869, the former being the year that Britain's most prestigious scientific prize, the Royal Medal, was bestowed upon him and the latter the year that saw the publication of his acclaimed memoir *The Malay Archipelago*. But 1869 was also the year that Wallace decided to go public with his personal heresies: perhaps no position embraced by him did more damage to his reputation in scientific circles than his very public endorsement of spiritualism coupled with declaration of the inadequacy of natural selection to account for the human mind (see, e.g., Kottler 1974; Schwartz 1984; Smith 1992; Fichman 2001; 2004, 150–157; Slotten 2004, 268–270; Nelson 2008; Gross 2010).

Wallace may have been "more Darwinian than Darwin" in that a strict adherence to gradualism seemed to lead him to this view: he argued that human cognition could not have evolved gradually since these abilities clearly lie latent, unused, even in so-called savages. He eventually saw divine agency in some sense guiding human mental evolution (and, concomitantly, physical attributes such as the organs of speech, manual dexterity, etc.) once natural selection had brought human physical evolution to a certain point (Fichman 2001). By and large his scientific colleagues had little time for spiritualism, and some felt his newly declared position on human evolution absurd. "I hope you have not murdered too completely your own and my child," a distressed Darwin wrote him in March of 1869 (Darwin Correspondence Project, letter 6684).

Certain controversial movements like mesmerism and phrenology were attractive to the iconoclastic Wallace even in his early days (Fichman 2004, 72–73, 157–165; Slotten 2004, 230–248), and his socialistic views and sense of social justice were certainly well developed before his Amazonian and Malay expeditions as well (e.g., Moore 1997; Jones 2002; Fichman 2004, 70–72). Mesmerism in particular may have fed naturally into Wallace's interest in spiritualism, which seemed to develop within two or three years of his return from Southeast Asia. He clearly had some spiritual (perhaps even deistic) sense before then. In later life he described himself as a skeptical materialist in his youth and early adulthood, for example, when he declared that "Up to the time when I first became acquainted with the facts of Spiritualism, I was a confirmed philosophical sceptic" and that he was so "thorough and confirmed a materialist" that he "could not at that time find a place in [his] mind for the conception of spiritual existence, or for any other agencies in the universe than matter and force" (Wallace 1875, vi–vii). Note, however, that some entries in the Species Notebook seem to have spiritual overtones, such as the passage by Ernest Renan copied from the *Revue des deux Mondes,* a periodical dedicated to literary and art criticism: "All religions & all philosophies alike teach us that man should have a higher aim than his physical enjoyment & interests" (Species Notebook, p. 154). That Wallace may have had an early but inchoate spirituality is consistent with the analysis of Smith (1992) and Fichman (2001), who argued that Wallace's view of humans as revealed in 1869 may not have represented an about-face at all: Wallace may have always harbored misgivings about extending natural selection to human intellectual and moral capacity, and seemed to become increasingly theistic in his view of humans over time (though certainly he was never religiously orthodox).

Wallace's later "heresies" have too often distracted from the fact that he was deeply admired by Darwin, Lyell, Huxley, and Hooker for his talent, perseverance, and the depth and breadth of his insights. Their views of the evolution of the human mind may have been unbridgeable, but in many other areas Wallace and Darwin maintained a long and fruitful correspondence exploring problems in geographical distribution, the action of natural selection, sexual selection, and other topics. Darwin often lauded his younger colleague for setting him straight: "I have long recognised how much clearer and deeper your insight into matters is than mine," he wrote in 1867 (Darwin Correspondence Project, letter 5528), and he took seriously Wallace's arguments where they disagreed: "I grieve to differ from you, and it actually terrifies me, and makes me constantly distrust myself" (Darwin Correspondence Project, letter 6386).

Wallace's growing embrace and often public defense of causes deemed dubious by most leading scientific lights of his day may have strained his relationship with some of these eminent men of science somewhat, but their ties were never severed. Darwin and the others even successfully petitioned the government to grant Wallace a Civil List Pension later in life (Darwin 1880; Colp 1992). Lyell, who so perceptively took notice of Wallace's talents at the first reading of his Sarawak Law paper in 1855, was perhaps the first among the giants of the scientific establishment to unequivocally laud Wallace both publicly and privately in the years to come. Indeed, in the tenth (1868) edition of the *Principles*, Lyell prominently cited Wallace's contributions to the development of evolutionary thought and quoted from the Sarawak Law paper, finding that "there are some points laid down more clearly than I can find in the work of Darwin himself, in regard to the bearing of the geological and zoological evidence of species" (Lyell to Wallace, 4 April 1867; Marchant 1916, 2: 21; WCP2086). By then Lyell had finally embraced the idea of transmutation, if not natural selection as its primary mechanism, and with an extensive citation in the latest *Principles* Wallace had come home, finding himself quoted and praised in the very treatise whose earlier arguments against transmutation had inspired his plan for a book to prove Lyell wrong.

On the Organic Law of Change was not to be, but Wallace (and Darwin) ultimately prevailed in convincing the old master, Lyell. Wallace had, remarkably, succeeded in his bold undertaking to solve the mystery of species origins, and in little more than a decade after setting out to do so. His subsequent contributions toward the understanding of geographical distribution, animal coloration, sexual selection, and the species concept further cemented his place in the pantheon of great naturalists and scientific travelers, but he should be best remembered, perhaps, for this most astonishing achievement: the perseverance and creativity to look deeply at nature, to see the underlying patterns where others only saw unintelligible chaos. In *A Narrative of Travels on the Amazon and Rio Negro,* Wallace described how "in all works of Natural History, we constantly find details of the marvellous adaptation of animals to their food, their habits, and the localities in which they are found." He must have been describing himself when in the next sentence he wrote of naturalists "now beginning to look beyond this, and to see that there must be some other principle regulating the infinitely varied forms of life" (1853, 83–84). He found that principle.

TWO

The Consilient Mr. Wallace

Transmutation and Related Themes of Wallace's Species Notebook

HOW DO WE KNOW when we are on the right track in elucidating a scientific principle or explaining some natural phenomenon or observation? Although scientific methods and conventions used today have their roots in the empirical sciences of astronomy and physics of the seventeenth century, the self-conscious study of science as a process—of *how* we study the natural world—had only begun to be formalized in the nineteenth century. In Britain the leading natural philosophers of that century charting the course for what we now call scientific methodology include the English astronomer Sir John Herschel (1791–1872), in his influential *A Preliminary Discourse on the Study of Natural Philosophy* (1830), and the polymath Welshman William Whewell (1794–1866), in a pair of important volumes: *History of the Inductive Sciences* (1837) and *The Philosophy of the Inductive Sciences, Founded Upon Their History* (1840). In the latter work Whewell introduced his concept of the "consilience of inductions"—the combining or merging of independent inductive strands. In Whewell's words, *"the Consilience of Inductions* takes place when an Induction, obtained from one class of facts, coincides with an Induction, obtained from another different class. This Consilience is a test of the truth of the Theory in which it occurs" (1840, 1:xxxix; emphasis in original). As more and more such inductive strands point to a common explanation, the greater the confidence that one is on the right track: "The cases in which inductions from classes of facts altogether different have thus *jumped together*, belong only to the best established theories which the history of science contains. . . . I will take the liberty of describing it by a particular phrase; and will term it the *Consilience of Inductions*"

(1840, 2:230). It is not clear if Wallace read Herschel, but he certainly read Whewell's *History* as he quoted a passage on the mutability of species from it in his 1854 field notebook (LINSOC MS 179). The essence of Whewellian consilience thinking is at the heart of his pro-transmutation arguments in the Species Notebook, and surely would have informed the structure of his planned book on the subject.

In this chapter we will see the disparate lines of evidence for transmutation pursued by Wallace in his Species Notebook (Costa 2013a), lines that reinforce one another in a philosophical sense, collectively forging a strong inductive argument. For convenience, I will group the relevant observations, discussions, and arguments from the notebook under the following thematic headings, though there is some overlap between them:

1. *Designedness, Balance of Nature, and Ethical Considerations* (entries on approximately thirty-two notebook pages) includes critiques of claims of "designedness" and the supposed balance or harmony of nature. It also includes entries that have an ethical or spiritual dimension.

2. *Geographical Distribution: Pattern and Process* (entries on approximately twenty pages) includes arguments bearing on island species and entries on comparative species richness of different regions.

3. *Morphology and Classification* (entries on approximately twenty-nine pages) includes entries relating to the significance of "affinities"—morphological homology and analogy, embryology, and classification.

4. *Instinct and Experience* (entries on approximately forty pages), explored in several contexts, includes birds' nests, construction of honeycomb by bees, and a critique of claims of certain instinctual behaviors in humans.

5. *Humans: Variation and Relationships* (entries on approximately fourteen pages) includes scattered entries describing the ethnic and "racial" features and behaviors of some of the peoples that Wallace encountered in his travels.

6. *Transmutation* (entries on approximately fifty-six pages) contains widely ranging entries that include evidence and arguments for the reality of transmutation, and is where Wallace's planned book on the "Organic Law of Change" is mentioned. Extracts from Lyell's *Principles* with Wallace's rebuttals form the centerpiece of this

category, with some twenty-four pages dedicated to Lyell alone. Topics include the limits (or not) of variability of species and varieties, a glimpse into Wallace's grasp of "tree thinking," and lessons from islands.

These themes were obviously interrelated for Wallace, and his observations, critiques, and musings for them are all of a piece. In exploring the "consilient" lines of argument of the Species Notebook, I will point out where Wallace's views intersect or, less often, diverge from those of Darwin, underscoring as I do throughout this book the similarities in their respective intellectual pathways to understanding and arguing for transmutation.

Designedness, Balance of Nature, and Ethical Considerations

Wallace's first entries relating to "designedness" are disparaging notes on so-called proofs of design from Charles Knight's *Cyclopedia of Natural History* (Species Notebook, p. 12). The issue for Wallace turns on the claim that animals are morphologically adapted in relation to their "necessities." In other words, the bat *needs* the bones of its forelimb to be elongated to support wings, and so it is. In the natural theology tradition prevalent in England at the time, all structure is adaptive to and reflective of the designed needs of the organism: necessity and structure/function go hand in hand. Wallace finds this nonsensical—animals could not have necessities before they existed, he says (though theologians might say they could, existing as ideas in the mind of the creator). Of course structure enables the organism to live as it does, but Wallace points out, "If the bat had not wings it would of course do without them & would have no more necessity for them than any other animal." In other words, it is possible for animals to adapt to different conditions; members of such a diverse group could "do without" such structures yet still be a member of the group taxonomically. There are no flightless bats, but plenty of examples of birds with reduced or lost wings (not Wallace's example, but consonant with his point). A similar criticism is later leveled against claims that different vertebrates were supposedly created with different numbers of cervical vertebrae, each with a particular number for its needs. "Here are several gratuitous statements & references. The writer seems to have been behind the scenes at the creation," Wallace says incredulously. How can the writer pretend to know the motives of the creator, he asks. How can anyone say that such-and-such species have a particular number of cervical vertebrae because their habits "require it"? Is it not possible that an omnipotent creator could fashion various species to their diverse niches making do with the same number of vertebrae? An analogy might be the pentadactyl limb, the bones of which are modified for so many ends (swimming, flying, running, burrowing, brachiating, etc.) with the same set of bones in evidence. And so, Wallace suggests, rather than simply declare that "their particular habits require it," consider instead the possibility that "some totally different causes absolutely hidden from us determined the form & structure of animals, and that their wants and habits resulted from that structure." The number of vertebrae varies for reasons that are unclear as yet, but rather than simply declare the number of vertebrae designed as such, look for the cause. In the very interesting passage that follows, Wallace says, "We are like children looking at a complicated machine of the reasons of whose construction they are ignorant, and like them we constantly impute as cause what is really effect in our vain attempts to explain what we will not confess that we cannot understand."

He later (p. 53) presents another example: "The Hornbills of Africa feed on reptiles, insects, such as grasshoppers lizards &c. & even small mammals whereas those of India eat only fruit. Yet both have exactly the same general structure & forms of bill . . . feet tail wings & stomach!" (See Figure 2.1 for an illustration of a female hornbill and chick Wallace collected in Sumatra.) To Wallace, this is clear proof that the structure of birds does not correlate absolutely with habitat, but rather "they are necessitated to adopt certain habits in order to obtain a subsistence in accordance with the peculiar circumstances by which they are surrounded." Trogons, too, vary in this manner, he notes, and inland versus coastal kingfishers. Wallace's idea parallels that of Darwin in citing the woodpeckers of the plains of La Plata (Darwin 1859, 184) and other species with "habits and structure not in accordance," and in a sense they are making the same point: structure or form is not simply correlated with mode of lifestyle, as natural theology would suggest, and variance in behavior despite having the same structure is proof of this. It should be borne in mind, however, that in other respects this variation means different things for Darwin and Wallace. When habits and structure are "not in accordance" Darwin sees this as evidence of transitional habits, while Wallace sees it as reinforcing his belief in the primacy of experience and environment over instinct (this is discussed in some detail later in this chapter).

Figure 2.1. Female Great Pied or Great Indian Hornbill (*Buceros bicornis*) and plump chick collected by Wallace in Sumatra, from *The Malay Archipelago* (Wallace 1869, 146). Hornbills (Bucerotidae) are found in the tropics and subtropics of the Old World from Africa to the Melanesian south Pacific. In this and related species the female is sealed in a tree cavity nest with the chick, and is fed by the male through a small hole. Wallace was not correct to generalize that the hornbills of India "eat only fruit"; many species are omnivorous, but some do specialize on fruit, including figs as shown here being fed to the chick. To Wallace variable diets of otherwise structurally identical species belied the assumption that structure was matched to diet by design. Courtesy of the Staatsbibliothek zu Berlin.

In another vein, Wallace is impatient with what he takes to be trivial claims of design and divine wisdom, as with the example of the three scars of coconuts. One of them is a soft spot through which the cotyledon extends in germination. Cited as a "wise contrivance," Wallace seems to find it pathetic to point to such trivia as evidence for an omniscient and omnipotent being, as if this would be a proof of "the superior wisdom of some philosopher, [if] it was pointed out that in building a house he had made a door to it." He is scathing: "Yet this is the kind and degree of design imputed to the Deity as a proof of his infinite wisdom. Could the lowest savage have a more degrading idea of his God." Wallace sounds quite scornful of religion, but at the time he was most likely not the atheist or strict materialist he seemed to be in his youth, judging from various comments he made about his early convictions. For example, in an 1861 letter he confided to his brother-in-law Thomas Sims:

> In my early youth I heard, as ninety-nine-hundredths of the world do, only the evidence on one side, and became impressed with a veneration for religion which has left some traces even to this day. I have since heard and read much on both sides, and pondered much upon the matter in all its bearings. . . . I have since wandered among men of many races and many religions. I have studied man, and nature in all its aspects, and I have sought after truth. In my solitude I have pondered much on the incomprehensible subjects of <u>space</u>, <u>eternity</u>, <u>life</u> and <u>death</u>. I think I have fairly heard and fairly weighed the evidence on both sides, and I remain an <u>utter disbeliever</u> in almost all that you consider the most sacred truths. I will pass over as utterly contemptible the oft-repeated accusation that <u>sceptics</u> shut out evidence because they will not be governed by the morality of Christianity. You I know will not believe that in <u>my case</u>, and <u>I</u> know its falsehood as a general rule. . . . To the mass of mankind religion of some kind is a necessity. But whether there be a God and whatever be His nature; whether we have an immortal soul or not, or whatever may be our state after death, I can have no fear of having to suffer for the study of nature and the search for truth, or believe that those will be better off in a future state who have lived in the belief of doctrines inculcated from childhood, and which are to them rather a matter of blind faith than intelligent conviction. (Marchant 1916, 1:82–83; WCP3351; emphases Wallace's)

Note that Wallace maintained that he was an unbeliever "in almost all that you consider the most sacred truths"—conventional Christianity—but did not assert atheism. He acknowledged himself a skeptic, and later, in his book *On Miracles and Modern Spiritualism* (1875), was more explicit about this early skepticism: "up to the time when I first became acquainted with the facts of Spiritualism [mid-1860s], I was a confirmed philosophical sceptic, rejoicing in the works of Voltaire, Strauss, and Carl Vogt, and an ardent admirer (as I am still) of Herbert Spencer. I was so thorough and confirmed a materialist that I could not at that time find a place in my mind for the conception of spiritual existence, or for any other agencies in the universe than matter and force" (Wallace 1875, vi–vii).

While it is not clear precisely what Wallace believed during the years of the Species Notebook, by the mid-1860s he had serious doubts about the universal applicability of natural selection (see, for example, Fichman 2001; 2004, 150–157; Slotten 2004, 268–270). Around this time he began exploring spiritualism as part of an effort to develop a more holistic conception of naturalistic processes such as natural selection operating in the material world while spiritual processes operated in the realm of human cognition and consciousness. Smith (1992, 2008) and Fichman (2001) among others have argued that the seeds of Wallace's dissatisfaction with a purely materialistic conception of organisms and their evolution germinated some years before his turn toward spiritualism in the 1860s. The Species Notebook bears this out. A few passages, copied out from various sources, seem to speak to a spiritual sense, as this one quoting French philosopher Ernest Renan (p. 154): "A gross materialism, valuing things only for their immediate utility tends more & more to seize upon humanity. . . . All religions & all philosophies alike teach us that man should have a higher aim than his physical enjoyment & interests."

His criticism was leveled against those who invoked poorly examined religious tropes to explain the natural world. In this respect he almost seems deistic, with his core of spiritual belief but rationally seeing the creator as working though natural law. It is in this spirit that he attacks those writers like Lyell who saw the world through the lens of natural theology, whether seeing structure designed just so or seeing only balance and harmony in nature. He notes that Lyell, in volume 3 of the *Principles,* saw balance in the rapacious destruction by swarming locusts or explosions in sugar ant populations. Wallace derides this idea—far from examples of balance, they are examples of *struggle.* On pages 49 and 50 he attacks Lyell's claim for the "balance of species being preserved by plants insects, & mammalia & birds all adapted to the purpose." To the contrary, "This phrase is utterly without meaning," Wallace wrote. "Where is the balance? Some species exclude all others in particular tracts. Where is the balance. When the locust devastates vast regions, & causes the death of animals & man what is the meaning of saying the balance is preserved." There is no "balance" but struggle: "To human apprehension this is no balance but a struggle in which one often exterminates another. When animals or plants become extinct where is the balance."

The prevalence of struggle is consistent, too, with the fact that introduced species tend to overrun native ones—European plants introduced to America, for example. On pages 146–147 he cites data from Harvard zoologist Louis Agassiz (1807–1873) on the prodigious number of such introductions. "What becomes of the 'Harmony of distribution,' the 'balance of species'; the 'proofs of intelligence in the [natural distribution] of species' &c. &c. Did the 'wonderful order' Agassiz speaks of exist before the country was overrun by these strange plants,—or does it exist now?" When Agassiz (the last great naturalist embracing natural theology post-1858–1859, as it happened) goes on about balance and harmony, Wallace is impatient. "What are the normal proportions & harmony spoken of," he continues. "The proportions have continually varied & are varying. Are the horses in S. America harmonious or not? In the tertiary period there were horses,—then none now they are again." He ends on an insightful note: "Whatever exists must be in harmony or it could exist no longer. The proportions of all animals are self regulating, & constantly varying, it has not been maintained unchanged for any great period as Agassiz well knows" (all emphases Wallace's).

Wallace had a sense of the ecological context for species interrelationships and struggles, as when he proposed what we can recognize as an ecological explanation for the tropical grassy plains of Gilolo or the llanos of Orinoco, Brazil (pp. 108–109). Why are these areas dominated by grasses, rather than the more common dense tropical forest? Invoking geological uplift, he posits that newly exposed flats become more readily colonized by grasses, seeds of which are more readily dispersed. Once established, such grasses grow densely and prevent the germination of the tardily arriving tree seeds. This is essentially a version of the modern principle of competitive exclusion.

These entries on the theme of "design" show Wallace skeptical of simplistic natural-theological explanations (or as he would characterize them, nonexplanations) for the structure, habits, and distribution of species. To paraphrase language he uses on page 33 of the notebook, for all of these things we are like children trying to comprehend complexities that far exceed our capacity to fully fathom, yet we insist on "explanations" that really say nothing. Cause and effect, natural law, material explanation is what Wallace seems to be striving to grasp, but there are also elements of Wallace being more reflective of the nature of religion.

Near the conclusion of the recto side of the notebook, on page 178, he quotes from Herbert Spencer's *First Principles of a New System of Philosophy* (1862) on the idea of religion as just another aspect of natural history to be explained—albeit human natural history. (Judging from the date of publication of the book, this entry perhaps postdates his return home.) Spencer first stated, "All general beliefs have some basis of truth." He then suggests that religious ideas either are themselves specially created or have arisen by a process of evolution (perhaps because it is in "some way conducive to human welfare"). This is concordant with Darwin's view in *The Descent of Man* (1871), which anticipated some lines of inquiry in modern evolutionary anthropology and evolutionary biology that consider the evolution of a tendency to belief as adaptive.

Geographical Distribution: Pattern and Process

Geographical distribution was an early and abiding interest of Wallace's, something that he felt held the key to understanding species origins. Accordingly, entries bearing on distribution are found throughout the notebook. Some are tallies of species numbers from different regions, taken from various sources. Other, more interesting entries bear on geographical distribution in relation to species relationships. The earliest entry in this regard pertains to the Galápagos Islands, likely taken from Darwin's *Journal of Researches* (1845), which is quoted elsewhere. "In a small group of islands not very distant from the main land, like the Galapagos, we find animals & plants different from those of any other country but resembling those of the nearest land. If they are special creations why should they resemble those of the nearest land? Does not that fact point to an origin from that land" (p. 46). Wallace was willing to draw the obvious conclusion that Darwin would not, using instead rather suggestive language:

> Reviewing the facts here given, one is astonished at the amount of creative force, if such an expression may be used, displayed on these small, barren, and rocky islands; and still more so, at its diverse yet analogous action on

points so near each other. I have said that the Galapagos Archipelago might
be called a satellite attached to America, but it should rather be called a
group of satellites, physically similar, organically distinct, yet intimately related to each other, and all related in a marked, though much lesser degree,
to the great American continent. (Darwin 1845, 398)

Wallace, long a transmutationist by the writing of the Species Notebook,
declares, "Here we must suppose special creations in each island of peculiar
species though the islands are all exactly similar in structure," but then concludes that "we can hardly suppose that islands would be left for ages to
become stocked in this manner." This commentary is part of the lengthy
section critiquing Lyell. As will be evident by the end of this chapter, the
interpretation of distribution is just one of Wallace's many points of dis-
agreement with Lyell. Some of his entries on the Canary Islands relate to
the same Galápagos observation, as in his compilation of flora of the Canaries listing "species peculiar" (endemic), "found elsewhere," "imported,"
and so on (p. 151), and his citation of Christian Leopold von Buch (1774–
1853), the German naturalist, who observed that the Canary Islands "have
risen separately from the sea, & never to have been joined together nor to
the continent of Africa."

But the Canary Islands also provided a case study in finer-scale distribution that seemed to bear on the species question: again looking to von Buch
(*Flora of the Canary Islands,* or *Physikalische Beschreibung der Kanarische
Inseln,* 1825, in the original), Wallace in the Species Notebook zeroed in on
a fascinating passage: "On continents the individuals of one kind of plant
disperse themselves very far, and by the difference of stations of nourishment & of soil produce <u>varieties</u>, which at such a distance not being crossed
by other <u>varieties</u> & thus brought back to the primitive type, become at
length permanent & distinct <u>species</u>." "Then," he continued, "if by chance
in other directions they meet with another <u>variety</u> equally changed in its
march, the two are [have become] very distinct <u>species</u> & [are] no longer
susceptible of intermixture" (p. 90, underscores Wallace's). Wallace further
comments that von Buch "then shows that plants on the exposed peak of
Teneriffe where they can meet & cross do not form varieties & species,
while others such as Pyrethrum or Cineraria living in sheltered vallies &
low grounds often have closely allied species confined to one valley or one
island." These passages are remarkable for their avowal of transmutation.
Plants colonizing islands like the Canaries produce new varieties by virtue

of their new environment, which "become at length permanent & distinct species," again with his emphasis. Note that the ability or not to interbreed is the criterion that defines species here—an early articulation of the modern biological species concept. This species definition has even earlier antecedents: the naturalists John Ray (1627–1705) in seventeenth-century England and the Comte de Buffon (1707–1788) in eighteenth-century France, for example, both had a concept of species based on the ability to interbreed and produce offspring which perpetuate their "type" (e.g., Corsi 1978, 7–8). Wallace was later to elaborate this concept of species through his analysis of the papilionid butterflies of the Malay Archipelago (Wallace 1865b; Mallet 2008, 2009).

In modern terms, the observation that "plants on the exposed peak of Teneriffe where they can meet & cross do not form varieties & species" might be understood as selection reinforcing species distinctness, incompatibility perhaps being favored by selection against production of hybrids. Or, it might refer to the homogenizing effect of interbreeding, preventing divergence. The further observation that species living in "sheltered vallies & low grounds" often have closely allied species confined to one valley (or one island) speaks to allopatric speciation: diversification among isolated valleys. On page 151 Wallace mentions that for plant genera naturalized in the Canaries, there is but a single species. Native *Pyrethrum* and *Cineraria*, however, have "many allied species."

A modern reading of these intriguing passages would suggest that the Canary observations from von Buch must have been key for Wallace, but it is not clear just how important they were: von Buch and his passage on "change of species" in the Canary Islands is listed in Wallace's index to the Species Notebook, but neither von Buch nor the Canaries are mentioned in many papers or other publications by Wallace (perhaps the most extensive reference was made in an 1887 address to the American Geographical Society, but this particular observation of species change was not recounted). When was this notebook entry made? On the same page is found a reference to pigeons collected at Macassar and Lombock, and on the following page there is a reference to ducks of the Aru Islands. Wallace first visited Macassar, from Lombock, September through December 1856, arriving in the Aru Islands (via the Ké Islands) in January 1857. He may have extracted von Buch in this time period, or once back in Macassar the following July where he wrote several papers exploring the species-variety relationship, though none of these discuss the Canary Islands or von Buch. This may be, then, an intriguing one-off entry, along with most others that bear on distribution pertaining to numbers and relationships of species in different geographical locales. This is the gist of entries on pages 145–147 (flora of North America and Asia), page 151 (flora of the Canaries, St. Helena, St. Michel, and different African regions), and page 177 (northern Africa vs. southern Europe). As a final comment regarding von Buch, it is noteworthy that this passage on species change did not pass unnoticed by Darwin, who made notes on two passages from the *Flora of the Canary Islands* in his Transmutation Notebook B (July 1837–February 1838) (Barrett et al. 1987, 209–210):

> Von Buch—Canary Islands: French Edit.—Flora of Islands very poor. . . . Analogous to nearest continent; poorness in exact proportion to distance (?) and similarity of type" (B Notebook, p. 156 [B156]), and
>
> Von Buch distinctly states that permanent varieties become species . . . not being crossed with others.—Compare it to languages. But how do plants cross?—Admirable discussion. (B158)

In late 1860 Wallace apparently copied out the key passage from von Buch's *Flora* and sent it to Darwin:

> On continents the individuals of one kind of plant disperse themselves very far, and by the difference of stations of nourishment & of soil produce *varieties* which at such a distance not being crossed by other varieties and so brought back to the primitive type, become at length permanent and distinct *species*. Then if by chance in other directions they meet with another *variety* equally changed in its march, the two are become *very distinct species* and are no longer susceptible of intermixture. (Darwin Correspondence Project, letter 2627; emphasis in original)

Wallace may have done so in reference to an earlier letter from Darwin mentioning people who had come forward after the *Origin's* publication claiming priority in conceiving the theory. If so, Wallace considered von Buch as one who had indeed suggested that varieties become species. Darwin already knew this, as evidenced by his citation of the very same passage in his B Notebook from 1838, but despite this and Wallace's having again drawn his attention to it, for some reason Darwin did not give von Buch credit in his "historical sketch" of the idea of transmutation included in the third edition of the *Origin* the following year.

A final geographical distribution-related observation in the Species Notebook worthy of mention is found on pages 132–133, a "Note for determining species population of Globe." This Humboldtian exercise was a method to "determine an approximate law for the total number of species common to two countries when only a portion of each is known," not the entire globe itself. Wallace's approach was based on statistical resampling. If the total species richness in two areas is known, then the ratio of species common to both locales is also known. The "law" he sought was some kind of adjustment factor applied to partial collections from both locales. Repeating this for, say, each pairwise comparison of all locales, it may be possible to gauge "the species population of a region from partial collections made at two or more localities in it, or of a group of islands only partially explored." This has a practical application for a working collector like Wallace—collecting in a new area, is he likely to obtain many more specimens per unit effort? But there is also the suggestion of a deeper interest: "A series of results of collecting in different countries compared in every possible way will shew whether a greater proportion of the common species of each country (which will be those first obtained) or of the rarer ones (which will enter only in a more complete collection) have a wider distribution." Do rare or common species tend to have a wider distribution? He does not elaborate here, but this statement resonates with his idea about geographical distribution of older, more established species and newer, more recently arisen species (see Chapter 1). In his Ternate essay (1858e) Wallace stated that commonness or rarity is determined by selection, but it is of interest to map the distribution of common and rare species in order to better understand how selection may be operating over a geographical area. This entry is not dated, but it is bookended (albeit a couple of pages away on either side) by entries dated May and September 1858, respectively.

Morphology and Classification

Wallace discussed a number of topics relating to structure, comparative anatomy, and classification in the context of transmutation. In the earliest of these (p. 54) he commented on a paper by comparative anatomist Richard Owen, who became superintendent of the natural history department of the British Museum in 1856 and was a star in the scientific firmament of Britain by the time of the Species Notebook. In the paper read by Wallace, Owen discussed the "law" of the prevalence of more generalized structure of extinct reptiles compared with the more specialized structures of living species. The earlier members of a group were recognized to be more "generalized" in structure than later members. There are two implications of this. First, Wallace seems to be noting that the adaptations of the extinct groups (fishlike features of ichthyosaurs, birdlike features of pterodactyls, etc.) are more generalized than those of later groups—the swimming and flying adaptations of modern fish and birds, say—from which "type" they have "borrowed." The bird's wing structure is "higher" and unique to birds, while the pterodactyl's wing is a more generalized saurian version. More importantly and with great insight, Wallace recognizes that this is just what one would expect with a "constant change of species," and those nearer the base of a "complicated many branching series" exhibit a *combination* of characters. Wallace appreciates that earlier groups exhibiting a combination of the characters of descendant groups is precisely what is predicted in an evolutionary context—they are linking forms, in the manner that common ancestors combine and link features of their descendant groups.

The relationship between structure and classification, or the respective "affinity" of groups, is a prevalent theme. This reflects the struggle at the time to develop a so-called natural system of classification that was supposed to reflect true affinities. External similarities, what we would call analogies, were not only seen as evidence of affinity but also often arrayed to illustrate supposed links of affinity between divergent groups. Thus bats were a linking form between mammals and birds, cetaceans between mammals and fish, and so on. One of Owen's significant contributions was clarifying the distinction between analogy and homology (terms he coined), the latter reflecting a deeper level of affinity based on anatomical analysis (Wood 1995). (He interpreted patterns of homology-based affinity in typological terms, but these same patterns would be recognized by Wallace and Darwin as consistent with transmutation too.)

In one entry (p. 76) Wallace says that the clear mammalian affinities of cetaceans show that they are mammals modified for an aquatic life, not some link connecting mammals with fish. Cetaceans and seals, he points out, "are both aberrant developments of their respective orders, not the foundation of the whole class which both of them can not be." On the next page, while Wallace is approving that the important distinction "between affinity & analogy is now generally recognised by Naturalists" as a result of which "we have no such absurdities as considering the Hummers to connect birds with Insects or the Bats to form a natural transition from Mammalia to Birds," still he feels that "the <u>principles</u> of the distinction are still often lost sight of" (his emphasis). The principles he alludes to are fundamental affinity through common descent.

Another passage even more explicit about the lessons of affinity is found on pages 98–99. Musing on the "import of the doctrine of Morphology of plants," Wallace describes the homologies of stamens, petals, and sepals, carpels and leaves, and other characteristics. "By these laws all the countless modification of flower & fruit can be reduced to a common type & most of the excentricities of vegetation explained & accounted for," he says. He then asks:

> Now what does all this beautiful law mean, what does it teach us? Is it a substance or a shadow, a truth or a fallacy?—For if we are to believe that each & every species is an absolute & distinct creation independent of any of its closest allies, & has come into existence at a different time & in a different place from them, if those having the highest developed parts may have had their origin before those others of which we suppose them to be a more complete development, then all this doctrine of morphology is meaningless & leads to error.

This passage is noteworthy. If "those having the highest developed parts" have had their origin—were found in the fossil record, say—"before those others of which we suppose them to be a more complete development," the doctrine of morphology would be meaningless because simple structure should not come before complex structure. The "doctrine of morphology" was a phrase then in common use, referring to fundamental affinity of the structural parts of an organism. In the botanical world the idea that stamens and pistils could be thought of as modified leaves gained acceptance slowly, as reflected in this editorial note from *The Gardener's Monthly and Horticulturist* (Meehan 1879; 21:278):

The Doctrine of Morphology
It is almost wonderful that the doctrine which teaches that all parts of a flower are modified primary leaves, should have such universal assent, when but a comparatively few years ago it was laughed at by the most intelligent men of the day. Speaking of the theory, an editorial article in *Paxton's Magazine* for 1844, says: "There is something so monstrous, so degrading in the idea, that the mind which contemplates all things as beautiful and perfect in their creation, revolts at it."

Owen and Louis Agassiz, following the great French comparative anatomist Georges Cuvier, may have seen homology as nothing more than creation on a morphological theme, but Wallace declared that the doctrine of
morphology—the true meaning of homology—is meaningless only in their
view, since it is nonsensical to use language like "petals are modified leaves"
in the context of special creation. He saw that homology makes sense only
on the supposition that it reveals *true* affinity, meaning affinity stemming
from shared ancestry: "for if stamens & petals & carpels have been in every
case independently created as such, it is absurd to say they are modifications or developments of any thing else." Worse, if they do not reveal true
history of affinity, they are positively misleading—much like fossils which, if
mere imitations of organic life, would be misleading if we mistook them for
remains of once-living beings. Wallace puts this in powerful terms: "In
that case all the beautiful facts of morphology are a delusion & a snare, as
much so as fossils would be were they really not the remains of living beings
but chance imitations of them" (p. 98). He goes on to declare that we can and
must learn from homology, as from fossils:

> The natural inference of an unprejudiced person however would be that
> both are true records of the progress of the organic world. Nature seems to
> tell us that as organs are occasionally changed & modified now, in individ-
> ual plants, we may learn how the actual changes have taken place in the
> species of plants. A key is offered us to a mystery we could otherwise never
> have laid open, why should we refuse to use it? (pp. 99–100)

Darwin was similarly impassioned on this count, declaring in *On the Origin of Species* (Darwin 1859, 167) that he would sooner believe with the
old-fashioned "cosmogonists" that fossils have a nonbiological origin than
that the patterns he was uncovering did not say something about transmutation and common descent.

Taking insights from homology a step further, Wallace explores how embryology can inform classification (p. 144). His insight here is clearest where
with emphasis he points out that "Raptores [or raptors, birds of prey are] only
the highest development of raptorial aquatics. Embryo [of passerine, or common perching] birds have all hooked bills. Raptores are therefore lower than
Passeres." In other words, birds of prey have hooked beaks while perching
birds, which include the common songbirds, do not. As embryos, however,
the beaks of perching birds are temporarily hooked. This trait seen in the
development of the perching birds indicates that raptors are "lower than"

(originated earlier than) perching birds, since they retain a "primitive" trait lost by the perching birds. He gives a few other examples in this vein, such as with insects. Insofar as groups with adult mouthparts that are "haustillate" (piercing and sucking, like certain flies) have mandibulate (chewing) mouthparts when immature, he concludes that the mandibulate insects are "lowest" (basal, in modern terms), with beetles (Coleoptera) below butterflies and moths (Lepidoptera).

In brief comments, finally, Wallace shows that he appreciates the pitfalls of carelessly invoking adaptive or design-based explanations for certain structures or functions, as where on page 112 he notes from the popular 1851 zoology treatise of French naturalist Henri Milne-Edwards (1800–1885) that the large air cells found in ostrich leg bones prove that air cells in the bones of birds are not a designed or adaptive feature to render birds' bodies lighter for flight. How could it be, since ostriches are flightless? Again we see a parallel with Darwin's insight, as where he points out in the *Origin* (p. 437) that the skull sutures of fetal mammals cannot be, as was often maintained, designed to permit parturition, since birds and reptiles have such sutures too when neonates, and these groups merely have to break out of their egg.

Instinct and Experience

Instinct, experience, and habit were interesting to Wallace in several respects, and his perspective often differed from Darwin's on the subject. Wallace tended to emphasize learning over instinct; he did not so much reject the existence of instinct as call into question the certainty of naturalists that behaviors commonly viewed as instinctive were not in reality subtly learned behaviors. Wallace seems to have downplayed instinct wherever he could make a case for the effects of environment, learning, and experience, and some historians have speculated that he did so in part owing to his commitment to Owenite socialism. Reformer and socialist philosopher Robert Owen (1771–1858) had a vision of social transformation based on a core belief in the perfectibility of humanity; he believed that improved environment and opportunity were key to lifting the poor out of lives of vice and misery, while Britain's class-ridden social structure with its vast inequities in wealth, privilege, and opportunity was a major hindrance to achieving social improvement. The question of social improvement was one of plasticity versus predetermination of behavior, character, and circumstance. When Wallace was frequenting the London mechanics' institutes as a teenager, he heard lectures on all manner of new and incendiary subjects, perhaps foremost among them Owen's humanistic philosophy; even late in life he cited Owen as one of the most important figures of the nineteenth century (Fichman 2004, 70–72; Slotten 2004, 10–13; Claeys 2008).

Jones (2002, 80–81) argued that Wallace's view of animal and human behavior—learning and instinct—was seen through an Owenite lens. The Species Notebook offers the earliest expressions of Wallace's position on this subject. First, there are entries relating to *variation* in behavior, meaning behaviors that vary geographically in the same species or, perhaps more importantly, expressions of behavior that changes over time or under new circumstances. Wallace's extensive notes (pp. 112–119) on nest-building in birds gleaned from Rennie (1839, 1844) include records of building materials and sites of nest construction, and comments on the relationship of structure and function. He states on page 117, that "The delicacy & perfection of a birds nest is generally in exact proportion to its size activity & to the perfect structure of its bill & feet." Web-footed birds, for example, make rude nests, while perching birds are capable of more elaborate domiciles. A few entries hint at Wallace's underlying interest. On page 116 he notes, "Birds nests said to be built by instinct because they don't improve. But they vary according to circumstances & does man do more." Note the two-part key to his interest: these behaviors *vary according to circumstances*, and *does man do more* than this. Compare this with the opening sentences to his paper "The Philosophy of Birds' Nests" (Wallace 1867, 413): "Birds, we are told, build their nests by *instinct*, while man constructs his dwelling by the exercise of reason. Birds never change, but continue to build for ever on the self-same plan; man alters and improves his houses continually. Reason advances; instinct is stationary." Wallace seems to be setting up a commonly accepted dichotomy only to demolish it. "This doctrine is so very general that it may almost be said to be universally adopted," he writes, continuing, "Philosophers and poets, metaphysicians and divines, naturalists and the general public, not only agree in believing this to be probable, but even adopt it as a sort of axiom that is so self-evident as to need no proof." And then the shoe drops:

> Yet I have come to the conclusion that not only is it very doubtful, but absolutely erroneous; that it not only deviates widely from the truth, but is in

almost every particular exactly opposed to it. I believe, in short, that birds do *not* build their nests by instinct; that man does *not* construct his dwelling by reason; that birds do change and improve when affected by the same causes that make men do so; and that mankind neither alter nor improve when they exist under conditions similar to those which are almost universal among birds. (Wallace 1867, 413)

In the Species Notebook (p. 116) he argues that both birds and humans alter their behavior over time in response to circumstances, facilitated by communication and information-sharing: "It is only by communication, by the mingling of different races with their different customs, that improvements arise & then, how slowly! A race remaining isolated will ever remain stationary, & this is the case with birds." This is part of a larger argument against the occurrence of instinctive behavior in humans. Wallace saw human "progress" or improvement coming from effort and collaboration, not contingent upon innate traits (and therefore traits subject to natural selection; I will explore this further in the final section of the chapter).

The distinction between instinct and reason is important to Wallace. Instinct is not infallible, but rather is nearly always modified or tempered by experience (implying malleability, and suggesting the possibility of gradual, transitional steps between expressions of even the most complex behaviors). This is clear on page 149 where Lyell is criticized for drawing what Wallace takes to be a false dichotomy between the passage from "an irrational to a rational being" (the difference between brutes and humans) and the passage from "the more simple to the more perfect forms of animal organization & instinct." This is part of Lyell's systematic attack on transmutation: "To pretend that such a step or rather leap can be part of a regular series of changes in the animal world is to strain analogy beyond all reasonable bounds," he wrote in the *Principles*. In other words, the gap between reasoning man and nonreasoning animals is as great as that between simple life and life in all of its complexity, including instinct. Wallace rallies, dismissing Lyell's argument as semantic: "Here the absolute distinctness of <u>reason</u> & <u>instinct</u> is assumed; the argument depends on the terms <u>rational</u> & <u>irrational</u> which imply no gradation" (emphasis in original).

In two places (pp. 143 and 177) Wallace paraphrased Alexander Pope on the subject from the *Essay on Man* (1734): "Whether with reason, or with instinct blest | Know, all enjoy that power which suits them best. . . . And reason raise o'er instinct as you can | In this 'tis God directs, in that 'tis man." Wallace writes (p. 143; emphasis his here and below), "On Instinct & Reason, a poet says 'In this 'tis God that acts, in that 'tis Man.' But errors of instinct show this can not be the case." Insofar as instinct was assumed to represent an innate, fixed, and infallible capacity endowed by the creator, Wallace puts his finger on the idea of anomaly and variation in instinct and on what this teaches. *Errors* of instinct give lie to the idea that "in this God acts," the deity presumably being infallible and not capable of errors. (Insofar as instinct was, like domesticated varieties, cited by anti-transmutationists as evidence against the possibility of change, Wallace is undermining the common conception of instinct for precisely the same reason that he argued that domestic varieties were irrelevant to the question of transmutation.) He repeats this on page 177 at the conclusion of a long discussion about honeybees and their honeycomb, the cells of which were taken as a supreme example of highly complex instinctive behavior. "Honey comb, is said to be constructed of such form as to combine with mathematical precision the greatest strength with the least materials," he writes, setting up an argument: "This is erroneous on two points." Darwin too addressed the evolution of bees' cells in the *Origin,* prompting an attack by the Rev. Samuel Haughton, the distinguished Irish physician, scientific writer, and clergyman. Wallace, often as much of a bulldog for Darwin as Thomas Henry Huxley, rose to the defense drawing in part on these entries in the Species Notebook. His spirited "Remarks on the Rev. S. Haughton's Paper on the Bee's Cell, And on the Origin of Species" appeared in the *Annals and Magazine of Natural History* in October of 1863 (see Wallace 1863b).

In a series of entries beginning on page 166, Wallace explores the concept of instinct in a more philosophical manner. It is curious that he opens by noting that "in investigating instinct we proceed by degrees, from the easy & near to the difficult & remote"—the degrees of difference consciously or unconsciously paralleling evolutionary relationships. This is reiterated on the next page, pointing out that Cuvier, Richard Owen, and T. H. Huxley gained insight into homologies by analyzing structural similarities and differences in minute step-by-step comparisons of a series, not by simply comparing two forms at opposite ends of a morphological spectrum. (Wallace says their method is by "tracing modifications step by step"—although he would take the word "modifications" rather more literally than Cuvier and Owen.) The point is to draw an analogy with instinct: that, too, must be compared carefully and step-by-step across a series of species to gain insight into how the most complex expressions may have arisen from the less complex. This point is made in chapter 7 of the *Origin*, where Darwin maintained that transitional series of expressions of behaviors (and expressions by "collateral relatives") can give insight into the evolution of even the most extreme or well-developed instincts such as honeybee cells, brood parasitism in cuckoos, and slave-making ants.

Wallace's entries on instinct served as draft notes for an essay that appeared much later, in 1870. "On Instinct in Man and Animals" is found in *Contributions to the Theory of Natural Selection* (pp. 91–97). Several important points are argued, the most important being that much of what we carelessly term "instinct" in nature actually involves some degree of experience and learning, and the assertion that humans do not have instinctive behavior. He defines instinct as "the performance by an animal of complex acts, absolutely without instruction or previously-acquired knowledge," so the key point is that birds are commonly viewed as being able to build their nests and honeybees their waxen cells "without ever having seen such acts performed by others, and without any knowledge of why they perform them themselves." Wallace disputes this, pointing out that no one had yet done the experiment of raising birds or bees in isolation to determine if they instinctively know how to build. Learning from nestmates may be essential, he concludes.

When it comes to humans, Wallace argued strenuously against the possibility of any instinctive behavior. Oft-cited cases of human instinct like the suckling response of newborns he dismisses as mechanical as breathing and swallowing—topics also mentioned in the Species Notebook. "What are very commonly called <u>instincts</u> in man are only habits," he declares. Another putative case of human instinct that Wallace undermines in both the Species Notebook and his later essay is the purported ability of Indians to instinctively find their way unaided in trackless wilderness. Wallace argues that learned skill is involved, reading the lay of the land—land that is home. "But take these same men into another country with other streams & hills & soil & vegetation, & after bringing them by a circuitous course from a given point tell them to return through a forest of some extent, & they will certainly decline the attempt. Their instinct will not act out of their own country" (Species Notebook, p. 172). He ended his 1870 essay on the same point, concluding: "It appears to me, therefore, that to call in the aid of a new and mysterious power to account for savages being able to do that which, under similar conditions, we could almost all of us perform, although perhaps less perfectly, is almost ludicrously unnecessary" (Wallace 1870, 210).

It is instructive to consider in more detail *why* Wallace objected to the existence of instinct in humans. In the Species Notebook as in the 1870 essay "On Instinct in Man and Animals," he opened the discussion with reference to insects, the instinctive faculties of which are claimed by some to differ in kind and not merely degree from that of humans: they are considered "conclusive as to the existence of some power or intelligence very different from that which we derive from our senses or from our reason" (Wallace 1870, 201). Such claims are provocative to Wallace, and he sets out to define instinct and undermine this argument by pointing out (as on pp. 166–167 of the Species Notebook) that we hardly understand the senses of insects, let alone tested their capacity for learning or gaining from experience. When he draws an analogy with Owen and Huxley identifying homologies by minute step-by-step comparison, he seems to argue for *continuity,* a spectrum of mental capacities whereby humans differ from other animals by degree and not of kind. Seeing humans as part of an organic continuum is in keeping with his evolutionary convictions, certainly (though he eventually, in the mid-1860s, comes to the conclusion of discontinuity when it comes to the evolution of the human mind). But by the end of the essay (and the latter entries on the subject in the Species Notebook, pp. 168–172), why is it necessary to deny that humans have *any* instinctive behaviors at all? Perhaps he is not so much denying this as pointing out that commonly cited examples of human instinct are flawed; he may remain agnostic on the question, but he seems critical of the "upholders of the instinctive theory" who maintain that humans have instincts "exactly of the same nature as those animals."

In the Species Notebook and the 1870 essay we have seen that he presents and discards cases like suckling infants and "navigating" Indians. Is he arguing after all that humans do differ in kind from other animals? I do not think that we are seeing an inconsistency that presages, say, his later spiritualist beliefs and denial of a role for natural selection in the evolution of the human mind; rather, Wallace believes that humans differ merely in degree from other animals, but rather than degree of instinct per se it is degree of "faculties of observation, memory, and imitation" and some amount of reason, as he puts it at the close of the 1870 essay. This argument is pursued in "The Philosophy of Birds' Nests" (1867), which also draws on observations recorded in the Species Notebook (pp. 112–119). The relationship and relative balance of *instinct* and *reason* are the crux of the issue for Wallace, but he sees a far greater role for noninstinctive faculties in both people and other animals. It is with this point that he concludes his essay on birds' nests:

> The mental faculties exhibited by birds in the construction of their nests are the same in kind as those manifested by mankind in the formation of their dwellings. These are, essentially, imitation, and a slow and partial adaptation to new conditions. . . . I do not maintain that birds are gifted with reasoning faculties at all approaching in variety and extent to those of man. I simply hold that the phenomena presented by their mode of building their nests, when fairly compared with those exhibited by the great mass of mankind in building their houses, indicate no essential difference in the kind or nature of the mental faculties employed. (Wallace 1867, 420)

Wallace repeated this position in *Darwinism:* "Much of the mystery of instinct arises from the persistent refusal to recognise the agency of imitation, memory, observation, and reason as often forming a part of it" (Wallace 1889, 442), but here he is more explicit in attributing "perfection of instinct" to "the extreme severity of selection" in the deeper evolutionary history of humans.

Humans: Variation and Relationships

It is unfortunate that Wallace did not make any notebook entries connected with his discovery of the principle of natural selection. The temporally closest entry is found on pages 108–109, where on 20 January 1858 he ventured an explanation for the occurrence of grassy tropical plains like those of Gilolo. His explanation is essentially one of competitive exclusion, and only in this tenuous way do we have evidence of Wallace thinking about a phenomenon concerning populations and competition about the time his musings on Malthus led to the discovery of natural selection. Some scholars—McKinney (1972) in particular—have argued that Wallace was thinking about human populations around this time, stemming from his recent first encounter with Papuans, and that it was his continued interest in the origins of human races that in turn brought to mind Malthus, competition and conflict, and natural selection. There is no doubt that the origins of humans generally and human races in particular were of interest to Wallace.

Recall from Chapter 1 that Wallace's keen interest in human racial origins focused on distinction and transition between the Malay and Papuan peoples. Recall too that Wallace hoped to find evidence of transitional forms between these races in the Moluccas. Wallace first encountered Papuans in Aru, and just prior to traveling there he spent several months (September to December 1856) in Macassar, southern Celebes, where he took detailed notes on the "Malay races" (Species Notebook, p. 65). He observed some people whom he thought were of mixed race with Papuan characteristics, but he did not encounter Papuans per se until reaching Aru in January 1857. Indeed, the very next notebook entry (p. 66) duly reports observations on the physical features of Papuans as well as Malays. The locale of greatest interest to him in this regard, however, was Gilolo (Halmahera), the island where his friend Robert Gordon Latham suggested he might find clues to the transition between the Malay and "negrito" (Papuan) races (Brooks 1984, 163–168). It is on Gilolo, right around the time of his "eureka moment" about natural selection, that he states most fully his interest in the native people of the island, not in the Species Notebook but in his narrative journal:

> The natives of this large & almost unknown island were examined by me with much interest, as they would help to determine whether, independent of mixed races, there is any transition from the Malay to the papuan type: I was soon satisfied by the first half dozen I saw that they were of genuine papuan race, lighter in colour indeed than usual but still presenting the marked characters of the type in features & stature. They are scarcely darker than dark Malays & even lighter than many of the coast malays who have some mixture of papuan blood. Neither is their hair frizzly or wooly, but merely crisp or waved, yet it has a roughness or slight woolliness of appearance produced I think by the individual hair not laying parallel & close together, which is very different from the smooth & glossy though coarse tresses, every where found in the unmixed malayan race. The stature alone marks them as distinct being decidedly above the average malay height, while the features are as palpably unmalay as those of the European or the negro. (second *Malay Journal;* LINSOC MS 178b, entry 127; emphasis in original)

Note that Wallace was keen to know "whether, independent of mixed races, there is any transition from the Malay to the papuan type." This is the last journal entry before his return to Ternate in February 1858, but in the Species Notebook the only reference to Gilolo around this time (dated 20 January 1858) is his hypothesis for the origin of the grassy plains of the island (Species Notebook, pp. 108–109), described earlier in this chapter. Not long before this entry, however, there are several others relating to human races, including (p. 100) observations of Malays in Java, and (p. 104) the peoples of Flores, Bali, Timor, and Batchian. He also copied out information from French naturalist René Lesson's narrative of the voyage of the *Coquille* (Lesson 1826–1830), in particular accounts of the geographical distribution of Malays and Papuans (Species Notebook, pp. 105–106).

Wallace left Gilolo to return to Ternate on 1 March 1858 and was not to return to Gilolo again for six months. Upon returning he seemed to waste no time resuming his observations of the native people. On page 134 of the Species Notebook are recorded observations of the Alfures (or Alfuros) in the village of Sahoe, a group that he later concluded represented an exact intermediate between the Malay and Papuan races. In the third *Malay Journal* (LINSOC MS 178c, entry 154) he noted his uncertainty of the status of these Alfuros at the time: "I was much interested in the indigenes or alfuros of this part of Gilolo, of which a large population are settled in the neighbouring interior & numbers are daily seen in the village, either bringing their produce for sale or engaged by the Chineese [*sic*] or Ternate traders. A careful examination has strengthened my previous idea that they are a mixed race." But a few years later, in a paper read at the 26 January 1864 meeting of the Ethnological Society of London (published to acclaim the following year), he seemed convinced that he had found the "contact zone" between the races:

> The northern peninsula [of Gilolo], however, is inhabited by a native race, whose principal tribes are the so-called Alfurus of Sahoe and Galela. These people are quite distinct from the Malays, and almost equally so from the Papuan. . . . The people of the coasts of all these islands between Celebes and New Guinea are constantly changing about. The one fact, however, that I consider indisputable is, that here the Malay race comes in contact with two other races, which are non-Malay—the Alfurus of Gilolo and Ceram, and the Papuans of New Guinea, Waigiou, Mysol, and the Arru and Ké Islands. (Wallace 1865a, 207–208)

The material in this paper was largely incorporated into chapter 40 of *The Malay Archipelago,* where he is even more definitive that the people of Gilolo are indeed "radically distinct" from the Malay groups, not merely a mixed race: "here then I had discovered the exact boundary-line between the Malay and Papuan races, and at a spot where no other writer had expected it. I was very much pleased at this determination, as it gave me a clue to one of the most difficult problems in ethnology, and enabled me in many other places to separate the two races, and to unravel their intermixtures" (Wallace 1869, 323). The range of phenotypes that Wallace observed at Sahoe and Galela (described on p. 134 of the Species Notebook) are precisely what would be expected of a zone of contact between two races, and he concluded that "A mixture of Dyak or Celebes Malay with a true Papuan race would produce such variety." Taking this a step further in the Species Notebook, he then suggested "if the men were the former [i.e., Malay] & the women the latter [i.e., Papuan] the light colour would be probably the result."

The remaining notebook entries bearing on humans consist of scattered observations suggestive of the affinity of people with other animals, such as the two anecdotes of humans with taillike structures (pp. 64, 91). I include in the "Humans" category Wallace's extensive entries bearing on orangutans (Figure 2.2), though it should be noted that these primarily consist of information from the native Dyaks of orang habits (pp. 7, 9), accounts of Wallace's orang hunts (pp. 10–11, 13–19, 28–30), and a moving record of the orphan infant orang that he tried unsuccessfully to rear after having killed its mother (pp. 20–27). Although Wallace surely believed that humans evolved from an apelike ancestor probably very like the orangutan—indeed, he playfully hints at the relationship in "On the Habits of the Orang-Utan of Borneo" (Wallace 1856a), where he refers to orangs as creatures "which at once resemble and mock the 'human form divine'"—he does not discuss this idea in either the Species Notebook or his journals. The orangutan entries are nevertheless instructive, showing how intimately Wallace observed these great apes and how he came to appreciate their humanlike sensibilities. In this there is another parallel with Darwin, who spent several weeks in the spring of 1838 making observations on a young orangutan named Jenny at the London Zoo. Jenny was the first orangutan to be displayed at the zoo and as such was both a public and a scientific sensation. Darwin, who was a transmutationist by then and keenly interested in human-primate relationships, recorded in his notebook Jenny's childlike emotional behavior (Barrett et al. 1987, 545, 551, 554).

Figure 2.2. A young female orangutan, from *The Malay Archipelago* (Wallace 1869, 56). Wallace and Darwin were among many observers struck by the humanlike qualities of orangutans. Some were more favorably struck than others: in 1842 Queen Victoria commented that the young orang she visited in London Zoo was "frightfully, and painfully, and disagreeably human." Courtesy of the Staatsbibliothek zu Berlin.

Transmutation

The entries considered under this heading, finally, largely relate to Wallace's extended critique of Lyell, whose anti-transmutation arguments were taken as the definitive statement on the subject. A detailed treatment of these arguments are given in appendix 2 of Costa (2013a); here I will summarize the main themes pertaining to Lyell and conclude with an overview of the remaining transmutation-relevant entries of the notebook. The fourth (1835) edition of Lyell's *Principles of Geology*, issued in four volumes, was a part of Wallace's traveling library in Southeast Asia. Wallace's critique of Lyell

Table 2.1. Species Notebook entries for Wallace's critique of Lyell's
anti-transmutation arguments in the *Principles of Geology*
(fourth edition, 1835). Adapted from Costa (2013a, Appendix 2).

Species Notebook	Location in Costa (2013a)	*Principles* (volume:page[s])
34	96	1:35–42, 69
35	98	—
36	100	1:226
37–38	102–104	1:231–234
39–40	106–108	2:435
41–42	110–112	2:437, 446–448, 452
43	114	2:414, 464
44	116	2:443
45–48	118–124	3:22–28
49–50	126–128	3:115–116, 154
51–52	130–132	3:161–162
53	134	3:172–173
149–150	326–328	1:239; 3:21

largely follows this *Principles* edition sequentially (see Table 2.1): entries
on pages 34–38 of the Species Notebook correspond to statements found in
volume 1, those on pages 39–44 correspond to volume 2, and those on pages
45–53 to volume 3. One later entry from Lyell (pp. 149–150) draws on statements found in volumes 1 and 3.

Wallace's key points and arguments are summarized as follows:

A. It is inconsistent to maintain gradualism and change
for the earth but not its inhabitants (pp. 36, 38–40)

Wallace cites Lyell's account of the French naturalist Georges Leclerc, the
Comte de Buffon, from the *Principles* section entitled "Historical Sketch of
the Progress of Geology." This sketch is based upon the accounts of Italian
geologist Giovanni Battista Brocchi (1772–1826) (McCartney 1976). Lyell recounted how ecclesiastical authorities deemed Buffon's opinions on the organic origins of fossils "contrary to scripture" and how the authorities pres-
sured Buffon to recant. This story served Lyell's rhetorical purposes well: his
encapsulated history identified lone figures like Buffon heroically advancing
revolutionary ideas once considered heretical but now accepted. Lyell

wrote of the opposition to Buffon's claim that "secondary" (natural) causes are responsible for shaping the landscape of the earth, in so doing setting himself up as a Buffon-like figure who was similarly ahead of his time. Wallace found Lyell inconsistent in his forceful advocacy of gradual changes in the earth by natural causes but refusal to apply this to the organic world.

Wallace may have intended to take a page from Lyell's playbook, using this approach as part of his own argument to anticipate and defuse opposition to the pro-transmutation argument of his planned book. On the page headed "Note for Organic Law of Change" (p. 36), Wallace forcefully applies to the organic world the implications of the Lyellian vision of change in the inorganic world: "we must in the first place assume that the regular course of nature from early Geological Epochs to the present time has produced the present state of things & still continues to act in still further changing it." He sees Lyell's inconsistency in noting that "while the inorganic world has been strictly shown to be the result of a series of changes from the earliest periods produced by causes still acting, it would be most unphilosophical to conclude without the strongest evidence that the organic world so intimately connected with it, had been subject to other laws which have now ceased to act, & that the extinctions & production of species and genera had at some late period suddenly ceased." Herein lies another parallel with Darwin, who made extensive use of Lyell in this way in *On the Origin of Species.* In a section on the "Lapse of Time" (p. 282) Darwin drew a comparison between the reception of Lyell's ideas and his own ideas: naturalists who once resisted Lyell's vision of earth history now accepted it, and by extension those whose initial response might be to dismiss Darwin's ideas about transmutation should give the notion more careful consideration.

B. The fossil record is a record of progressive change; lineages branch from lineages over time (pp. 37–38)

Wallace examined Lyell's claim of "non-progression" in the fossil record—the supposed lack of progressive change of fossil groups over time, which Lyell claimed was strong evidence against transmutation. "Geological succession," as the pattern of groups succeeding groups over time was called, was not inconsistent with the idea of special creation and an unfolding divine plan—indeed, most naturalists of the day, including Lyell's fellow

anti-transmutationist geologists, saw progressive change in the fossil record (Haber 1968; Rudwick 1972; Corsi 1978; Burchfield 1998). But geological succession fit nicely with a transmutational explanation too, uncomfortably so for Lyell, who attacked the notion and thereby aimed to undermine
one of the strongest arguments in favor of transmutation (see Rudwick 1998,
11–12).

Wallace gathered evidence to controvert Lyell from the *Principles* itself: Lyell had stated that "some of the more ancient Saurians approximated more nearly in their organisation to the types of living Mammalia
than do any of our existing reptiles." "Which?" Wallace asked; "just what
I want" (p. 37). He has an intuitive grasp of tree-thinking: in modern
terms, the more ancient "saurians" are structurally closer to certain types
of mammals than living saurians, which is what would be expected of
groups representing basal lineages, closer to groups sharing a common
ancestor. Tree-thinking is further in evidence with Wallace's critique of
Lyell's interpretation of an early mammalian fossil. Lyell maintained that
Jurassic-age fossils of *Didelphis*, a marsupial mammal, are "fatal to the
theory of progressive development." Wallace rejected this: "if low organized mammalia branched out of <u>low</u> reptiles [and] fishes[,] all that is
required for the progression is that <u>some</u> reptiles should appear before
Mammalia & birds or even that they should appear together. In the same
manner reptiles should not appear <u>before fishes</u> but it matters not how
soon after them." He continues: "all that the development theory requires
is that <u>some</u> specimens of the lower organized group should appear earlier than any of the group of higher organization" (emphases in original).
This shows an understanding of a genealogical pattern of common descent, as opposed to the largely linear (within-lineage) sequence of
change posited by Lamarck. Lyell's criticism is valid for a Lamarckian
concept of a more or less linear succession of species, where the "highest" form of one group gives rise to the "lowest" form of the next. Wallace has a branching pattern of relationship in mind: "the supposed contradictions all arise from considering it necessary that the highest forms
of one group appear before the lowest of the next succeeding, not considering that each group goes on progressing after other groups have
branched from it. They then go on in parallel or diverging series & may
obtain their max[imum] together."

C. Domestic varieties. There is no reason to believe there are
limits to variability; Domesticated varieties as evidence
of transmutation; A "thought experiment" with dog breeds
developed from dog breeds; Reversion unlikely when change
is slow, hand-in-hand with changing geological and
climatic conditions (pp. 39–45)

In the *Principles* Lyell devoted much space to showing how domesticated
varieties undermine transmutationism. In the most important of these arguments Lyell maintained that domestic varieties teach us (1) that species
and varieties vary only within limits, since no domestic variety was known
to have been transmuted into a new species, and (2) that domesticated
varieties "revert" to parental type, showing their limited natural capacity
for change. Lyell further pointed to mummified animals (including domestic species like cats and dogs) found in ancient Egyptian tombs; though
thousands of years old they appear morphologically identical to living species of their group—no evidence of transmutation.

Wallace first rhetorically asks what evidence we have that species vary
within limits. He offers a "thought experiment," a scenario where all dog
breeds but one become extinct, and that remaining one is then spread
around the world and used as stock to develop new breeds. He then sup-
poses that all breeds but one of *those* ("farthest removed from the original") become extinct, and the process begins again. "Does it not seem
probable that again new varieties would be produced," he asked, "and have
we any evidence to show that at length a check would be placed on any
further change & ever after the species remain perfectly invariable." He
argues that there is no good reason for assuming that variability would
simply halt. Indeed, domestic varieties are themselves evidence of transmutation, Wallace argues, pointing to the diverse varieties of dog: "is not
the change of one original animal to two such different animals as the
Greyhound & the bulldog a transmutation?" (p. 41). Similarly, the varieties
of primrose that Lyell discusses as proof against transmutation are, to Wallace, proof positive *for* transmutation. "It only shows the impossibility of
convincing a person against his will," he commented with scorn. "Where
an instance of the transmutation is produced, he [Lyell] turns round & says
'You see they are not species they are only varieties.'" His frustration with
such arguments is evident in another entry on the case of the goatgrass
Aegelops ovata, reported (incorrectly, as it happens) by Frenchman Esprit

Fabre in 1854 to have been transformed into wheat. The suggestion caused a minor sensation at the time; even John Stevens Henslow, Darwin's Cambridge mentor, tried to replicate Fabre's results. Wallace commented on the case in both his 1854 notebook (LINSOC MS 179) and the Species Notebook where, sounding a bit irritated, he anticipates the response of anti-transmutationists: even when a member of one genus seems to be transformed into another genus, this would be dismissed by claiming the two are merely varieties of a single species.

Wallace concludes his entries relating to domestication with a pronouncement that echoes the main themes of the Sarawak Law and Ternate papers. On page 44 he commented that while "many of Lamarck's views are quite untenable & it is easy to controvert them, but not so the simple question of a species being produced in time from a closely allied distinct species"—the language of the Sarawak Law. The next entry, arguing against "reversion," states the main thesis of the Ternate essay (though that paper was still two or more years later than these entries):

> Changes which we bring about artificially in short periods may have a tendency to revert to the parent stock. This is considered a grand test of a variety. But when the Change has been produced by nature during a long series of generations, as gradual as the changes of Geology, it by no means follows that it may not be permanent & thus true species be produced. (p. 45)

D. Island species' relationship to those of the nearest mainland suggests an origin from that mainland; Islands of great antiquity have the most "peculiar productions" [endemic species], showing that much time is necessary for such species to arise (pp. 47–49)

Geographical distribution was one of Wallace's main interests, and his entries on this topic in the Species Notebook show that he interpreted biogeographical relationships in transmutational terms. He wondered why species found in "distant countries of similar climate" should differ, and commented upon the relationship between unique island species (endemics, to use the modern term) with species of the nearest mainland. "If they are special creations why should they resemble those of the nearest land? Does not that fact point to an origin from that land[?]" He drew on Darwin's Galápagos Islands observations in the *Journal of Researches* (Darwin 1845): "we find species peculiar to each island, & not one of them containing all the species found in the others as would be the case had one been peopled with new creations & the others left to become peopled by winds currents &c. from it." Anticipating Lyell's explanation for island-mainland relationships, he comments that "here we must suppose special creations in each island of peculiar species though the islands are all exactly similar in structure soil & climate & some of them within sight of each other" (p. 46). Such a view only introduces difficulties: "it may be said it is a mystery which we cannot explain, but do we not thus make unnecessary mysteries & difficulties by supposing special creations contrary to the present course of nature[?]" (p. 47). Wallace also points out that *recent* volcanic islands contain species from the nearest land, and "nothing peculiar!" (his emphasis, where "peculiar" = endemic). Very old islands, in contrast, are "peculiarly inhabited," leading him to conclude that "a long succession of generations appears therefore to have been requisite, to produce those peculiar productions found no where [*sic*] else but allied to those of the nearest land" (pp. 48–49).

Wallace argues that it would be strange if islands were "stocked" or "peopled" by change colonization by extant species for ages, but then all of a sudden, upon reaching a certain age, have "new & peculiar" creations appear. Making a point that resonates with modern ecological thinking, he says that this would be just when new species were *not* wanted, being well stocked for a long time: the newly created species "would hardly be able to hold their own against the previous occupiers of the soil & there would have to be a special extermination of them to make room for the new & peculiar species" (pp. 47–48). Invoking special creation of island endemics requires an increasingly complicated and implausible scenario of colonization, creation, and extinction, highlighting the inadequacies and inconsistencies of this view.

E. The "balance of nature"; Rarity versus abundance,
insect irruptions, and extinction: Wallace asks,
"Where is the balance?" (pp. 49–50)

In the natural theology tradition nature was seen to be in balance or harmony, its checks and balances like predator-prey interactions and disease seen as elements of an ultimately benevolent design. Wallace's engagement with the supposed "balance of nature" was discussed earlier in this chapter,

in the section "Designedness, Balance of Nature, and Ethical Considerations." In the Species Notebook both Lyell and Agassiz are taken to task.
Here, in his extended critique of the *Principles,* Wallace takes issue with
Lyell's pronouncement that all species are "adapted to the purpose" of preserving balance, charging that "this phrase is utterly without meaning" (p.
49). To Wallace the struggle that is in abundant evidence in nature gives lie
to the balance-and-harmony view: species displace other species, some are
common and some rare, some are driven extinct while others thrive. Lyell
may see the destruction brought about by irruptions of locusts or ants as
balance-preserving, but to Wallace this is simply evidence of struggle: "To
human apprehension this is no balance but a struggle in which one often
exterminates another," he states on page 50.

> F. Climate and earth gradually change, new species
> replace existing ones; Would not new species be modified
> forms of the previously existing ones? Rapid change
> may force migration or extinction, but gradual change
> should lead to modification (pp. 50–53)

As seen in the Sarawak Law paper, Wallace at this time understands that
the patterns of biogeography speak to species change. He returns to this
point toward the end of the Lyell critique. Gradual geological changes go
hand-in-hand with change in environment, but while Lyell asserts that as
landscape changes—the Sahara giving way to a chain of lofty mountains,
say—its species disappear and are gradually replaced by others "perfectly
dissimilar in their forms habits & organization," Wallace strongly disagrees.
"But have we not reason to believe they would be modified forms of the
previously existing Northern African species," he protests. "The climate
might then more resemble that of the W. Indies, but we know the productions would not resemble them" (p. 50). With a consistency that Lyell was
then incapable of, Wallace points out, "It would be an extraordinary thing if
while the modification of the surface took [place] by natural causes now in
operation & the extinction of species was the natural result of the same
causes, yet the reproduction & introduction of new species required special acts of creation, or some process which does not present itself in the
ordinary course of nature" (pp. 50–51). Lyell asserted that one species cannot change into another by a change of environment because other species
already well suited to those new conditions would preclude the new species

from becoming established. Wallace does not think this likely if new species come about very gradually. It was regarding this point that Wallace
inserted his memo to "Introduce this and disprove all Lyells [*sic*] arguments first at the commencement of my last chapter" on page 51.

G. Wallace challenges Lyell on the gulf between "rational" and "irrational" beings, the question of limited capacity for change, and the claim that change beyond a certain point is fatal to the organism (pp. 149–150)

In this final extract from the *Principles* that Wallace takes exception to,
Lyell puts forth an analogy between change from an irrational to a rational
being and change from simpler to more complex organisms (p. 149). He sees
discontinuity, not a gradual series with many intermediate steps—reflecting
the main reason for his resistance to transmutation, namely, its implications
for human origins (Bartholomew 1973). Wallace argues, on the other hand,
that discontinuity assumes an absolute distinction between reason and
instinct, and points out that Lyell's argument is semantic, depending on
the terms themselves. He further quotes Lyell on the reality of species
(pp. 149–150), where Lyell argues in his anti-transmutation "recapitulation"
that species can vary only so much and that varying beyond certain bounds
is fatal to the individual. Wallace says that Lyell assumes that "only change of
circumstances produce variety" and demands to know how, then, varieties
are constantly produced in the same place under the same conditions. Furthermore, Lyell cannot prove that variation will not continue "at a rate commensurate with Geological changes." "How can man's hasty experiments
settle this," Wallace asks in the final comment in this passage. He then points
out that Lyell's "'though ever so gradually' is a gratuitous assumption. What
are 'the defined limits',—he assumes that they exist."

The experiments he refers to are found immediately prior to Lyell's recapitulation: German physiologist and anatomist Friedrich Tiedemann (1781–
1861), who studied under Georges Cuvier, published a study of fetal brain
development in 1816 (English translation 1826) in which he argued that the
brain appears to pass, developmentally, through fish, reptilian, avian, and
other stages. Lyell grasped the transmutational interpretation of Tiedemann's ideas (Corsi 1978, 17) and attacked Tiedemann on this point in the
Principles, arguing that "on the contrary, were it not for the sterility im-
posed on monsters, as well as on hybrids in general, the argument to be

derived from Tiedemann's discovery, like that deducible from experiments respecting hybridity, would be in favour of the successive degeneracy, rather than the perfectibility, in the course of ages, of certain classes of organic beings" (Lyell 1835, 3:20). Wallace felt that the weighty question of limits of variation in the grand sweep of time cannot be settled by mere "hasty experiments," implying that Lyell is too quick to marshal any observations at all as ammunition in his anti-transmutation arsenal.

Wallace addressed other transmutation arguments throughout the notebook beyond the long and informative critique of Lyell. A recurring theme is the nature of species and varieties. For example, on pages 57–58 Wallace cites an account of the formation of a new rice variety, "Yu-mi" or "imperial rice," discovered as a "sport" (in modern terms) in the reign of the Chinese emperor Khang-Hi (1662–1723). This variety was found to differ from its parent species in color and growth period, which was shorter. Owing to its rapid ripening it was the only rice variety that could be grown north of the Great Wall, where the growing season is short. Wallace muses that if some grains were carried away by birds and propagated "in a country where the other kinds were not found," it would be seen as a new species given its differences in form and growth habit—a species endemic to the new country. This entry is followed by a note on a putative change (over many years) in the mint *Zizyphora dasyantha* in the Berlin Botanical Garden, which was seemingly altered so markedly that a German botanist (Heinrich Friedrich Link, the garden's director) said it might be a new species, which was soon dubbed *Zizyphora intermedia*. A few pages later (p. 62) zoologist Edward Blyth on species and varieties is quoted at length (Blyth 1835). Blyth, who was curator of the museum of the Royal Asiatic Society of Bengal, shared Wallace and Darwin's keen interest in the subject and had even conceived of a process of selection apparently unknown to the two naturalists. He does not appear to have been in contact with Wallace at this time, but was in contact with Darwin and had written in glowing terms to him about the Sarawak Law paper, recognizing it as a tour de force: "What think you of Wallace's paper in the Ann. M. N. H.? Good! Upon the whole!" (Darwin Correspondence Project, letter 1792). In the Species Notebook Wallace does not comment in detail on Blyth's classification of varieties, but in brief notes he questions whether Blyth's second category, individuals with "acquired variations," are ever propagated, and later penciled in an emphatic "yes –." Beddall (1972) discussed the influence of Blyth's ideas about species and varieties on both Wallace and Darwin.

A second recurring transmutational theme is geological. Entries on this theme fall into two general categories: entries bearing on geological phenomena and evidence of geological change such as recent uplift (e.g., pp. 92–93, 147, 153), and entries bearing on the fossil record and its inherent patterns. Two important patterns in the fossil record that concerned nineteenth-century geologists are (1) the so-called law of succession where extinct and living species of a given type are found in the same location—extant forms "succeeding" extinct forms of a given group in the same area (for example, extinct and living armadillos being found in South America), and (2) the "law of generalized to specialized structure," the idea that general form is found in older, lower fossil strata, while more specialized forms occur in younger, more recent strata. A third law, combining fossil data with morphological development, was the "law of parallelism"—the idea that there exists a correspondence between the morphological stages an organism exhibits through its development and the sequence of those stages, as forms or taxa, in the fossil record. This view of development was first articulated by the German-Estonian embryologist Karl-Ernst von Baer (1792–1876) in his 1828 treatise *Über Entwickelungsgeschichte der Thiere* ("On the development of animals"). He held that animal embryos appear similar early in development and become less similar, or more differentiated, over time as successively more specialized structures become manifested. This means that the closer species are taxonomically, the longer they will resemble one another in development. The "laws" pertaining to fossils could be (and were) interpreted in terms of successive creations and extinctions over time, implicitly or explicitly according to a divine plan. Neither Lyell nor Richard Owen, the eminent paleontologist and comparative anatomist, saw a divine hand *directly* involved in this process. Lyell had a commitment to changes in the earth (both nonorganic and organic) according to natural law, not miraculous interposition (e.g., Rudwick 1970, 10–11), and Owen, though often (and erroneously) pegged as a strictly anti-transmutation typologist, had his own complex view of species change inspired by von Baer (e.g., Ospovat 1976; Richards 1987; Camardi 2001, 496–501).

There are several references to the geological record in the Lyell critique, and further entries on this head are found throughout the notebook. A few of these pertain to papers by Owen. Although he was later to become bitterly opposed to Wallace and Darwin's transmutation ideas and adhered to the end to the "archetype" concept (a term he coined), Owen did as much as anyone to help establish the reality of the so-called law of succession and the "law of generalized to specialized structure"—the patterns of sequential change in the fossil record and the affinities between extinct and extant groups that Wallace and Darwin would see as not only consistent with their conception of transmutation, but necessitated by it. On pages 54–55 Wallace perceptively picks up on just such an observation by Owen, in a paper from the May 1855 *Proceedings of the Geological Society of London*. Extinct reptiles, Owen states, exemplify the "law of the prevalence of a more generalized structure as compared with the more specialized structures of existing species." *Rhynchosaurus* is a saurian with turtlelike characters, labyrinthidonts are saurians with froglike characters, *Icthyosaurus* "had modifications borrowed from the class of fishes" and pterodactyls "thus borrowed from the type of Birds & Bats," and so on. What does this mean?

To Wallace the meaning was clear: "The above is what might be expected, if there has been a constant change of species, by the modifications of their various organs, producing a complicated many branching series." He continues: "Those nearer the base must exhibit to some degree a combination of those characters, which in a higher developed condition are characteristic each of one group of animals which have since come into existence." Note that Wallace is articulating the concept of the intermediate nature of common ancestors here. This is another example of an observation by Wallace paralleling Darwin's insights in *On the Origin of Species*. The same goes for the long entries on pages 92–97 regarding gaps in the fossil record. The sudden appearance of novel forms of life in a series of strata suggests special creation, but Wallace points out how this is an artifact of the sedimentation and fossilization process (which varies locally as land is elevated and subsides) as well as the effect of destruction of exposed strata by erosion. "Interruptions therefore in the series of organic remains sometimes for small sometimes for immense periods must be expected. Continuity must be the exception not the rule," he forcefully argues. Like Darwin in his *Origin* chapter on geological succession, Wallace concludes that "the changes in the organic as well as in the inorganic would have been continuous & gradual in the record"—the *complete* record—but as it is, the record is "discontinuous and incomplete." He further elaborates on the nature of geological and species change and the patchy and discontinuous nature of fossilization, giving a remarkably modern explanation for the gaps in the fossil record.

Geological succession also comes up in some of the notes Wallace made on Louis Agassiz's book *Lake Superior* (Agassiz and Cabot 1850). Agassiz was a believer in fixity of species, but his scientific work was nonetheless seen by Wallace and Darwin as supporting species change. Owen, Agassiz, and others articulated the "law of parallelism," which described the concordance between the pattern of succession in the fossil record and the embryological stages that vertebrates pass through in development. That is the context for the entry on page 145 (with Wallace's emphases): "In Molluscs the naked gasteropoda [*sic*] & Cephalopoda are higher than the testaceous. This agrees with Geolog[ical] succession <u>Ammonites</u> before <u>Belemnites</u>. In crustacea <u>crabs</u> in early stages have a lobster form—[therefore] Brachyarea [a crab] higher than Macroura [lobsters] and are geologically later."

In his fine book *In Darwin's Shadow: The Life and Science of Alfred Russel Wallace* (2002), Michael Shermer pointed out the irony in William Whewell's rejection of the Wallace-Darwin theory: that the brilliant philosopher who coined the term "consilience" and defined it as a powerful mode of inductive argument should reject what is "arguably the most consilient theory ever generated" (Shermer 2002, 204), even banning *On the Origin of Species* from the Trinity College library at Cambridge. If the idea behind Whewellian consilience is that confidence in an explanation increases in proportion to the separate lines of evidence supporting that explanation, then are we not all the more confident when different individuals independently adduce much the same consilient evidentiary lines? We need Whewell himself, inveterate coiner of terms, to provide a resonant name for this situation; for want of something better we might consider Wallace and Darwin's case one of superconsilience: "supersilience."

THREE

Wallace and Darwin

Parallels, Intersections, and Departures on the Evolutionary Road

WALLACE'S PAPERS, notably but not exclusively the Sarawak Law and Ternate papers, show him assembling the key building blocks of a cogent argument for species change: (1) evidence supporting the contention that transmutation occurs, (2) a model for how this occurs in the context of Lyellian
change in earth and climate, and (3) a plausible natural mechanism for species change. The Species Notebook fills in many "behind the papers" details
about Wallace's thinking, in much the same way that the Transmutation
Notebooks do for Darwin. This chapter explores the commonalities and differences in the sorts of information drawn upon by the two in their arguments for transmutation—the lines of evidence they used and their sources
of inspiration, observations, and arguments. I will do this largely through
the lens of the Species Notebook in Wallace's case and *On the Origin of
Species* and pre-*Origin* notebooks and manuscripts in the case of Darwin.
Natural selection itself will be considered in Chapter 5. Consider the analysis here an extension or elaboration of Chapter 2 on Wallace's "consilience"
argument, pursuing his ideas in the Species Notebook in greater detail. I
will also discuss Darwin in comparison to Wallace to a greater degree than
I did in Chapter 2. This analysis should not be mistaken as an effort to measure Wallace's worthiness as a scientific thinker in relation to Darwin.
Rather, since his early views are less widely known and appreciated than
those of Darwin, the only other person to have both adduced diverse forms
of evidence in support of the idea of transmutation and offered a solution
to its mechanism, we benefit all the more by understanding Wallace's path
in relation to that taken by his more famous colleague.

Comparisons have their pitfalls, however. Although the thinking of Wallace and Darwin may be thought to have converged at the point in 1858 when both had a more or less complete theory—a branching model of transmutation plus the principle of natural selection as a mechanism of change—it is important to bear in mind that they have very different starting points and trajectories of thinking. Darwin had discovered the mechanism of species change by late 1838, less than two years after becoming a transmutationist, while Wallace was a transmutationist for nearly thirteen years before making his own discovery of the mechanism. This means that, at least as reckoned by time, a comparison of their views as of 1858 is not entirely valid or fair insofar as Darwin had far longer to think through the implications of and evidence for transmutation *and* natural selection. In regard to natural selection in particular Wallace's paper on the subject was written rather quickly following his sudden insight. Wallace's explorations pre-1858 thus consist of the construction of a pro-transmutation argument without a mechanism, while Darwin labored longer to adduce lines of evidence for both the reality of transmutation and the mechanism of natural selection. Indeed, evidence for the two often go hand-in-hand for Darwin: in a number of places in *On the Origin of Species* he discusses observations that, he argues, make sense only if natural selection occurs, and by extension this is used to further his transmutation argument.

With this caveat, the parallels in Wallace's and Darwin's thinking are well worth exploring as a way of mapping more precisely the similarities and dissimilarities in the early transmutational thinking of these two naturalists. I will reveal at the outset the main lesson from this chapter: Wallace and Darwin labored along strikingly similar paths as they gathered evidence for transmutation, often making the same sorts of observations, consulting many of the same authorities, and crafting many of the same pro-transmutation arguments: variation in structure and habit; the significance of islands; homology-based affinity; geological succession; the gradual nature of change in both inorganic and organic worlds; the nature of species, varieties, and variation; competition, population pressure, and a want of real harmony in nature; the highly complex but nonetheless imperfect nature of bees' cells; variation in and errors of instinct and behavior; and more. They even both had opportunities to observe closely the behavior of a young orangutan, with an eye to its humanlike characteristics. Differences will become more evident too, both in their interpretation of certain observations and in some of the evidentiary paths explored.

Simultaneity and Similitude

It is a widely held view that Wallace and Darwin discovered a natural mechanism of transmutation simultaneously. There is only an element of truth in this, an element that depends on taking a long view. Insofar as the idea of species change was considered, more or less seriously, by at least some naturalists for several centuries, one might view the embrace of the concept and the discovery of a naturalistic mechanism for it by two Englishmen more or less in the middle of the nineteenth century as having occurred simultaneously. But could this send investigators off on a snipe hunt, looking for all manner of sociocultural reasons for the putatively simultaneous discovery, or even suggesting that Wallace and Darwin's accomplishment was not very remarkable, as evolution was "in the air" and about that time almost any investigator could have been the lucky one to pluck it from the zeitgeist? But Wallace and Darwin were not just anyone, and there are good reasons why profoundly intelligent naturalists who thought long and hard about whether or how species change might occur were consistently far off base, incapable of envisioning a messy branching and rebranching manner of change based on a populational process driven by external forces (Bowler 2003, 139–140).

The "simultaneity" of Wallace's and Darwin's discovery is relative, rather like the apparently sudden appearance of complex life forms in the lower Paleozoic—a problem that Darwin defuses near the end of chapter 9 of the *Origin* by pointing out that "we do not make allowance for the enormous intervals of time, which have probably elapsed between our consecutive formations." The time between the respective evolutionary discoveries of these naturalists is not great even in human terms, but it is not trivial either. Wallace and Darwin were converted to the idea of transmutation eight years apart; Wallace was twenty-two years old at the time he became convinced of the reality of transmutation in 1845, while Darwin was twenty-seven years old when he had his revelation in late 1836 or early 1837. They discovered the principle of natural selection twenty-one years apart: Darwin's eureka moment came little more than a year after becoming a transmutationist, while Wallace's insight came after a dozen years. From the vantage point of the early twenty-first century, there is a tendency to compress the Wallace-Darwin timeline and to see Wallace's and Darwin's insights as nearly coincident when in fact the two naturalists were on different temporal and in some measure intellectual trajectories of discovery. Appreciating this fact is an important first step toward realizing the independent nature of their respective insights.

If simultaneity can be seen as a relative thing dependent on time scale, so can identity of views be seen as a matter of degree. The evolutionary thinking of Wallace and Darwin bear some striking similarities but also important differences. The similarities stem from their shared sources of inspiration and information, as well as parallel experiences. As young men both were profoundly influenced by Alexander von Humboldt, and both were keen botanists and insect enthusiasts reveling especially in beetles, a group that could only have heightened their awareness and appreciation of variation (Berry 2008; Berry and Browne 2008). Each traveled to exotic locales, Darwin before he became a transmutationist and Wallace afterward. And when it came to natural selection, both men cited Malthus's Enlightenment-era *Essay on Population* as a touchstone inspiring insight into the broader significance of struggle and population pressure.

It is to be expected that broadly contemporaneous individuals interested in the same subject would have been reading much the same literature, but in the case of Wallace and Darwin in those critical years leading up to 1858 this is only partially true. Wallace cited, discussed, extracted, and critiqued thirty-nine authors in the Species Notebook. Darwin, in his Transmutation Notebooks of 1837–1844 (Barrett et al. 1987) and *Natural Selection* manuscript (precursor to *On the Origin of Species;* Stauffer 1975), cited just over half of these very same authors, and in eight (21 percent) of the cases he cited the same work as Wallace (see Appendixes 1 and 2). This does not mean Darwin and Wallace zeroed in on precisely the same information in these works, however. Of the eight works cited by both Wallace and Darwin, they refer to the same information in only three instances, and then only broadly (von Buch on islands and Lyell on a number of topics) or without comment (D'Urville on fern distribution). As an aside, Wallace cited Darwin in the Species Notebook (*Journal of Researches,* regarding observations of the Tucotuco), and Darwin in turn cited Wallace in his *Natural Selection* manuscript (pp. 339–380, in the chapter "Difficulties on Theory").

In all other cases Wallace and Darwin picked up on different things in the works they read in common. For example, Wallace cited French naturalist René Lesson (1794–1849) in connection with human races and variation, while Darwin was concerned with Lesson's observations on the geographical distribution of certain mammals, and some of his geological observations. Wallace cited adventurer William Dampier (1651–1715) on volcanoes, while Darwin noted the adventurer's observations of shoals and flats, and a curious account of a dissected shark found to have a rather fresh-looking hippo head in its stomach. Évariste Huc (1813–1860), a French missionary who journeyed to China, gave an intriguing account of the origin of "imperial rice" (in essence the origin of a new variety) that caught Wallace's attention. Darwin did not note this, but instead cited Huc regarding a night-flowering wheat variety in a section on outcrossing in his *Natural Selection* manuscript.

On occasion Wallace and Darwin noted the same information, but from different sources. A good example is the report by the noted German botanist Heinrich Friedrich Link (1767–1851), director of the Berlin Botanical Garden. Link maintained that a species of the mint *Ziziphora* in his botanical garden "changed"—transmutated, or otherwise gave rise to—a new form which John Lindley dubbed *Z. intermedia,* a name still recognized today. Wallace extracted this information from Lindley's *Introduction to Botany* (see p. 59 of the Species Notebook), while Darwin recorded the same observation attributed to Link from Bronn's *Handbuch einer Geschichte der Natur* (Bronn 1842–1843; Stauffer 1975, 127).

Great Minds

To gain an appreciation of the commonalities and differences in Wallace's and Darwin's interests and sources, it is useful to step back and consider broadly the major topics, themes, and observations of the Species Notebook that have a direct correspondence with Darwin's writings (notably the Transmutation Notebooks, *Natural Selection,* and *On the Origin of Species*). Identifying discrete topics has an element of subjectivity, but for the sake of discussion I consider Wallace's "evolutionary" Species Notebook entries in relation to Darwin in six thematic areas (largely but not precisely mirroring the notebook themes discussed in Chapter 1): Geology and Paleontology, Morphology and Affinity, Geographical Distribution, Instinct and Habit, Human-Primate Relationship and Human Variation, and Arguments and Observations for Transmutation. These are summarized in Appendix 3.

1. Geology and Paleontology

Wallace and Darwin shared the Lyellian view of a continuously and gradually changing earth, one dominated by slow uplift and subsidence of different regions at different times. They both saw, too, Lyell's inconsistency
in articulating a vision of continuous change in the earth even while arguing strenuously for the impossibility of organic change. Wallace says as
much on page 35 when he charges Lyell with being "unphilosophical" on
this point. Darwin never criticized his friend in such strong terms, prefer-
ring a long-term approach to wearing down Lyell's resistance to the idea
of transmutation.

Both also addressed the intermediate nature of common ancestors, gaps
in fossil record, geological succession and its continuous and gradual nature, and the parallelism between development and the fossil record (on this
last point, both cited Agassiz's work—Agassiz, like Lyell, elucidated phenomena that Wallace and Darwin saw as supporting the idea of transmutation,
though he himself did not). The points they make in all of these cases are
quite similar. Perhaps the most insightful of these observations concerns the
"intermediate" morphology of putative common ancestors of two species or
species groups: "Those [species] nearer the base must exhibit to some degree
a combination of those characters, which in a higher developed condition are
characteristic each of one group of animals which have since come into existence," Wallace wrote on page 54 following a passage extracted from Richard Owen. The idea of common ancestors as linking forms that have features
in common with descendant groups is a crucial departure from earlier conceptions of transmutation—a key aspect of this model of change is its branching pattern. (That Wallace had a branching model of transmutation is clear
from several notebook entries, discussed in the next section.) In making the
same point as Wallace about intermediate forms, Darwin also cites Owen,
albeit a different work. In the *Natural Selection* manuscript (p. 384) and
again in the *Origin* (Darwin 1859, 329), Darwin made references to a paper
by Richard Owen published in 1848 describing the teeth of extinct pachy-
derm relatives that reveal a relationship with certain ruminants. The discovery of these Eocene-age linking forms compelled a taxonomic reorgani-
zation recognizing the relationship of pachyderms and ruminants, groups
that were formerly considered to be widely divergent and unrelated.

Even on the one geological topic where Wallace and Darwin seem to
disagree—namely, on theories of the origin of erratic blocks and "drift"

deposits—the two can be seen as ultimately having the same position. The relevant passage in the Species Notebook is on pages 142–143, headed "Erratic drift theories." Lyell had long maintained that erratic blocks (boulders, sometimes of immense size and number, that occur in localities far from the nearest source of that particular rock type) were deposited in their current location by icebergs in a distant and colder epoch when the sea encroached upon the land. Indeed, there is abundant evidence that icebergs do sometimes carry rocks and other debris, and Darwin himself published an observation of such in 1839. The ice sheet theory of Agassiz and Charpentier suggested an alternative explanation: rather than ice transport via oceangoing icebergs, continental ice sheets may have accomplished the same thing on dry land. Lyell eventually came to appreciate that both processes operated, but in the mid-1830s the idea of widespread ice sheets had not yet been proposed (the theory did not come on the scene until the 1840s, while the edition of Lyell's *Principles* used by Wallace at the time of the Species Notebook was dated 1835). In that edition of the *Principles* Lyell was still stuck in his iceberg mode, a model requiring significant incursions of the ocean onto the land in order to account for erratic blocks. Under the iceberg transport theory these blocks were carried over the present land by floating ice during cold periods when the sea extended far inland. Wallace thought of a fatal argument undermining the iceberg theory which appeared to have eluded both Lyell and Agassiz: if the erratic blocks around Lake Superior were carried by icebergs, the depression forming the great lake would have been beneath the sea, leaving a salt lake behind when the ocean receded. That the lake is freshwater proves, according to Wallace, that it was formed *since* the last elevation of the region above sea level. Continental ice sheets would seem to be the better explanation, given that erratic blocks were accepted by everyone to be of relatively recent origin. Darwin eventually came to accept the glacial theory too, although like Lyell never completely gave up a role for iceberg transport.

2. Morphology and Affinity

With regard to various subjects relating to morphology and affinity—structural relationships and transitional forms, the utility of embryology in informing classification, and a clear conception of an arborescent (as opposed to ladderlike) pattern of evolutionary relationship—Wallace and Darwin were broadly in agreement. One topic of special interest was the significance of the relationship between embryology and classification. This was an area of interest to Agassiz, following Karl Ernst von Baer and Henri Milne Edwards, who showed that traits in evidence during successive stages of development reveal affinities of relationship that can help resolve classification problems based on adult characters (Ospovat 1981, 117–129). None of these men interpreted this in an evolutionary context, but Wallace and Darwin were quick to see the significance for transmutation. They made use of different works by Agassiz: Wallace's observations (pp. 144–145 of the Species Notebook) come from *Lake Superior* (Agassiz and Cabot 1850), while Darwin drew his information from Agassiz's later writings such as the 1857 *Essay on Classification*.

Another topic of interest is the articulation of the branching or treelike pattern of evolutionary relationship, which is found relatively early in the Species Notebook. On page 37, in noting Lyell's claim that the "Didelphys of the Oolite"—a fossil marsupial found in Mesozoic formations—"is fatal to the theory of progressive development," Wallace says, "Not so." Lyell thought that transmutation, á la Lamarck, entails a model of progressive and wholesale replacement of groups one after another. This was an oversimplification of Lamarck, who after all did allow for limited evolutionary branching (e.g., Lamarck 1809, 179), but it is fair to say that even Lamarck might have paled at the shrubs and trees advanced by Wallace and Darwin (for one thing, Lamarck held that plants and animals arose and developed along separate lines). Lyell rejected any model of species change, branched or linear. Lyell's point in the passage cited by Wallace was that a member of a supposedly "higher" or more "advanced" group (in this case *Didelphys*) could not possibly be found sharing the same geological strata with earlier forms (the "Oolite" dates to the reptile-dominated Jurassic), as that would contradict the progressive, linear, stepwise mode of change that his opponents advocated. Wallace correctly points out that it matters not if members of so-called advanced groups co-occur with earlier groups, as long as they do not predate those groups altogether. Why? Because if those "advanced" groups branched out of early members of the "lower" groups, their lineages would exist alongside one another. He concludes: groups branch from other groups, and "they then go on in parallel or diverging series." Early in his own evolutionary musings Darwin came to precisely the same view, as exemplified, for example, by the tree diagrams of his B Notebook of late 1837–early 1838. For his part, Darwin's comments on *Didelphys* largely concerned the origins of groups like mammalia and their fossil record. In Transmutation Notebook B entry B87–88, for example, Darwin commented on *Didelphys* as "the father of all Mammalia in ages long gone past," in the context of the incompleteness of the fossil record and the lack of complete intermediate series for many groups. Nuanced discussion of branching order seen in Wallace is not found in Darwin's notebooks, at least in connection with this early marsupial.

As a side note, *Didelphys* also led to an interesting comment by Darwin in Transmutation Notebook E. In entry E128 (Barrett et al. 1987, 434) Darwin noted that Cambridge philosopher and mathematician William Whewell discussed the "controversy on Didelphys" in his presidential address to the Geological Society for 1838. The controversy had to do with the identification of the fossils as a marsupial (see Whewell 1839, 86–89)—Owen, following Cuvier, had identified it as such, but some other naturalists questioned the identification, suggesting that given the age of the Oolite it was a reptile of some kind. The issue was eventually decided in Owen and Cuvier's favor. Whewell dedicated so much time in his address to the issue because, he said, it involves "considerations . . . of the most vital importance." He continued: "The battle was concerning the foundations of our philosophical constitution; concerning the validity of the great Cuvierian maxim,—that from the fragment of a bone we can reconstruct the skeleton of the animal. This doctrine of final causes in animal structures, as it is the guiding principle of the zoologist's reasonings, is the basis of the geologist's views of the organic history of the world." Darwin quoted Whewell's next statement: "If we cannot reason from the analogies of the existing to the events of the past world, we have no foundation for our science." Darwin then remarked in his notebook, "It is only analogy, but experience has shown we can & that analogy is sure guide & my theory explains why it is sure guide"—a key indicator of Darwin's philosophical view at the time. Michael Ruse, in writing of "Darwin's debt to philosophy" in *The Darwinian Paradigm,* suggested that this notebook entry reflected Darwin's view of domestication as a powerful analogy for natural selection (Ruse 1989, 25). Thus *Didelphys* provoked noteworthy and in some ways rather different responses from Wallace and Darwin.

One final point on the topic of morphology concerns a case not of Wallace and Darwin converging on a common source per se, but an instance in the Species Notebook where Wallace quotes Darwin. This concerns the semiblind burrowing rodent *Ctenomys,* or tucotuco, which Darwin described in his *Journal of Researches* (1845). Darwin first commented on this curious South American rodent in the first edition of the *Journal* (1839), reporting that an Englishman residing in the area asserted that its frequent blindness stems from inflammation of the nictitating membrane of the eyes. In 1839 Darwin wrote, "Considering the subterranean habits of the tucotuco, the blindness, though so frequent, cannot be a very serious evil; yet it appears strange that any animal should possess an organ constantly subject to injury," and then passes on to a comment about moles. Perhaps feeling mischievous, in the second (1845) edition he inserted this comment following the passage quoted above: "Lamarck would have been delighted with this fact, had he known it, when speculating . . . on the gradually-*acquired* blindness of the Aspalax, a Gnawer living under ground, and of the Proteus, a reptile living in dark caverns filled with water" (emphasis Darwin's). Why did he insert this comment, and why did he emphasize the word "acquired?" Like his passages on the species of the Galápagos, pregnant with meaning, his mention of Lamarck is an indication, at least, that he himself was thinking in transmutational terms (though not tipping his hand). And so too was Wallace, who on page 60 of the Species Notebook copied out Darwin's passage on Lamarck. On the next page Wallace speculated that the eyes of nocturnal species are either very large and sensitive or small and imperfect; in the latter case, they become imperfect "for want of use."

3. Geographical Distribution

The distribution of species on the globe was, as we have seen, a subject of keen interest not only to Wallace and Darwin but also to nineteenth-century naturalists generally. An ever-more-detailed understanding on the number of distribution of species was widely believed to give insight into "creative force" if not the plan of the creator—a pursuit pioneered by Humboldt, who was greatly admired by both Wallace and Darwin. Wallace does not discuss species distribution at length in the Species Notebook, but his particular interest in the subject is revealed in the many entries consisting of brief notes, lists, or summaries of species numbers, often in a comparative context, taken from a diversity of sources (Mary Somerville's *Physical Geography*, Zollinger and Mousson on mollusks, von Buch on the flora of the Canary Islands, Pieter Bleeker on fish, etc.). Many of these entries note numbers of species that are common to different locales, or are "peculiar" (endemic) to particular ones. One of Wallace's entries reflects the quantitative approach Humboldt advocated in his "botanical arithmetic," computing ratios of species and genera of different groups and comparing these between regions, at different latitudes, and so on. On page 110 of the Species Notebook Wallace compares the richness of terrestrial mollusk species ("land shells" as they were called then) in Jamaica versus Java, calculating the ratio in relation to area and noting that "Jamaica only 1/10th area has more than 10 times as many shells," and "in proportion to area more than 100 to 1!!!" There are many instances of Darwin making the same kinds of calculations, for example, entries B156–157 in his Transmutation Notebook B, explicitly citing Humboldt and recording ratios of genera to species on different islands. Elsewhere (e.g., B204; Barrett et al. 1987, 222) he notes the differences in land shell species between islands, relating more directly to Wallace's mollusk entry.

Wallace and Darwin also comment on species distribution in relation to climate. This comes up in two interrelated contexts: distribution in response to changing climate (here both Wallace and Darwin take their cue from Lyell and his model of ever-changing climate, envisioned as a slow, gradual process) and the crucial observation that geography and not climate is the more reliable determinant of species relationship. Wallace, reading Agassiz, made note of the close relationship between the flora of eastern North America and eastern Asia on page 145, while Darwin investigated precisely the same phenomenon through his American botanist friend Asa Gray. Wallace was interested in Agassiz's argument that the eastern North American and eastern Asian regions are older than western North America and Europe, supporting this with the "extraordinary and unexpected" fact that fossil plants found in Europe resemble modern trees and shrubs of eastern North America. Darwin's interest, too, was in shifting geographical distribution over time. At the time he was investigating possible mountain "highways," as he put it, to explain the occurrence of Northern Hemisphere plants deep in the Southern Hemisphere as earth's climate slowly cooled and then warmed. He prevailed upon Gray to summarize relationships of the North American flora for him, one happy outcome of which was Gray's important paper "Statistics of the Flora of the Northern United States" (Gray 1856). While earlier observers knew something of the striking correspondence between eastern North American and eastern Asian plant groups, in this paper Gray took this to a whole new level with his comprehensive tabulation of the pattern. Darwin wrote to Gray in October 1856 upon reading this paper: "Nothing has surprised me more than the greater generic & specific affinity with E. Asia than with W. America. Can you tell me . . . whether climate explains this greater affinity? or it is one of the many utterly inexplicable problems in [Botanical] Geography?" (Darwin Correspondence Project, letter 1973).

Also in the vein of broad patterns of species relationship, on page 45 of the Species Notebook Wallace notes that Lyell goes on at length about accidental dispersal explaining species similarities between distant locales, while "the matter of wonder has always been that in distant countries of similar climate so many should be <u>different</u>" (Wallace's emphasis). This is a key aspect of Wallace's Sarawak Law: if it is true that all species have "come into existence coincident both in space and time with a pre-existing closely allied species," it follows that related species will be geographically proximate, at least on a broad regional scale. Wallace is even clearer on this point on pages 50–51: Lyell described how as climate changes in, say, North Africa, the region "would gradually become fitted for the reception of a population of species perfectly dissimilar in their forms habits & organization." Wallace disagreed, noting that the new species would be modified forms of species previously living there—and so would be related to them. "The climate might then more resemble that of the W. Indies," he writes, "but we know the productions would not resemble [those of the West Indies]." The best predictor of species relationships in these respective regions is not their tropical climate.

This is a theme echoed by Darwin in several places in the Transmutation Notebooks, but he had not quite gotten to that topic in his *Natural Selection* manuscript (having been interrupted in its writing partway through geographical distribution when Wallace's manuscript arrived in early 1858). His first geographical distribution chapter in the *Origin* opens with this very observation, however: "In considering the distribution of organic beings over the face of the globe, the first great fact which strikes us is, that neither the similarity nor the dissimilarity of the inhabitants of various regions can be accounted for by their climatal and other physical conditions" (Darwin 1859, 346). This is followed by an argument paralleled by one Wallace made earlier: Darwin noted in chapter 11 of the *Origin* how the similarity of climate in continental areas between 25° and 35° south latitude does not predict species affinities (while geography does). Similarly, along the north-south latitudinal and environmental gradient of, say, South America the diversity of environmental conditions does not predict broad species affinities, but their shared geography of the South American continent does. Wallace made precisely the same argument in his 1857

Annals paper on the Aru Islands (Wallace 1857c), contrasting the species *differences* between climatically and topographically similar New Guinea and Borneo with the species *similarity* of climatically and topographically divergent New Guinea and Australia.

Perhaps the most important area of common interest for Wallace and Darwin, at least with regard to geographical distribution, concerns islands. The Swiss botanist Augustin de Candolle, a follower of Humboldtian botanical arithmetic, had highlighted the special nature of islands in his *Géographie botanique* (de Candolle 1820), noting that the high genus-to-species ratio found on remote islands meant that the creative force had clearly been more active in such places. Islands had henceforth assumed a special place in the study of "centers of creation" as well as in regard to the mystery of species origins, and so it is unsurprising that Wallace and Darwin were especially interested in island flora and fauna. Both commented that the species of a given island or archipelago tend to be related to species of the nearest mainland, drawing the inference that these species stemmed from colonists originating on that mainland. We find Darwin commenting in his Transmutation Notebooks, for example, on island flora being "analogous to nearest continent," their poorness in species "in exact proportion to distance" (B156) or his excited note, "The creative power seems to be checked when islands are near continent: Compare Sicily & Galapagos!!" (B160; Barrett et al. 1987, 210).

Wallace was blunt on this point on page 46 of the Species Notebook: "In a small group of islands not very distant from the main land, like the Galapagos, we find animals & plants different from those of any other country but resembling those of the nearest land. If they are special creations why should they resemble those of the nearest land? Does not that fact point to an origin from that land." He takes this an important step further, too, noting that different island groups contain different sets of species: "Again in these islands we find species peculiar to each island, & not one of them containing all the species found in the others." This insight may have eluded Darwin early on, who to my knowledge did not comment in his notebooks on the phenomenon that modern biogeographers term "disharmony," the seemingly random composition of flora and fauna of archipelagos in relation to groups represented on the nearest mainland and to each other. Disharmony stems from the stochastic nature of island colonization, something that Wallace clearly grasped. Darwin later touched on disharmony and a host of other intriguing aspects of island biogeography in his section on oceanic islands in chapter 12 of *On the Origin of Species.*

Wallace also expressed a sense that geological antiquity of islands might somehow be related to the observed degree of species endemism. Near the very beginning of the Species Notebook, for example, he recorded the numbers of plant species of the Canary Islands and St. Helena, in both cases noting the number that were "peculiar," or endemic. Significantly, on the very next line he wrote, "? the Geological age of these islands." Later (pp. 48–49) he wrote that "none of the islands which we have any reason to believe have been formed, since a very late geological era, are inhabited by such peculiar species. They generally have not one species peculiar to themselves." In the margin Wallace noted that "this must be proved." He continued: "On the other hand islands which are thus peculiarly inhabited, appear to be of a considerable antiquity. A long succession of generations appears therefore to have been requisite, to produce those peculiar productions found no where else but allied to those of the nearest land." The idea is a good one, but Wallace does not name the young islands thought to have formed "since a very late geological era," and there are no further entries on the subject. The few corresponding musings on the significance of island age found in Darwin's writings are rather general. Transmutation Notebook entry B152 (Barrett et al. 1987, 207) reports that the French naturalist and explorer Jean-Baptiste Bory St. Vincent (1778–1846) "considers [species] in recent volcanic islets not well fixed." Later, in entry E135 (Barrett et al. 1987, 436) Darwin wondered whether geographical isolation is more efficient than time in making new species—mere antiquity, he means, may not be as important as isolation. This may be why neither Wallace nor Darwin addressed island antiquity much, namely, because the overwhelmingly important factor in determining endemism is island isolation or remoteness from continents, while there were few examples of very young *and* remote islands with which to make an adequate comparison.

Wallace made reference to the Galápagos Islands twice in the Species Notebook, and his source of information in both cases is almost certainly Darwin's *Journal of Researches* (1845). The first entry, concerning the relationship of Galápagos species to the South American mainland, has already been mentioned. The second (notebook p. 48) relates again to island age. Wallace cites the Galápagos as part of an argument about islands being "stocked with their peculiar species immediately on their being raised from the ocean," and makes an interesting ecological argument suggesting that well-established resident species will have the best chance of resisting newcomers. There are other instances of Wallace advancing the idea of ecological or competitive exclusion in the Species Notebook, such as his hypothesis for the origin of tropical grasslands of Gilolo and the Orinoco (pp. 108–109). Darwin's view differs, at least in the context of islands, arguing in the *Origin* that island species will almost always yield before introduced continental species.

A final aspect of geographic distribution to consider brings us back to the Canary Islands. Recall that Wallace and Darwin were impressed with the African affinity of the flora of this archipelago, and with Leopold von Buch's assertion that permanent varieties change into species (see Chapter 2). Recall, too, that Wallace was impressed with von Buch's report that plants on the peak of Teneriffe (Teide, elevation 3,718 m) freely meet and cross, and as a result "do not form varieties & species," while those species confined to the low-elevation valleys "often have closely allied species confined to one valley or one island"—an observation that can be interpreted in terms of allopatric speciation. Darwin, in contrast, took note of an 1837 paper by Barker-Webb and Bertholet, who commented that the vegetation of the peak region of Teide "is altogether original." In his Transmutation Notebook C, entry C184 (Barrett et al. 1987, 296), he attributed this to the peak vegetation "being oldest & having undergone changes." The observations that Wallace and Darwin focused on can be seen as different facets of the same phenomenon: von Buch's suggestion that there are no or few varieties and species may reflect extremely low species diversity, as would be expected at high elevation on an island. At the same time, the few plant species that are found there include quite a few "altogether original" endemics, which is indeed the case.

4. Instinct and Habit

Wallace and Darwin were interested in several aspects of instinct and habit, as well as behavior generally. As with other areas of correspondence, in some cases they have coincident interests and in other cases they had different uses or came to different conclusions. The subjects of ants and aphids and of bees' cells are treated rather differently by the two, for example. Wallace simply reported observations of ants tending aphids, while Darwin discusses these insects in the context of natural selection and the evolution of their relationship. Wallace's interest in how honeybees construct their hexagonal cells lay in attacking the idea that bees are exemplars of the "supremacy of instinct"—he takes issue with their "perfection" and alleged solution of an "abstruse mathematical problem" (p. 177). Comb-building by bees was considered the epitome of complex instinctive behavior—bees were often cited as veritable mathematicians of the insect world, their structures evidence of design. Wallace attacks this idea on two fronts. First, he suggests in one passage (p. 168) that the cell-building behavior of bees may not be instinctive. They may build from experience, for all we know, as no one has reared bees apart from a hive from the time they are larvae to see if they can still build their cells. Second, he attacks the idea that bees' cells are in any way "perfect," arguing that on the contrary they are wasteful and do not, despite claims to the contrary, employ the best solution to their problem of honey storage.

In Transmutation Notebook N, Darwin too questioned whether the bees' cell-building behavior is truly instinctive, or if experience plays some role. He takes a different tack, however, in *Natural Selection* and the *Origin*. In the *Origin* he grants that bees exhibit a wonderful behavior, using language similar to Wallace's in stating that "we hear from mathematicians that bees have practically solved a recondite problem" (Darwin 1859, 224). He endeavors to show how natural selection could result in this by favoring economy of wax, and traces out a transitional series with bumblebees and *Melipona* bees. Darwin's ultimate aim was to argue that regardless of how marvelous bees may be, "all this beautiful work can be shown . . . to follow from a few very simple instincts." As an aside, note that when Wallace rallied to Darwin's defense in his paper "Remarks on the Rev. S. Haughton's Paper on the Bee's Cell, and on the Origin of Species" (Wallace 1863b), he employed his argument from the Species Notebook: "The same mathematical knowledge that enables us to see the beauty and economy of the form of the individual cells, as surely points out the great waste of material in building the upper and lower portions of the comb of the same thickness and strength. We have here, I think, a conclusive argument against the notion that the bees are guided by any supernatural impulse to construct their cells on the best mathematical principles, so as to economize, in the highest degree, labour, space, and material." In this paper Wallace also supports Darwin's idea that different and often simpler versions of the honeybees' hexagonal comb are evident in other bee species. "Some of these steps do actually occur in the *Melipona domestica*," he points out, agreeing with Darwin that selection favoring the economy of wax is the primary force: "every step in this direction would tend to the well-being of the community, what was at first done under the pressure of necessity would at length become a regular practice, and finally settle into that class of hereditary habits which we call instinct." This paper is post-*Origin,* but it is of interest nonetheless to note that Wallace was thinking about bees and their remarkable behaviors just as Darwin was.

Darwin's scenario for the evolution of bees' cell-building instinct is just one example of several where he suggested that insight could be gained through a consideration of transitional series. Wallace did not discuss transitional series as such, but alludes to this idea on page 166 of the Species Notebook where he states that "in investigating instinct we proceed by degrees, from the easy & near to the difficult & remote." While it is inaccurate to say Wallace rejected the idea of instinctive behavior altogether, he seems to demote its significance relative to learning and experience. In the Species Notebook most entries on instinct (there are twenty-one of them) are devoted to refuting its occurrence, or at least in criticizing commonly cited instances of instinct by pointing out what he sees to be unwarranted assumptions and lack of experimental rigor in demonstrating instinct. At this stage he is, like Darwin, critical of the idea of instinct "implanted by the deity," but he is also more vocal than Darwin in pointing out that insofar as instinct is defined as "complicated acts" performed "absolutely without previous instruction or knowledge," the experiment has never been done to disentangle instinct from experience. In other cases, he argues, what we mistakenly call instinct is merely habit. These are themes that Wallace continued to explore for years (e.g., Wallace 1867, 1870, 1873) and are further discussed later in this chapter.

Darwin was not so concerned with debunking instances or claims of instinct, though at least one of his notebook entries (C70; Barrett et al. 1987, 261) implicitly acknowledged problems with two putative examples that were also criticized by Wallace (suckling of human infants [Species Notebook p. 168] and the behavior of newly hatched ducks and chicks near water [p. 170]), while citing the beeline that hatchling sea turtles make for the sea as a good instance of heritable instinct. We find Darwin grappling with the definition of instinct versus habit in several places, as in the N Notebook where he critiques a paper by Algernon Wells on the subject. He comments that "pure instinct is not imitative," while "imitations seem invariably associated with reason." He later inserted a note to himself putting his finger, like Wallace, on the experimental acid test of instinct: "NB.

insects which have never seen their parents offer best cases of instincts" (N69; Barrett et al. 1987, 583).

Turning to arguments for transmutation based on observations of behavior (instinct, habit), two effective approaches are found in the Species Notebook and Darwin's writings. The first is to highlight the idea that structure and habits are often not in accordance. In the natural theology tradition it was often argued that each species is specially adapted for its own particular "role" in nature, and therefore structure was assumed to be specially designed to match habit, mode of feeding, and so on. Thus Wallace makes much of the hornbills of Africa and India. The former "feed on reptiles, insects, such as grasshoppers lizards &c. & even small mammals," while the latter feed only on fruit. "Yet both," says Wallace, "have exactly the same general structure & forms of bill . . . feet tail wings & stomach! Here is the most palpable proof that the structure of Birds is not varied in accordance with their habits" (p. 53). Earlier in the Species Notebook (pp. 32–33) Wallace criticizes the several "gratuitous statements & references" made in the *Cyclopedia of Natural History,* in an article purporting to explain the skeletal structure of different bird groups in terms of design on the grounds that "their peculiar habits require it." He had little patience for such reasoning, and he goes on at length chastising the author and concluding that "We are like children looking at a complicated machine of the reasons of whose construction they are ignorant, and like them we constantly impute as cause what is really effect in our vain attempts to explain what we will not confess that we cannot understand." Wallace later inserted a note on this page to "see p. 53"—referring to his argument about the hornbills.

Darwin surely would have agreed with these points made by Wallace, recounting his own examples of species "having habits and structure not at all in agreement" such as the woodpeckers of La Plata that see no trees, and the "water ouzel" songbird that forages underwater in rivers and streams, to cite just two of several examples from chapter 6 of the *Origin.* He also had little patience with the closed arguments of the natural theology adherents. "He who believes that each being has been created as we now see it," he wrote, "must occasionally have felt surprise when he has met with an animal having habits and structure not at all in agreement." Those who invoke special creation as an explanation "will say, that in these cases it has pleased the Creator to cause a being of one type to take the place of one of another type; but this," he says dismissively, "seems to me only restating the fact in dignified language."

Then, there is the matter of "diversified habits in the same species," as Darwin put it. Wallace and Darwin each addressed birds' nest-building behavior in this regard, but they had rather different objectives. Wallace's treatment is part of his attack on claims of instinct in birds (and other animals). On pages 112–119 of the Species Notebook Wallace provides a long list of modes of nest construction in birds. Why he does so is hinted at on pages 166–168, under the heading "Instinct—Birds' Nests," where the comment, discussed earlier, that "we proceed by degrees, from the easy & near to the difficult & remote" in investigating instinct can be found. Wallace attacks the assumption that we have any clear idea of the senses of other animals (insects in particular), and then segues into the entry that defines instinct. Here at last is the crux of the issue for Wallace: "Thus it is said & repeated, that birds & insects build nests gather & store food & provide for the future wants without any instructions for the fellows & without even knowing that such acts have been performed by others. This however is assumed." His larger point is presented far more fully a decade later in "The Philosophy of Birds' Nests" (1867), where he argued that birds, like people, build by imitation and with materials at hand. He does not claim that birds have no instinct at all, but rather argues strenuously against the claim that their nest-building behavior is *purely* instinctive.

It is important to understand that the ultimate object of Wallace's essay, the foundation of which is evident in the Species Notebook, was to weigh in on an important contemporary debate: what is the source of human knowledge? Instinct, habit, experience, learning, imitation—precisely how people *know* things was a question of great philosophical significance, and still is. By extension, how other species "know" things bears on the question. Were certain species specially endowed with certain knowledge? What are the nature and extent of innate or hereditary knowledge in, say, birds (like how to build a nest or sing a courtship song)? The final sentences of "The Philosophy of Birds' Nests" make reference to philosopher John Stuart Mill's "sensationalism," and how this and "all the modern philosophy of experience" would be overthrown if true instinct (defined as "the capacity to perform some complex act without teaching or experience") could be demonstrated. Wallace says that while it is not improbable that "the existence of true instinct may be established in other ways," in the case of birds' nests ("usually considered one of its strongholds") he finds no evidence at all.

Darwin was far more inclined than Wallace to accept the idea of instinct, even in humans. We have already seen how he took it as a given that the construction of bees' cells was instinctive. In a fragment published with his Transmutation Notebooks Darwin also commented on the very same instinct versus experience issue as Wallace in connection with Mill (1840), who commented on the works of poet Samuel Taylor Coleridge. There Darwin is clear that instinct surely exists in animals, and he seems to entertain the possibility of instinct in humans: in response to Mill's statement "We see no ground for believing that anything can be the object of our knowledge except our experience," Darwin asks, "Is this not almost a question whether we have any instincts, or rather the amount of our instincts—surely in animals according to usual definition, there is much knowledge without experience . . . so there *may* be [instinct] in men" (Barrett et al. 1987, 610).

It makes sense that Darwin, given his extensive reading and other experience with domestic animal breeding, would assume that animals, including humans, have some degree of instinctive behavior—how else could the striking behavioral traits of working dogs like pointers be selected for and improved if such traits were not in some measure hereditary? And if they are hereditary such that even young animals have the capacity to express the behavior (however imperfectly) without experience, is that not instinct? For Darwin, then, the more interesting line to pursue was *variation* in the expression of instinctive behavior. Many animals can modify their behavior through experience, of course, but more important for Darwin were cases of heritable within-species variation in the expression of instinctive traits— such variation must exist, he felt, and provides raw material for selection to act upon. Here Darwin and Wallace had rather different perspectives on birds' nests. In *Natural Selection* under the heading "Nidification & habitation" (Stauffer 1975, 498–505) Darwin discussed at length the "complex instinct" of birds' nests, to see "whether there is any variation in [this] important instinct." This was amplified in the *Origin:*

> I can only assert, that instincts certainly do vary—for instance, the migratory instinct, both in extent and direction, and in its total loss. So it is with the nests of birds, which vary partly in dependence on the situations chosen, and on the nature and temperature of the country inhabited, but often from causes wholly unknown to us: Audubon has given several remarkable cases of differences in nests of the same species in the northern and southern United States. (Darwin 1859, 211)

At least one of Darwin's many examples was also cited by Wallace, and so provides another case of how Wallace and Darwin sometimes had different perspectives on the same topic. In the discussion in *Natural Selection* on variation in instinct, Darwin cited William B. O. Peabody (1799–1847), American clergyman and naturalist, on differences in the nests made by northern versus southern populations of the Baltimore oriole, *Icterus galbula:* "in the South [Baltimore orioles] make their nests of light moss, which allows the air to pass through, & complete it without lining; while in the cool climate of New England they make their nests of soft substances closely woven, with a warm lining" (Peabody 1840). In contrast to this, Wallace, in "The Philosophy of Birds' Nests," commented that this is "an excellent example of a bird which modifies his nest according to circumstances," continuing, "it has been observed that the nests built in the warm Southern states are much slighter and more porous in texture than those in the colder regions of the north." Instinct versus experience or necessity? From a modern perspective, both Darwin and Wallace were correct in part.

There is a final noteworthy facet to Wallace's and Darwin's respective approaches to the question of instinct as it pertains to birds. I have already noted that Darwin thought in terms of heritable variations in behavior subject to selection: "the nest of each bird, wherever placed & however constructed be good for that species under its own conditions of life; and if the nesting-instinct varies ever so little, when a bird is placed under new conditions, & the variations can be inherited . . . then natural selection in the course of ages might modify and perfect almost to any degree the nest of a bird in comparison with that of its progenitors in long past ages" (Stauffer 1975, 500). Wallace, in contrast, focused on learning and experience, questioning whether the cases often cited as instinct really were cases of learning. On pages 116–117 of the Species Notebook he discussed what we might describe today as cultural transmission of knowledge as a means of birds' "improving" (or at least altering) their nest building. He opens by stating that birds' nests are "said to be built by instinct because they don't improve. But they vary according to circumstances & does man do more." He then suggests, "It is only by communication, by the mingling of different races with their different customs, that improvements arise & then, how slowly!" He draws a curious parallel between the way humans and birds improve their domiciles: "A race remaining isolated will ever remain stationary, & this is the case with birds.

Each species is generally confined to a limited district in which the circumstances are similar & give rise to no diversity of habits." Wallace seems to be applying the same kind of cultural exchange found between peoples to the way in which birds improve nest architecture. With his focus on experience and environment as an impetus for behavioral change, he does not address the question of whether or to what extent such behaviors might change through selection, despite the likelihood that the bird entries, beginning with "definition of birds" (p. 111) and continuing through the extensive treatment of bird architecture (pp. 112–119), apparently followed soon after his discovery of the principle of natural selection. This is not to say that Wallace did not appreciate the nature of variation and varieties or that change of circumstance is not necessary for the occurrence of such variation—this is seen in his critique of Lyell, for example, as when he points out that varieties "constantly occur in the same place & under the same circumstances as the original species" (p. 150). But Wallace did have a rather different perspective on instinct than Darwin, at least at the stage of the Species Notebook, which resulted in his not drawing on variations or transitions in behavior and instinct in his arguments for transmutation. In his Sarawak Law and Ternate papers, for example, the words "habit" and "instinct" are not found at all in the former and only the word "habit" occurs in the latter, in the context of habits changing along with structure.

5. Human-Primate Relationship and Human Variation

Human descent from primates was clear to Wallace and Darwin, and in their respective writings they addressed this in more or less straightforward terms. Wallace took note of accounts of tails or taillike structures occurring in humans (pp. 64, 91) and studied the behavior of his orphaned orangutan. Most of Darwin's commentary on the human-primate relationship pre–*Descent of Man* is found in his notebooks, as he famously avoided the subject in the *Origin*. The notebooks include a range of observations and notes concerning the nature of humans, from philosophical to morphological and behavioral observations much like Wallace's. He also comments on "tails" in humans, for example (E89, in this case referring to the *os coccyx* itself, not freakish taillike extensions of it), as well as other structural curiosities such as male nipples (their presence in orangs as well as humans: D61).

In a parallel briefly mentioned earlier, both Wallace and Darwin had opportunities to observe a young orangutan for an extended period. On Species Notebook pages 20 through 27 Wallace describes the "Habits of a Young Mias," an infant orangutan ("Mias" being the Dyak name for these primates, "Orang" the Malay name) orphaned when Wallace killed its mother in early May of 1856. He attempted to raise the young orang but it survived only a few months. In that time he recorded details of its behavior and development, later published as "Some Account of an Infant 'Orang-Utan'" and "A New Kind of Baby" (both in 1856). Wallace's observations make it clear he found the humanlike qualities of his young Mias quite interesting, drawing explicit comparisons with humans, such as in a passage describing how the orang would "scream & kick about violently exactly like a baby in a passion," or another where the orang's behavior is described "like a young baby." His paper "A New Kind of Baby" was written in a mischievous tongue-in-cheek style for the popular *Chambers's Journal*, drawing comparisons with "common" (human) babies and only revealing at the end of the piece that the infant he describes is an orangutan. Twenty years earlier, in early spring of 1838, the London Zoo had acquired a young orang named Jenny. The timing could not have been better for Darwin, who had just become convinced of transmutation the previous year and was then deeply immersed in all manner of evolutionary speculations, including the implication for human origins. He made several visits to the zoo to spend time with Jenny. Beyond noting its similarities in behavior with humans (such as his M Notebook entries describing Jenny's pouting or crying "like [a] naughty child"), Darwin was interested in Jenny's emotions and cognitive capacities. Thus several of his notebook entries speak more to such feelings as guilt, shame, fear, and interest—Jenny would look to her keeper "as if for approval" before taking bread offered by a visitor, and "when she knows she has done wrong will hide herself." He described Jenny's "expression and whine," sulkiness, and passionate outbursts like a human child's meltdown, and has no doubts that he sees the human in Jenny:

> Let man visit Ouranoutang in domestication, hear its expressive whine, see its intelligence when spoken to; as if it understands every word said—see its affection.—to those it knew.—see its passion & rage, sulkiness, & very actions of despair; . . . and then let him boast of his proud pre-eminence.
> (C Notebook, p. 79; Barrett et al. 1987, 264)

The similarity was not lost on Queen Victoria, who, upon seeing Jenny, remarked that she was "disagreeably human." Observing Jenny confirmed for Darwin the fundamental relationship between humans and other animals: "Man in his arrogance thinks himself a great work, worthy the interposition of a deity. More humble and I believe true to consider him created from animals" (C Notebook, pp. 196–197; Barrett et al. 1987, 300). Darwin took to observing the behavior of his infant children a few years later, and wrote questions to friends and family encouraging them to send him observations of their own young children. His "Queries About Expression" (1867), which had their origin in a set of questions headed "Natural History of Babies" penned in the inside back cover of the M Notebook, were one of many of Darwin's investigations into human origins and diversification, culminating in *The Descent of Man* (1871) and *On the Expression of the Emotions in Man and Animals* (1872).

These investigations do not appear to have any parallel in Wallace's writings, but there are two additional human-related topics addressed by both naturalists. The first relates to the question of instinct in humans, and the second to human racial diversity. Earlier in this chapter I discussed how Wallace was dubious of the existence of true instinct in humans, while Darwin seemed to believe that people had some instinctive behaviors but also acknowledged the importance of reason and experience. I have also discussed Wallace's interest in racial diversity in chapters 1 and 2. Recall that observations on race are scattered throughout the Species Notebook, some of which give better clues than others to his overarching interest in the subject. It was in the Moluccas that Wallace first encountered Papuans, and soon after he believed he found a line of demarcation between the Papuan and Malay "races" which originated from peoples of Pacific and Asian stock, respectively. This is the context for notes such as that on page 63 of the Species Notebook: "Papuan races use bow & arrow, Malays not . . . query is this a universal difference? If so, good proof of diversity of origin." Wallace's entries on race are of a practical, immediate nature: in the field, he was recording observations that he thought might shed light on the origin of the Papuan and Malay races, and by extension human races more generally. Darwin did not address origins or distribution of particular races in his notebooks and other early writings, but rather was interested at a more general level in the human-animal (especially primate) relationship. Like Wallace he was a monogenist—one who believed that human races constituted "varieties" of one species and therefore have a single origin, as opposed to the polygenists who held that different races represented separate created species. Of several references to the idea of a unitary origin for human races in the Transmutation Notebooks, Darwin gave one argument (entry C234) based on lice: "Why if louse created should not new genus have been made[?]" he asked. The fact that all humans are afflicted with the same species of lice is a "good argument for origin of man one" he concludes, albeit in an awkward construction.

Wallace was deeply impressed by William Lawrence's *Lectures on Man* (1822), as the book was known, and James Cowles Prichard's *Researches into the Physical History of Mankind* (1851) (see Chapter 1). Darwin, too, read these authors closely—both are cited in his Transmutation Notebooks, and he was personally acquainted with Prichard as the two served on a committee appointed in 1839 by the British Association for the Advancement of Science, the charge of which was to draw up a questionnaire on various aspects of "race" for travelers. The opening paragraph of the resulting document, *Queries Respecting the Human Race, to Be Addressed to Travellers and Others* (Darwin et al. 1841), explains the rationale:

> At the meeting of the British Association held at Birmingham, Dr. Prichard read a paper "On the Extinction of some varieties of the Human Race." He pointed out instances in which this extinction had already taken place to a great extent, and showed that many races now existing are likely, at no distant period, to be annihilated. He pointed out the irretrievable loss which science must sustain, if so large a portion of the human race, counting by tribes instead of individuals, is suffered to perish, before many interesting questions of a psychological, physiological and philological character, as well as many historical facts in relation to them, have been investigated.

"At the suggestion of the Natural Historical Section, to which Dr. Prichard's paper was read," the report continues, "the Association voted the sum of £5 to be expended in printing a set of queries to be addressed to those who may travel or reside in parts of the globe inhabited by the threatened races." The questionnaire consisted of eighty-nine questions divided into ten categories: Physical Characters; Language; Individual and Family Life; Buildings and Monuments; Works of Art; Domestic Animals; Government and Laws; Geography and Statistics; Social Relations; and Religion,

Superstitions, &c. The questionnaire ultimately failed to yield much new information, and in his later writings on human evolution (*Descent* and *Expression*) Darwin did not refer to it. Instead he cited Lawrence, Prichard, and other authorities and observers for information regarding human races and their practices. Wallace's later disavowal of the applicability of selection in human cognitive evolution (discussed briefly in Chapter 1; see Fichman 2001; Gross 2010) is clear, though it should be borne in mind that Wallace never denied the fundamental kinship between humans and other species. Note too that Wallace showed a deeper interest than Darwin in details of racial and ethnic diversity; the relatively few human racial entries in the Species Notebook do not reflect his extensive research documenting the human diversity of the region, as demonstrated by the remarkable comparative linguistic database he appended to *The Malay Archipelago*. This interest is also manifest in the abundant ethnological information Wallace included in his 1853 narrative of his travels on the Amazon and Rio Negro rivers.

To conclude this section, it is clear that at the time period of the Species Notebook Wallace and Darwin shared essentially the same views of the human-primate relationship and human races. The one subject that they perhaps viewed from different perspectives was the question of human instinct. It is possible that Wallace was already questioning the idea of a material origin for the human brain and its cognitive and moral faculties (recall his extracts from Ernest Renan on the "higher objects of existence"—Species Notebook, pp. 154–155), but all things considered both he and Darwin saw humans as a part of organic nature, as the parallels between their writings and sources demonstrate.

6. Observations and Arguments for Transmutation

Wallace's pro-transmutation arguments in the Species Notebook include observations favoring a transmutational explanation and attacks on anti-transmutation positions, such as the claims of "design" in nature. These arguments and their relationship to Darwin's thinking deserve special treatment. I discuss them in connection with morphology, the supposed balance of nature, domestication and the nature of varieties, and geographical distribution and isolation.

Morphology

Wallace's and Darwin's conceptions of transmutation as an ever-branching process were similar, and in relation to this both evinced an appreciation of the concept of history inherent in organisms in terms of morphology. Thus on page 112 of the Species Notebook Wallace notes that the presence of "large air cellules" in the femurs of ostriches offer proof that the air cells of birds' bones are not an adaptation to render the body lighter to facilitate flight. He does not elaborate, but he surely has a sense that the air cells are there for other reasons and that these are then "taken advantage of" or happen to be beneficial later, in the context of flight. Darwin makes precisely this point about the skull sutures of newborn mammals. The sutures and skull malleability they afford should not be seen as an adaptation for parturition, as reptiles and birds have these sutures too and they need only escape from an egg. Like the air cells in the ostrich bones, these sutures are there for other reasons, maybe correlated with something quite different developmentally, but can then be "seized upon" by selection in another, newer context (Darwin 1859, 197). Other examples are given by Darwin in *Natural Selection.*

Rhetorically speaking, the tone, language, and angle of attack seen in many of Wallace's entries are echoed by Darwin. A good example is Wallace's choice of language on notebook page 99 in discussing homology and the "import of the doctrine of Morphology of plants." Wallace points out that it is absurd for naturalists to refer to stamens as modified petals or carpels as fused leaves, while they consider each part of the plant as being specially designed and created for its function. Are "all the beautiful facts of morphology . . . a delusion & a snare, as much so as fossils would be were they really not the remains of living but chance imitations of them[?]" There is a tone of mixed impatience, incredulity, and condescension in Wallace's statement. There is a parallel in context as well as tone in chapter 5 of *On the Origin of Species,* where Darwin summarized the striping patterns of various equine species, arguing that this empirical observation speaks of transmutation and makes little sense if each species was a special creation:

> He who believes that each equine species was independently created, will, I presume, assert that each species has been created with a tendency to vary, both under nature and under domestication, in this particular manner, so as often to become striped like other species of the genus; and that

each has been created with a strong tendency, when crossed with species inhabiting distant quarters of the world, to produce hybrids resembling in their stripes, not their own parents, but other species of the genus. (Darwin 1859, 167)

The rhetorical parallel with Wallace comes next: "To admit this view . . . makes the works of God a mere mockery and deception," says Darwin, echoing Wallace's "delusion & snare." He continues: "I would almost as soon believe with the old and ignorant cosmogonists, that fossil shells had never lived, but had been created in stone so as to mock the shells now living on the sea-shore"—paralleling Wallace's comment about fossils not being the remains of once-living organisms but merely chance imitations of them. Another example along these lines is found in Wallace's and Darwin's invocation of Lyell to make much the same point in rather different but strikingly parallel ways. Wallace noted Lyell's account in the *Principles* of how some thinkers of an earlier generation (he mentions Buffon) were ridiculed for suggesting the earth was ancient and were forced to retract their statements as "contrary to scripture, though they are now universally admitted to have been correct" (Species Notebook, p. 34). A few pages later (p. 39) he questions Lyell's stance that we should have confidence in the stability of species. But why should we have that confidence, asks Wallace; "Is it not a mere . . . prejudice like that in favour of the stability of the earth?"—a position which Lyell has argued against so convincingly, no less. Darwin too points to the past prejudice of those insistent on an unchanging earth, but in his case he draws a parallel between himself and Lyell in chapter 4 of the *Origin:* "I am well aware that this doctrine of natural selection . . . is open to the same objections which were at first urged against Sir Charles Lyell's noble views on 'the modern changes of the earth, as illustrative of geology.'" Pushing the point further, Darwin urges that just as modern geology "has almost banished such views as the excavation of a great valley by a single diluvial wave, so will natural selection . . . banish the belief of the continued creation of new organic beings" (Darwin 1859, 95–96).

Part of Wallace and Darwin's rhetorical strategy, too, is to attack claims of "designedness" in relation to morphology. Wallace underscores the absurdity of speaking of adaptations of structure reflecting design to an organism's needs, as if it could have needs before coming into existence, or of citing the wisdom of the creator for "designing" the obvious—like the "wise

contrivance" of the soft scar of the coconut without which the embryonic
plant could not escape. "Is not this absurd?" Wallace asks. "To impute to
the supreme Being a degree of intelligence only equal to that of the stupid-
est human beings." He points out that this is tantamount to lauding the
great wisdom of one who, in building a house, made a door to it—hardly
evidence for good design, but rather something to be expected. Likewise,
Wallace suggests that it is presumptuous to explain the variable number of
cervical vertebrae in birds to design for "the wants which the peculiar habits of particular birds require." A "humbler mortal" might have supposed
that the same deity that enabled mammals as disparate as elephant, giraffe,
whale, and camel to live with seven vertebrae could easily have created
birds with this number too, even those that make their living in varied ways.
Darwin is rather philosophical in making much the same points. In his
notes on John Macculloch's *Proofs and Illustrations of the Attributes of
God* (1837), Darwin takes the author to task for suggesting that plants
growing in deltaic areas were specially created to hold mud or sediment in
place. Referring to Macculloch's "long rigmarole," Darwin opined that if
we presume that the deity created these plants simply to hold sediment in
place, "we lower the creator to the standard of one of his weak creations."
Similarly, when Macculloch compares vertebrate skeletal anatomy and explains "abortive bones" found in some groups by claiming that the creator
determined to stick with one plan once adopted, Darwin writes, "What
bosch!! . . . the designs of an omnipotent creator, exhausted & abandoned.
Such is Man's philosophy when he argues about his Creator!" (Barrett et al.
1987, 633–634).

Balance and Harmony

Wallace's and Darwin's approaches to attacking the idea of designedness
extends to another fundamental tenet of natural theologians: the idea of
balance and harmony in nature. Wallace first took aim at Lyell's discussion
of balance in the *Principles,* where Augustin de Candolle, the Swiss botanist, is quoted as saying that "all the plants of a given country are at war
with one another." Lyell agreed, but saw this as part of a checks-and-
balances system designed to preserve a harmonious balance in nature. Wallace disagreed: "Lyell talks of the 'balance of species being preserved by
plants insects, & mammalia & birds all adapted to the purpose,'" he wrote
in the Species Notebook (p. 49), continuing, "This phrase is utterly without
meaning. Some species are very rare & others very abundant. Where is the

balance? Some species exclude all others in particular tracts. Where is the
balance[?] When the locust devastates vast regions, & causes the death of
animals & man what is the meaning of saying the balance is preserved[?]"
It is noteworthy that Darwin opened his discussion of the struggle for existence with reference to the very same statement by de Candolle, also taken
from Lyell. He used this to evoke one of his memorable images: "We behold the face of nature bright with gladness," as he put it in the *Origin*
(Darwin 1859, 62). Darwin went on to show that this sense of balance and
harmony is illusory.

Later in the Species Notebook Wallace (pp. 146–147) turned to Agassiz
on balance and harmony. In *Lake Superior*, Agassiz described how many
native plants have been displaced, replaced by many of European origin.
"What becomes of the 'Harmony of distribution', the 'balance of species';
the 'proofs of intelligence in the nat[ural] distrib[ution] of species' &c. &c.,"
Wallace asks. "Did the 'wonderful order' Agassiz speaks of exist before the
country was overrun by these strange plants,—or does it exist now?" Darwin picked up on the same phenomenon, though he did not cite this example from Agassiz. Rather, in arguing that species are not perfectly adapted
to their locale, he pointed out how some species transported to foreign locales thrive, even displacing the natives. In both *Natural Selection* and the
Origin Darwin pointed out that no country could be named in which the
native species had not been displaced by invaders. "And as foreigners have
thus everywhere beaten some of the natives," as he put it in the *Origin* on
page 83, "we may safely conclude that the natives might have been modified with advantage, so as to have better resisted such intruders." This is a
good argument against perfect adaptation, undermining the claim that
species are created to be perfectly suited to their home turf, so to speak.
Imperfect adaptation calls into question the assumption of special creation
itself. Incidentally, Darwin makes an additional and (to him) very important point concerning naturalized plants:

> It might have been expected that the plants which have succeeded in be-
> coming naturalised in any land would generally have been closely allied to
> the indigenes; for these are commonly looked at as specially created and
> adapted for their own country. It might, also, perhaps have been expected
> that naturalised plants would have belonged to a few groups more especially
> adapted to certain stations in their new homes. But the case is very different;
> and Alph. De Candolle has well remarked in his great and admirable work,

that floras gain by naturalisation, proportionally with the number of the native genera and species, far more in new genera than in new species. (Darwin 1859, 115)

This observation relates to Darwin's "principle of divergence," a subject
taken up in Chapters 4 and 5. Darwin made an ecological argument about
divergence in niche occupancy, to put it in modern terms, leading to coex-
istence. This is an additional twist on the observation about naturalized
plants that Darwin made much of, but it is not something seen in the Species Notebook or other writings by Wallace.

As a final note on the topic of balance and harmony (or lack thereof),
recall that Wallace and Darwin drew inspiration from Malthus's *Essay on
Population,* which played the same inspirational role for both naturalists in
their discovery of the principle of natural selection. Malthus is not mentioned in the Species Notebook, while he is mentioned a half-dozen times
in Darwin's Transmutation Notebooks. The earliest mention of Darwin's
reading of Malthus is near the end of the C Notebook, in a list of books he
examined "with reference to Species." His discovery of natural selection
dates to about this time, in the autumn of 1838. The seeming absence of
Malthus from the Species Notebook perhaps reflects the timing of Wallace's discovery of natural selection, which was relatively late in the notebook as reckoned by its rough chronology—and even then Wallace reported
in his Ternate essay that he recollected key ideas from his reading of Malthus some dozen years earlier, or the mid-1840s.

Wallace likely encountered Malthus even earlier than this, perhaps
first in 1837 when living in London, attending lectures and reading at the
Mechanics' Institute. Though Malthusianism may have been contrary to
Owenite socialist thinking—to which Wallace was an adherent—Wallace
seemed able to reconcile this by distinguishing between the operation of
population pressure in nature (where it is very important) versus human
populations (where it is of little or no importance) (Moore 1997; Jones 2002).
Regardless of the correctness or not of the suggestion of McKinney (1972)
and other scholars that Wallace's Malthusian insights were partly or wholly
inspired by reflections on the indigenous peoples he had encountered in
the East (see the treatment in Chapter 1), he later seemed to downplay the
applicability of Malthusianism to humans. This reading is supported by the
subject of one of Wallace's last letters to Darwin, written in July of 1881, in
which Wallace heartily recommended to Darwin the book *Progress and*

Poverty by Henry George. In it, Wallace said, Darwin would find "an elaborate discussion of Malthus's 'Principles of Population,' to which both you and I have acknowledged ourselves indebted. . . . Mr. George, while admitting the main principle as self-evident and as actually operating in the case of animals and plants, denies that it ever has operated or can operate in the case of man" (Marchant 1916, 1:317; WCP1992).

Varieties, Variability, and Domestication

Another important topic of discussion by Wallace and Darwin in their arguments for transmutation focused on the nature of varieties—in particular their origin and fate. Note that this differs from the origin and fate of *variations;* neither Wallace nor Darwin knew where variations came from, of course, though they implicitly or explicitly held that variations were spontaneous and nondirectional and that only heritable ones were of any consequence to transmutation. When it came to varieties, both rejected the prevailing view of the day that there are "defined limits" to variation from the parental type, beyond which no further change was possible. Lyell took this view, and Wallace criticized him for it on pages 45 and 150 of the Species Notebook. The latter entry is particularly to the point: "How can [Lyell] prove that variation may not go on at a rate commensurate with Geological changes? . . . What are 'the defined limits',—he assumes that they exist." He continues: "Changes which we bring about artificially in short periods may have a tendency to revert to the parent stock though this in animals is not proved. This is considered a grand test of a variety. But when the Change has been produced by nature during a long series of generations, as gradual as the changes of Geology, it by no means follows that it may not be permanent & thus true species be produced."

Darwin did not take Lyell to task in strong terms, but clearly rejected his argument for reversion. For example, in Transmutation Notebook C (entry C176) Darwin made a telling entry: his friend William Lonsdale was ready, he says, "to admit, permanent small alterations in wild animals, & thinks Lyell has overlooked argument that domesticated animals change a little with external influences—& if those changes [are] permanent so would the change in animal be permanent.—It will be easy to prove persistent Varieties in wild animals." He further commented on reversion in the *Origin* (Darwin 1859, 15): "Having alluded to the subject of reversion, I may here refer to a statement often made by naturalists—namely, that our domestic varieties, when run wild, gradually but certainly revert in character to their aboriginal stocks. Hence it has been argued that no deductions can be drawn from domestic races to species in a state of nature. I have in vain endeavoured to discover on what decisive facts the above statement has so often and so boldly been made." In the next passage he made a point that echoes Wallace in the opening argument of the Ternate essay, namely, that arguing that domestic varieties always revert to parental type is in one sense moot since few domestic varieties could survive in nature to begin with: "we may safely conclude that very many of the most strongly-marked domestic varieties could not possibly live in a wild state. In many cases we do not know what the aboriginal stock was, and so could not tell whether or not nearly perfect reversion had ensued." Darwin, much like Wallace, concludes that "natural selection . . . will determine how far the new characters thus arising shall be preserved"—the persistence and continued divergence of natural varieties is the crux of the issue.

Darwin became convinced early on that domestic varieties held lessons for an understanding of transmutation, in particular by providing an analogy: development of domestic varieties by artificial selection as a microcosm of natural selection (Vorzimmer 1969; Wood 1973; Secord 1981; Bartley 1992). In the Ternate essay Wallace took a very different view, taking the position that since domestic varieties are "unnatural" their reputed reversion in a state of nature has no bearing on the possibility that natural varieties can "depart indefinitely from the original type," to borrow from the title of the essay. The supposed instability and limited variation of domestic varieties were much cited by anti-transmutationists, and Wallace sought to sidestep if not undermine this argument. The point I would like to make here, however, is that in the Species Notebook Wallace seems rather closer to Darwin's view than the Ternate essay would suggest. There, domestic varieties are discussed more as the microcosm of divergence that Darwin articulated, though his arguments are heuristic and not based on analysis of a particular group. On page 39 of the Species Notebook, for example, he gives a scenario using dogs: "Let us suppose that every variety of the Dog but one was to become extinct & that one say the spaniel, to be gradual spread over the whole world, subjected to every variety of climate & food, & domesticated by every variety of the human race. Have we any reason for supposing that in the course of ages a new series of varieties quite distinct from any now existing would not be developed." He continued the scenario with further and further divergence, making the point that there is no reason to think there are limits on the continued variability and divergence of these dog breeds. He is more explicit on page 41, where he declares, "In a few lines Lyell passes over the varieties of the Dog & says there is <u>no transmutation</u>—Is not the change of one original animal to two such different animals as the Greyhound & the bulldog a transmutation?" (emphases Wallace's, here and below).

Wallace also states his conviction that varieties can give rise to new varieties: "All varieties we know of are produced at <u>birth</u> the offspring differing from the parent. This offspring propagates its kind. Who can declare that it shall not produce a <u>variety</u> which process continued at intervals will account for all the facts" (p. 40)—in other words, varieties will "depart indefinitely from the original type" (becoming species in the process and giving rise to other such varieties). Domestication, then, was a phenomenon that was viewed by Darwin and Wallace in rather different ways. From a modern perspective it would be a mistake to see Wallace's argument about domestication in the Ternate essay as simply erroneous, suggesting that he did not fully understand the selection and diversification process. The opposite is true. First, the language he used in the dog evolution discussion in the Species Notebook shows that Wallace did see the generation of new domestic varieties as transmutation per se, and he took for granted that domestic varieties of a given species or species group are genealogically related. But second and more importantly, Wallace viewed domestication as a weak analogy for transmutation in the wild because human-mediated selection served to decrease, not increase, the fitness of organisms (as reckoned by how they would fare in a state of nature). The "limited and temporary" nature of change by human-mediated selective breeding, and the inefficacy of artificial selection in bringing about new species relative to natural selection, suggested to Wallace that artificial and natural selection differed fundamentally (Richards 1997). It is easy to see why Wallace would open his essay announcing a mechanism for transmutation in nature by dismissing the leading anti-transmutation argument of his day—that of Lyell, based on domestic varieties—on the basis that domestic varieties are misleading.

As might be expected, finally, both Wallace and Darwin were alert to possible examples of *new* varieties arising. Cases like the new *Ziziphora* variety (or species?) reported by Link at the Berlin Botanical Garden caught their attention—Wallace recording the account in Lindley's *Introduction to Botany* (p. 59 of the Species Notebook), and Darwin noting that given in Bronn's *Handbuch* (Stauffer 1975, 127). Both of them were also well aware of the puzzling case of William Herbert's primroses and cowslips, two common species of *Primula*. Herbert found that a single individual cowslip in his experimental garden produced a diverse array of offspring judged to be different varieties; indeed, some of the plants produced were even deemed to be true primroses, and everything in between. Darwin dedicated several notebook entries and a long (ten manuscript page) discussion in *Natural Selection* to these plants (Stauffer 1975, 128–133), ultimately viewing them as a good example of the innate variability of species, such that the two *Primula* species are "united by many intermediate links." Given the "overwhelming amount of experimental evidence, showing that they descend from common parents," Darwin concluded in the *Origin* (Darwin 1859, 50) that primroses and cowslips must be considered varieties of the same species. There he was merely presenting this as an especially well-documented example of "doubtful species," but he goes into more detail in *Natural Selection* where it is evident that the stakes are high in making sense of this case study: "An able Botanist has remarked that if the primrose and cowslip are proved to be specifically identical, 'we may question 20,000 other presumed species'" (Stauffer 1975, 133). Unsurprisingly, Lyell, in the *Principles* (1835, 2:447–448), declared that these plants "afford no ground for questioning the instability of species," maintaining that the remarkable variability that Herbert found is merely "part of the specific character." Wallace was having none of that: he saw Herbert's experiment as clear evidence for transmutation, and felt that Lyell was simply prejudiced in his refusal to accept the evidence: "The varieties of the <u>Primrose</u> adduced by <u>Lyell</u> [are] complete proof of the transmutation of species. It only shows the impossibility of convincing a person against his will. Where an instance of the transmutation is produced, he turns round & says 'You see they are not species they are only varieties'" (Species Notebook, p. 42; emphases Wallace's). For his part, Darwin may have cited the primrose example merely to help underscore how fuzzy species boundaries can be, but at the same time he certainly believed that the many varieties intermediate between primroses and cowslips produced by Herbert and others constituted "linking forms" proving that these species shared a recent common ancestor. Like Wallace, Darwin saw this as a welcome example of transmutation, but unlike Wallace he was uneasy about the idea of the primrose and cowslip changing into each other instantaneously, per saltum—a concern he expressed in all three Transmutation Notebook entries on the subject (E16, E113, and E141). This is an important difference between Wallace and Darwin; while both generally had a gradualistic notion of evolutionary change, Wallace seemed (at least in the 1850s) more ready than Darwin to accept rapid and even instant origin of new varieties and perhaps species.

There are other examples of the two naturalists' interest in varieties. Wallace copied out a long passage from Évariste Huc's *L'Empire Chinois* (Huc 1854) on the origin and maintenance of a new early-ripening variety of rice (Species Notebook, pp. 57–58), and an account from the horticultural literature of how "the miserable grass *Aegelops ovata* was sown year after year till it became wheat in no respect different from the common hard wheat of the South of France" (Fabre 1854). Darwin, on the other hand, noted in his "Questions and Experiments" notebook (Barrett et al. 1987, 506) a paper on another cereal grain: Weissenborn (1838), "On the Transformation of Oats into Rye." Weissenborn claimed that oats, sown late and cut back twice before flowering, can transform into rye when regenerated a third time and allowed to flower. Oats turning into rye was widely held to be true—an urban legend, to use a modern term—but was likely viewed as dubious by naturalists, reflected by the defensive tone of the paper and the fact that Darwin did not discuss it further. However, reports along these lines persisted, such as John Lindley's account (1844) of the plant-breeding experiments of Lord Arthur Charles Hervey, who was purported to have produced barley from oats—examples duly reported in *Vestiges of the Natural History of Creation* in support of the idea of transmutation. That Darwin took an interest in such accounts is seen in a comment by his friend the botanist Joseph Hooker in a letter from February 1845 telling Darwin, "The change of Wheat into Rye is here [in Paris] wholly disbelieved" (Darwin Correspondence Project, letter 832).

Geographical Distribution: Isolation and Islands

The question of the permanence of varieties is integral to the transmutation process, and these naturalists expressed rather similar ideas as to how varieties might remain distinct long enough to become permanent, and perhaps even new species. Their view found in their notebooks is not far from the modern one: varieties remain distinct to the degree that they do not intercross, and so isolation from other varieties (or the parental form), preventing intercrossing, plays a key role in facilitating divergence.

Wallace, on page 62 of the Species Notebook, took notes on Edward Blyth's paper on varieties in which isolation was emphasized (Blyth 1835). In fact, three of the four types of variety discussed by Blyth and recorded by Wallace in the Species Notebook are defined explicitly in terms of physical separation, of being "kept apart & propagated," "propagated or increased by isolation either natural or artificial," and "kept distinct from the original stock." Wallace's reading of Leopold von Buch was in part focused on this same issue. Wallace extracted von Buch's scenario for the origin of new varieties and eventually new species: a plant, say, is dispersed a long distance from the parent. It varies in its new environment (in part, because of the new conditions there), and being separated and therefore not able to cross with other varieties and "brought back to the primitive type" they become, he says, "at length permanent and distinct species." Moreover, von Buch noted that "if by chance in other directions they meet with another variety equally changed in its march, the two have become very distinct species & are no longer susceptible of intermixture" (p. 90; Wallace's emphases)—an early statement of the modern biological species concept.

Von Buch had an island colonization scenario in mind, writing of the Canary Islands (von Buch 1836), and given that Wallace was then living in the world's largest archipelago, islands certainly loomed large for him too in the 1850s. The word "island" appears more than thirty times in the Species Notebook, more than half of these in connection with transmutation. In Darwin's earlier writings (notably the Transmutation Notebooks) he, too, wrote of the significance of isolation in preventing intercrossing and enabling varieties to become more and more distinct. He also recognized the special significance of islands in this regard: "According to [my] view animals, on separate islands, ought to become very different if kept long enough apart" (B7); "the type [of species on an island] would be of the continent though all species different" (B11); "As I have said before isolate species & . . . especially with some change [in environment, they will] probably vary quicker" (B17) (Barrett et al. 1987, 172, 173, 175). He extends this to isolation by mountains and habitat as well: "The reason why there is not perfect gradation of change in species . . . [is that] if after isolation (seed blown into desert) or separation by mountain chains &c. the species have not been much altered they will cross" (B209); "Animals of same classes differ in different countries in exact proportion to the time they have been separated" (D23). These are just a few examples (Barrett et al. 1987, 223, 337; emphases in the original).

Darwin later downplayed the importance of isolation in favor of competition and his "principle of divergence" (discussed in Chapters 4 and 5), but he never abandoned it completely, as shown by Sulloway (1979). The key point to bear in mind is that both Wallace and Darwin had a concept of varieties as incipient species—both recognized that the transition from unstable to permanent varieties set the stage for the transition from permanent varieties to full species, despite their ignorance of the origin of variation. In Wallace's view, and initially Darwin's, this process of diversification was facilitated if not necessitated by physical separation to prevent intercrossing.

It has long been known that Wallace's and Darwin's formulations of natural selection were made in quite similar ways, but it has not been fully appreciated how remarkably concordant they were in the lines of evidence pursued to further their pro-transmutation arguments. The analysis of this chapter shows that the two traveled very similar paths, recording observations and constructing arguments on patterns of morphology (including embryology), geographical distrubution and island life, fossils and the branching of lineages, the capacity of species and varieties to vary indefinitely, and both marshaled arguments against design, "balance and harmony" in nature, and special creation. For Wallace's part, most of these evidentiary lines are found in the Species Notebook, making this the single most important document we have in shedding light on Wallace's thinking in the crucial years leading up to the events of 1858–1859. Additional lines of evidence were discussed by Wallace elsewhere, such as the significance of rudimentary structures in the Sarawak Law paper (which although misinterpreted in one sense also represented evidence of transmutation for him). Darwin also treated rudimentary structures in various writings, and both naturalists took note of or made the same kinds of observations in other areas too that fit into their pro-transmutation agenda: behavior of a young orangutan with an eye to its humanlike qualities, for example, and literature reports of new varieties arising. Important differences exist as well: they discussed and made use of domestic varieties in different ways (though the fact that they paid attention to domestication signals an awareness of its bearing on the nature of species and varieties), and they had their own perspectives on instinct and experience. In that regard Wallace did not discuss behavioral traits in connection with transmutational series or heritable variability as did

Darwin, but simply dissociated habit and structure as an argument against design.

What is the most important point to emerge from this analysis? Both Wallace and Darwin sought a consilience argument—indeed, converged on much the same one in good Whewellian inductive fashion, to the consternation of Whewell.

FOUR

Two Indefatigable Naturalists

Wallace and Darwin's Watershed Papers

WALLACE AND DARWIN were hard at work through the 1850s, each pursuing an impressive range of topics. Extracts from a steady stream of letters chronicling Wallace's observations and experiences afield in Amazonia and then Southeast Asia were duly published by Samuel Stevens, punctuated by a series of scientific papers on a multitude of topics: from entomology and ornithology to geology, geography, and ethnology. No slouch either, Darwin pursued mostly private investigations into such diverse subjects as seed dispersal, longevity, and viability in seawater, hybridization, fertilization by bees, and insectivorous plants—all undertaken with an eye toward gathering evidence for his still-secret theory of evolution by natural selection. Several papers resulted along the way, including one on the power of icebergs to make grooves on uneven surfaces on the seafloor. That decade was by any measure one of great intellectual ferment for the two of them, with some fifty-six published papers and letters by Wallace plus two books from his Amazonian expedition and thirty-two papers and letters by Darwin, plus four volumes on living and fossil barnacles, and, capping the decade, the epochal *On the Origin of Species* (Barrett 1977; van Wyhe 2009; Alfred Russel Wallace Page [people.wku.edu/charles.smith/index1.htm]; Darwin Online [darwin-online.org.uk/contents.html]).

Amid the sizable stack of Wallace and Darwin's collective papers of the pre-*Origin* years, Wallace's 1855 "Sarawak Law" paper and the Wallace and Darwin papers of 1858 stand out as watershed publications from our point of view today. The Sarawak Law paper made an argument (with transmutational overtones) for relationships of species in space and time, while the

1858 Wallace and Darwin papers represent the first announcement of natural selection (named as such by Darwin; Wallace left the mechanism unnamed), which took place at the Linnean Society of London meeting of 1 July 1858. At the time Darwin and Wallace, as well as Lyell and Hooker acting on behalf of these "two indefatigable naturalists," believed that their ideas on natural selection were virtually identical—Darwin called it a "striking coincidence," lamenting that "even [Wallace's] terms now stand as Heads of my Chapters." A closer reading reveals important distinctions that have relevance for, among other things, the idea that Darwin may have appropriated ideas from Wallace in formulating his ideas about transmutation, common descent, and natural selection. Accordingly, in this chapter I present a guide to the structure and content of these watershed papers. Each paper is presented in facsimile with accompanying annotations. In Chapter 5 I build upon this with an analysis of the similarities and differences in Wallace's and Darwin's ideas on transmutation by natural selection as reflected in the 1858 papers.

The "Sarawak Law" Paper of 1855

This paper was written in Sarawak, Borneo, in February 1855 and published in volume 16 (second series) of the *Annals and Magazine of Natural History* the following September. The paper was written while Wallace was waiting out the rainy season, bottled up in a bungalow at the foot of Santubong Mountain near the mouth of the Sarawak River. "I was quite alone," he wrote in his autobiography, "with one Malay boy as cook, and during the evenings and wet days I had nothing to do but to look over my books and ponder over the problem which was rarely absent from my thoughts" (1905, 1:354). The significance of this paper in reflecting the development of Wallace's thinking is evident from the effusive response it elicited from Henry Walter Bates, Wallace's Leicester friend and sometime Amazonian collecting partner. It is worth quoting Bates's November 1856 letter to Wallace from "Tunantins—Upper Amazon" at length:

> I received about 6 months ago a copy of your paper in the "Annals" on the "Laws which have governed the introduction of new species"—I was startled at first to see you already ripe for the enunciation of the theory."—You can imagine with what interest I read & studied it, & I must say that it is perfectly well done. The idea is like truth itself, so simple & obvious that

those who read & understand it will be struck by its simplicity: & yet it is perfectly original the reasoning is close & clear, & although so brief an essay it is quite complete—embraces the whole difficulty & anticipates & annihilates all objections—Few men will be in a condition to comprehend & appreciate the paper, but it will infallibly create for you a high & sound reputation—The theory I quite assent to & you know was conceived by me also, but I confess that I could not have propounded it with so much force & completeness. (WCP824)

Note the last sentence: "the theory" that Bates "quite assent[s] to" and that "was conceived by [him] also" is their idea that geological change—elevation and subsidence of the land, the creation of barriers—somehow induces transmutation. This is evident from an earlier passage in this same letter:

What you say about the similarity of the species between Malacca & several of the Islands of the Archipelago—compared with the great difference we find at different points & near on opposite sides of the Amazon—suggests the hypothesis that Central S. America is a region of elevation—formerly consisting of Islands long isolated & containing separate Faunas—whilst the Easter Archipelago is a region of depression with its opposite results. (WCP824)

Bates was excited by the new avenues for research that Wallace's paper pointed to:

Many details I could supply, in fact a great deal remains to be done to illustrate & confirm the theory—a new method of investigating & propounding Zoology & Botany inductively is necessitated, & new libraries will have to be written—in part of this task I hope to be a laborer for many happy & profitable years—What a noble subject would be that of a monograph of a group of being peculiar to one region but offering different species: in each province of it=tracing the laws which connect together the modifications of forms & colors with the local circumstances of a province or station—tracing as far as possible the actual affiliation of the species. (WCP824)

This letter is a reminder of Wallace and Bates's motivation to travel to the tropics to begin with, to solve the mystery of species origins (see Chapter 1), and their holistic vision of transmutation in both organic and inorganic worlds, taking Lyellian geology where Lyell would not go. Indeed, historian Martin Fichman noted that Lyellian geology "permeates" the Sarawak Law paper (Fichman 2004, 34), a fact that resonates with the lengthy critique of Lyell found early in Wallace's Species Notebook. Lyell as both inspiration and foil is in evidence in this paper, and Wallace later pointed to the *Principles* as inspiration for it: having been reading up on geographical distribution in his bungalow, he realized that "the great work of Lyell had furnished me with the main features of the succession of species in time, and by combining the two I thought that some valuable conclusions might be reached" (1905, 1:355).

The argument structure of the Sarawak Law paper takes an inductive, lines-of-evidence approach that reflects the breadth of topics Wallace discussed in support of transmutation in the notebook. Wallace builds up to his "law" by putting forth in logical sequence a series of observations, offered almost as axioms building to the grand conclusion of a mathematical proof. The paper says much about Wallace's understanding of the transmutation concept by 1855: it is clearly an evolutionary manifesto, as Lyell recognized, but argues for the fact of species change without mentioning transmutation at all. The key insights to look for are as follows:

- Transmutation of species over time is a reality.
- This transmutation takes place extremely gradually, in parallel with gradual changes in the earth (geology, climate).
- Transmutation occurs both within continuous lineages (what we call anagenesis) and through a process of branching (cladogenesis, in modern terms), the latter brought about by isolation of some members of a species geographically.
- The branching and rebranching process leads to an arborescent pattern of species relationships, or a "divergent series" of lineages—this at once explains the basis of the natural system of classification, as well as the observed patterns of biogeographical, paleontological, and morphological relationships (including anatomical oddities like rudimentary organs, though these are interpreted as progenitor rather than vestigial structures).
- Geographical isolation, as by separation by physical barriers such as mountain ranges or expanses of ocean, leads to the formation of new species (allopatric speciation, in modern terms).

- The dynamic of anagenetic and cladogenetic change over time offers one explanation for the occurrence of "analogous" groups in disparate parts of the world (they are descended from once-widespread ancestors) and for phenomena such as apparent "retrogression" where species of a given group seem to have become less complex over time.

The Sarawak Law paper puts forth, then, an inductive case for gradual transmutation, a general evolutionary argument, consistent with the word "law" in its title. There is no mention at this stage of nuts-and-bolts process: varieties, variations, or populations, key elements of the 1858 formulation of selection, are not discussed here (notwithstanding one reference to the human population). The word "varied" appears twice in the paper, in connection with observed modifications of form and structure. Similarly, the word "varying" appears once, in a similar context where a progenitor species with an extensive range might give rise to two or more descendant groups of species, "each *varying* from [the ancestral species] in a different manner" (my emphasis). The closest we get to the idea that the evolutionary drama is played out through individuals in populations is found on page 192, where Wallace discusses how conditions are sometimes conducive to the rapid growth and increase of individuals, as well as giving rise to the greatest "profusion of species and the greatest variety of forms," while other conditions might arise that are unfavorable to individuals, leading to extinction. This reflects a populational understanding of species—no conceptual breakthrough as expressed in this paper, but worth noting because such an understanding is necessary to grasp the process of natural selection. At this stage, Wallace was lacking only a mechanism for the pattern of species juxtaposition that he argued for. That mechanism, natural selection, occurred to him less than three years later.

XVIII.—*On the Law which has regulated the Introduction of New Species.* By ALFRED R. WALLACE, F.R.G.S.

EVERY naturalist who has directed his attention to the subject of the geographical distribution of animals and plants, must have been interested in the singular facts which it presents. Many of these facts are quite different from what would have been anticipated, and have hitherto been considered as highly curious, but quite inexplicable. None of the explanations attempted from the time of Linnæus are now considered at all satisfactory; none.of them have given a cause sufficient to account for the facts known at the time, or comprehensive enough to include all the new facts which have since been, and are daily being added. Of late years, however, a great light has been thrown upon the subject by geological investigations, which have shown that the present state.of the earth, and the organisms now inhabiting it, are but the last stage of a long and uninterrupted series of changes which it has undergone, and consequently, that to endeavour to explain and account for its present condition without any reference to those changes (as has frequently been done) must lead to very imperfect and erroneous conclusions.

The facts proved by geology are briefly these :—That during an immense, but unknown period, the surface of the earth has undergone successive changes ; land has sunk beneath the ocean, while fresh land has risen up from it ; mountain chains have been elevated ; islands have been formed into continents, and continents submergéd till they have become islands ; and these changes have taken place, not once merely, but perhaps hundreds, perhaps thousands of times :—That all these operations have been more or less continuous, but unequal in their progress, and during the whole series the organic life of the earth has undergone a corresponding alteration. This alteration also has been gradual, but complete ; after a certain interval not a single species existing which had lived at the commencement of the period. This complete renewal of the forms of life also appears to have occurred several times :—That from the last of the Geological epochs to the present or Historical epoch, the change of organic life has been gradual : the first appearance of animals now existing can in many cases be traced, their numbers gradually increasing in the more recent formations, while other species continually die out and disappear, so that the present condition of the organic world is clearly derived by a natural process of gradual extinction and creation of species from that of the latest geological periods. We may therefore safely infer a like gradation and natural sequence from one geological epoch to another.

1 The "Sarawak Law" paper is so called for the "law" mentioned here in the title (given on p. 186) and the fact that Wallace signed it "Sarawak, Borneo" (p. 196). Scientific "laws" describe outcomes or phenomena consistently observed or following inexorably from established premises. Or, as the *Oxford English Dictionary* defines the scientific and philosophical uses of the word, "In the sciences of observation, a theoretical principle deduced from particular facts, applicable to a defined group or class of phenomena, and expressible by the statement that a particular phenomenon always occurs if certain conditions be present." The paper was written in February 1855, likely while Wallace was stuck indoors during the rainy season in the bungalow provided by Sir James Brooke. He sent it to the *Annals and Magazine of Natural History*, where it was published in September of that year.

2 This opening paragraph is framed in Lyellian terms: the "singular facts" of geographical distribution are illuminated by "geological investigations" of recent years. Indeed, the then-young science of geology was something of a child prodigy, easily the most rapidly progressing and most exciting of the sciences in early to mid-nineteenth-century Britain.

3 The Lyellian vision continues. Note that this paragraph emphasizes gradual change in both the earth and the life upon it. Wallace fully embraces a Lyellian view of gradual uplift and subsidence in earth history, setting up the key points of this paragraph—continuous and gradual change in earth and climate, hand-in-hand with "complete renewal" of life forms on the earth over time. Note that these changes are seen as deriving from a natural process. These parallel arguments from the Species Notebook; for example, at the start of his critique of Lyell under the heading "Note for Organic law of change" we find this same Lyellian vision of earth and life gradually changing together, as a natural process:

> We must at the outset endeavour to ascertain if the present condition of the organic world, is now undergoing any changes—of what nature & to what amount, & we must in the first place assume that the regular course of nature from early Geological Epochs to the present time has produced the present state of things & still continues to act in still further changing it. While the inorganic world has been strictly shown to be the result of a series of changes from the earliest periods produced by causes still acting, it would be most unphilosophical to conclude without the strongest evidence that the organic world so intimately connected with it, had been subject to other laws which have now ceased to act." (Species Notebook, p. 35)

The "law" of the notebook may be the Sarwak Law, though the word "change" in the notebook signals a more explicit transmutationism than is found in the paper.

1   Wallace points out that the geological changes just described must affect the geographical distribution of species—indeed, they must result in the observable biogeographic patterns of species throughout the earth.

2   In fact, the insights gained over the previous two decades into the present and past history of the organic world (read: present species and their distribution, *and* fossil species and their history or distribution through time) suggests that it is possible to frame a comprehensive "law" explaining observed patterns. As a side note, in this paragraph Wallace states that it had been "about ten years" since the idea of such a law occurred to him; the timing corresponds to 1844–1845, the period of his "conversion" to transmutation stemming in large part from reading *Vestiges* (discussed in Chapter 1).

3   Here Wallace enumerates nine "propositions in Organic Geography and Geology," culminating in a tenth proposition, namely, his law: *"Every species has come into existence coincident both in space and time with a pre-existing closely allied species."*

Now, taking this as a fair statement of the results of geological inquiry, we see that the present geographical distribution of life upon the earth must be the result of all the previous changes, both of the surface of the earth itself and of its inhabitants. Many causes no doubt have operated of which we must ever remain in ignorance, and we may therefore expect to find many details very difficult of explanation, and in attempting to give one, must allow ourselves to call into our service geological changes which it is highly probable may have occurred, though we have no direct evidence of their individual operation.

The great increase of our knowledge within the last twenty years, both of the present and past history of the organic world, has accumulated a body of facts which should afford a sufficient foundation for a comprehensive law embracing and explaining them all, and giving a direction to new researches. It is about ten years since the idea of such a law suggested itself to the writer of this paper, and he has since taken every opportunity of testing it by all the newly ascertained facts with which he has become acquainted, or has been able to observe himself. These have all served to convince him of the correctness of his hypothesis. Fully to enter into such a subject would occupy much space, and it is only in consequence of some views having been lately promulgated, he believes in a wrong direction, that he now ventures to present his ideas to the public, with only such obvious illustrations of the arguments and results as occur to him in a place far removed from all means of reference and exact information.

The following propositions in Organic Geography and Geology give the main facts on which the hypothesis is founded.

Geography.

1. Large groups, such as classes and orders, are generally spread over the whole earth, while smaller ones, such as families and genera, are frequently confined to one portion, often to a very limited district.
2. In widely distributed families the genera are often limited in range; in widely distributed genera, well-marked groups of species are peculiar to each geographical district.
3. When a group is confined to one district, and is rich in species, it is almost invariably the case that the most closely allied species are found in the same locality or in closely adjoining localities, and that therefore the natural sequence of the species by affinity is also geographical.
4. In countries of a similar climate, but separated by a wide sea or lofty mountains, the families, genera and species of the

one are often represented by closely allied families, genera and species peculiar to the other.

Geology.

5. The distribution of the organic world in time is very similar to its present distribution in space.
6. Most of the larger and some small groups extend through several geological periods.
7. In each period, however, there are peculiar groups, found nowhere else, and extending through one or several formations.
8. Species of one genus, or genera of one family occurring in the same geological time are more closely allied than those separated in time.
9. As generally in geography no species or genus occurs in two very distant localities without being also found in intermediate places, so in geology the life of a species or genus has not been interrupted. In other words, no group or species has come into existence twice.
10. The following law may be deduced from these facts:—*Every species has come into existence coincident both in space and time with a pre-existing closely allied species.*

This law agrees with, explains and illustrates all the facts connected with the following branches of the subject:—1st. The system of natural affinities. 2nd. The distribution of animals and plants in space. 3rd. The same in time, including all the phænomena of representative groups, and those which Professor Forbes supposed to manifest polarity. 4th. The phænomena of rudimentary organs. We will briefly endeavour to show its bearing upon each of these.

If the law above enunciated be true, it follows that the natural series of affinities will also represent the order in which the several species came into existence, each one having had for its immediate antitype a closely allied species existing at the time of its origin. It is evidently possible that two or three distinct species may have had a common antitype, and that each of these may again have become the antitypes from which other closely allied species were created. The effect of this would be, that so long as each species has had but one new species formed on its model, the line of affinities will be simple, and may be represented by placing the several species in direct succession in a straight line. But if two or more species have been independently formed on the plan of a common antitype, then the series of affinities will be compound, and can only be represented by a forked or many-branched line. Now, all attempts at a Natural classification and arrangement of organic beings show, that both

1   This law, Wallace asserts, "agrees with, explains and illustrates" all the facts connected with (1) the "system of natural affinities" (= classification), (2) the geographical distribution of species on earth, (3) the distribution of species in time (fossils), and (4) the phenomena of rudimentary organs.

2   Edward Forbes (1815–1854) of Edinburgh, who had put forth a quasi-mystical theory of the unfolding plan of life on earth positing two periods of creation, where that the richness of created species was initially high (Forbes's "Paleozoic"), steadily decreased to a low point, and then steadily increased again ("Neozoic"). The net effect is an hourglass shape where width of the glass represents species richness, the symmetry of which gives "polarity" theory its name (Forbes 1854). Forbes's theory provoked Wallace to write the Sarawak Law paper—Wallace was astonished with its thesis, which he later referred to as an "ideal absurdity," when the facts of paleontology and biogeography so plainly spoke otherwise. Wallace did not know when he penned and mailed off his paper that Forbes had meanwhile died (see footnote on p. 192).

3   Wallace explains how his law is expected to give rise to a "natural series of affinities" (by which he means relationships). Branching order maps onto geological succession of species in the fossil record, giving a natural series. In modern terms, the phenomena of both anagenesis (succession of species within a lineage) and cladogenesis (new species originating by branching) are described, the latter conceived as "compound affinities" that are "represented by a forked or many-branched line." Note the curious term "antitype" for ancestral species. Wallace's term reflects the idea that species arise from (or somehow on the model of) predecessor species in an "antitype-type" or ancestor-descendent relationship.

1   Earlier in this paragraph Wallace described how species relationships can be represented by "a forked or many-branched line," which can be traced in "parallel or divergent series." Here he is even more explicit, invoking the tree metaphor to represent the "complicated branching of the lines of affinity" that is the history of life. These branching lines are, he says, "as intricate as the twigs of a gnarled oak or the vascular system of the human body." Wallace's concept of branching and rebranching relationships over time is also found in the Species Notebook. In one notebook entry Wallace noted that geological succession is consistent with branching, where "each group goes on progressing after other groups have branched from it. They then go on in parallel or diverging series" (Species Notebook, p. 38; note the very phrase "parallel or divergent series" is found later in this paragraph of the Sarawak Law paper). In another entry, after discussing a paper by Richard Owen on relationships of different fossil and extant groups, he comments that "the above is what might be expected, if there has been a constant change of species by the modifications of their various organs, producing a complicated many branching series" (Species Notebook, p. 54).

2   Wallace recognized that branching species relationships gives rise to the "Natural System" of classification, where groups are nested within groups. He states here that it is often difficult to properly relate groups when we are missing linking forms, as represented by extinct species of the "stem and main branches." His comment about having "only fragments of this vast system" echo a comment he made in the Species Notebook: "Systems of Nature, compared to fragments of dissected Map or picture or a mosaic.—approximation of fragments shew that all gaps have been filled up" (Species Notebook, p. 52).

3   Further in regard to classification, having identified the basis for the Natural System, on this basis Wallace rejects a priori classification systems such as the so-called quinarian (circular) system of William Macleay (see, e.g., Holland 1996) as "contrary to nature."

these plans have obtained in creation. Sometimes the series of affinities can be well represented for a space by a direct progression from species to species or from group to group, but it is generally found impossible so to continue. There constantly occur two or more modifications of an organ or modifications of two distinct organs, leading us on to two distinct series of species, which at length differ so much from each other as to form distinct genera or families. These are the parallel series or representative groups of naturalists, and they often occur in different countries, or are found fossil in different formations. They are said to have an analogy to each other when they are so far removed from their common antitype as to differ in many important points of structure, while they still preserve a family resemblance. We thus see how difficult it is to determine in every case whether a given relation is an analogy or an affinity, for it is evident that as we go back along the parallel or divergent series, towards the common antitype, the analogy which existed between the two groups becomes an affinity. We are also made aware of the difficulty of arriving at a true classification, even in a small and perfect group ;—in the actual state of nature it is almost impossible, the species being so numerous and the modifications of form and structure so varied, arising probably from the immense number of species which have served as antitypes for the existing species, and thus produced a complicated branching of the lines of affinity, as intricate as the twigs of a gnarled oak or the vascular system of the human body. Again, if we consider that we have only fragments of this vast system, the stem and main branches being represented by extinct species of which we have no knowledge, while a vast mass of limbs and boughs and minute twigs and scattered leaves is what we have to place in order, and determine the true position each originally occupied with regard to the others, the whole difficulty of the true Natural System of classification becomes apparent to us.

We shall thus find ourselves obliged to reject all those systems of classification which arrange species or groups in circles, as well as those which fix a definite number for the divisions of each group. The latter class have been very generally rejected by naturalists, as contrary to nature, notwithstanding the ability with which they have been advocated; but the circular system of affinities seems to have obtained a deeper hold, many eminent naturalists having to some extent adopted it. We have, however, never been able to find a case in which the circle has been closed by a direct and close affinity. In most cases a palpable analogy has been substituted, in others the affinity is very obscure or altogether doubtful. The complicated branching of the lines of affinities in extensive groups must also afford great

13*

facilities for giving a show of probability to any such purely artificial arrangements. Their death-blow was given by the admirable paper of the lamented Mr. Strickland, published in the 'Annals of Natural History,' in which he so clearly showed the true synthetical method of discovering the Natural System.

If we now consider the geographical distribution of animals and plants upon the earth, we shall find all the facts beautifully in accordance with, and readily explained by, the present hypothesis. A country having species, genera, and whole families peculiar to it, will be the necessary result of its having been isolated for a long period, sufficient for many series of species to have been created on the type of pre-existing ones, which, as well as many of the earlier-formed species, have become extinct, and thus made the groups appear isolated. If in any case the antitype had an extensive range, two or more groups of species might have been formed, each varying from it in a different manner, and thus producing several representative or analogous groups. The *Sylviadæ* of Europe and the *Sylvicolidæ* of North America, the *Heliconidæ* of South America and the *Euplœas* of the East, the group of *Trogons* inhabiting Asia, and that peculiar to South America, are examples that may be accounted for in this manner.

Such phænomena as are exhibited by the Galapagos Islands, which contain little groups of plants and animals peculiar to themselves, but most nearly allied to those of South America, have not hitherto received any, even a conjectural explanation. The Galapagos are a volcanic group of high antiquity, and have probably never been more closely connected with the continent than they are at present. They must have been first peopled, like other newly-formed islands, by the action of winds and currents, and at a period sufficiently remote to have had the original species die out, and the modified prototypes only remain. In the same way we can account for the separate islands having each their peculiar species, either on the supposition that the same original emigration peopled the whole of the islands with the same species from which differently modified prototypes were created, or that the islands were successively peopled from each other, but that new species have been created in each on the plan of the pre-existing ones. St. Helena is a similar case of a very ancient island having obtained an entirely peculiar, though limited, flora. On the other hand, no example is known of an island which can be proved geologically to be of very recent origin (late in the Tertiary, for instance), and yet possesses generic or family groups, or even many species peculiar to itself.

When a range of mountains has attained a great elevation, and has so remained during a long geological period, the species

1   The facts of geographical distribution, Wallace says, are "beautifully in accordance" with his hypothesis. Isolation—regional separation—naturally gives rise to a pattern of *intra*regional relatedness among species and, concomitantly, *inter*regional differences. "Peculiar" (or "endemic" in modern terms) species, genera, and even families are seen by Wallace as the "necessary result" of isolation. And not isolation generally, but isolation over long periods of time. Here Wallace offers one ingenious explanation for species convergence in different parts of the world: should a wide-ranging ancestral species or species group (early "antitypes") become separated and isolated in different quarters of the world, over time these can give rise to different but analogous species groups. He cites as examples the old- and new-world warblers, and trogon and butterfly species of Asia and South America.

2   This pattern is also in evidence with islands such as the Galápagos and St. Helena. These archipelagos and their ancestral colonization are also discussed in the Species Notebook (pp. 47–49). Islands of "high antiquity" and isolated will, once initially colonized, eventually hold only "modified prototypes" descended from the colonists that have since died out. In this way, Wallace argues, we can also understand why separate islands in an archipelago each have their own endemic species—either the various islands happened to be colonized at the same or different times by the same progenitor species or they were "successively peopled from each other," the new species arising having been "created in each on the plan of the pre-existing ones."

3   In this paragraph Wallace underscores the importance of both *isolation* and *time*. Mountain ranges can present barriers as effective as expanses of open ocean, but in either case the antiquity of the resulting isolation is key to whether many "peculiar" species occur. Here differing though related species on either side of the Andes or Rockies are mentioned, while Wallace made a related point in the Species Notebook, where, on p. 90, he cited von Buch on the Canary Islands: plant genera "living in sheltered vallies & low grounds often have closely allied species confined to one valley or one island." Note too his point here about the species of Malacca, Java, Sumatra, and Borneo—he recognizes that their similarity stems from the union of these areas into a common landmass at some point in the not-too-distant past, while their differences speak to their relative isolation since that landmass sank and created islands of the high-lying areas. Wallace's discovery of the east-west faunal discontinuity in the archipelago came a year and four months later with his May–June 1856 visit to Bali and Lombock, but here we see him already attuned to the Lyellian model of uplifting and subsiding landmasses and the resulting isolation and linkage of neighboring land areas.

1 Isolation (as on an island) in a "recent period" affords insufficient time to give rise to endemics. Wallace cites the British Isles as an example, having been connected with mainland Europe very recently in geological terms. He expresses this in the Species Notebook as well, pointing out that "islands which are . . . peculiarly inhabited [i.e., inhabited by endemics] appear to be of a considerable antiquity. A long succession of generations appears therefore to have been requisite, to produce those peculiar productions found no where else but allied to those of the nearest land" (pp. 48–49).

2 Here Wallace amplifies the point that "closely allied species in rich groups" are found geographically proximate to one another, a "most striking and important" fact. This pattern is seen in *Bulimi* snails, hummingbirds, toucans, various unnamed fishes, certain orchid and palm genera, different trogons, and macaws and cockatoos. Insects furnish many an example, he says. He cites a work by Lovell Augustus Reeve (1814–1865), an English mollusk specialist best known for his *Conchologia Systematica* (2 vols., 1841–1842) and *Conchologia Iconica* (20 vols., 1843–1878).

of the two sides at and near their bases will be often very different, representative species of some genera occurring, and even whole genera being peculiar to one side only, as is remarkably seen in the case of the Andes and Rocky Mountains. A similar phænomenon occurs when an island has been separated from a continent at a very early period. The shallow sea between the Peninsula of Malacca, Java, Sumatra and Borneo was probably a continent or large island at an early epoch, and may have become submerged as the volcanic ranges of Java and Sumatra were elevated. The organic results we see in the very considerable number of species of animals common to some or all of these countries, while at the same time a number of closely allied representative species exist peculiar to each, showing that a considerable period has elapsed since their separation. The facts of geographical distribution and of geology may thus mutually explain each other in doubtful cases, should the principles here advocated be clearly established.

In all those cases in which an island has been separated from a continent, or raised by volcanic or coralline action from the sea, or in which a mountain-chain has been elevated, in a recent geological epoch, the phænomena of peculiar groups or even of single representative species will not exist. Our own island is an example of this, its separation from the continent being geologically very recent, and we have consequently scarcely a species which is peculiar to it; while the Alpine range, one of the most recent mountain elevations, separates faunas and floras which scarcely differ more than may be due to climate and latitude alone.

The series of facts alluded to in Proposition 3, of closely allied species in rich groups being found geographically near each other, is most striking and important. Mr. Lovell Reeve has well exemplified it in his able and interesting paper on the Distribution of the *Bulimi*. It is also seen in the Humming-birds and Toucans, little groups of two or three closely allied species being often found in the same or closely adjoining districts, as we have had the good fortune of personally verifying. Fishes give evidence of a similar kind: each great river has its peculiar genera, and in more extensive genera its groups of closely allied species. But it is the same throughout Nature; every class and order of animals will contribute similar facts. Hitherto no attempt has been made to explain these singular phænomena, or to show how they have arisen. Why are the genera of Palms and of Orchids in almost every case confined to one hemisphere? Why are the closely allied species of brown-backed Trogons all found in the East, and the green-backed in the West? Why are the Macaws and the Cockatoos similarly restricted? Insects

190   Mr. A. R. Wallace *on the Law which has regulated*

furnish a countless number of analogous examples;—the *Go-liathi* of Africa, the *Ornithopteræ* of the Indian islands, the *Heli-conidæ* of South America, the *Danaidæ* of the East, and in all, the most closely allied species found in geographical proximity. The question forces itself upon every thinking mind,—why are these things so? They could not be as they are, had no law re-gulated their creation and dispersion. The law here enunciated not merely explains, but necessitates the facts we see to exist, while the vast and long-continued geological changes of the earth readily account for the exceptions and apparent discre-pancies that here and there occur. The writer's object in putting forward his views in the present imperfect manner is to submit them to the test of other minds, and to be made aware of all the facts supposed to be inconsistent with them. As his hypothesis is one which claims acceptance solely as explaining and connect-ing facts which exist in nature, he expects facts alone to be brought to disprove it; not *à-priori* arguments against its pro-bability.

The phænomena of geological distribution are exactly analo-gous to those of geography. Closely allied species are found associated in the same beds, and the change from species to spe-cies appears to have been as gradual in time as in space. Geo-logy, however, furnishes us with positive proof of the extinction and production of species, though it does not inform us how either has taken place. The extinction of species, however, offers but little difficulty, and the *modus operandi* has been well illustrated by Sir C. Lyell in his admirable ' Principles.' Geo-logical changes, however gradual, must occasionally have modi-fied external conditions to such an extent as to have rendered the existence of certain species impossible. The extinction would in most cases be effected by a gradual dying-out, but in some instances there might have been a sudden destruction of a species of limited range. To discover how the extinct species have from time to time been replaced by new ones down to the very latest geological period, is the most difficult, and at the same time the most interesting problem in the natural history of the earth. The present inquiry, which seeks to eliminate from known facts a law which has determined, to a certain de-gree, what species could and did appear at a given epoch, may, it is hoped, be considered as one step in the right direction towards a complete solution of it.

Much discussion has of late years taken place on the question, whether the succession of life upon the globe has been from a lower to a higher degree of organization? The admitted facts seem to show that there has been a general, but not a detailed progression. Mollusca and Radiata existed before Vertebrata,

1 "The phenomena of geological distribution are exactly analogous to those of geography." Wallace proceeds to describe how the same pattern of allied species being found in proximity holds in the fossil record, and slow, gradual geological change results in gradual extinction. Note his citation of "Sir C. Lyell in his admirable 'Principles.' " Note Wallace's point that the occurrence of extinction is not a difficulty to understand. Rather, just how extinct species are replaced one after another is "the most difficult, and at the same time the most interesting problem." Wallace sees that identifying as close relatives those succeeding species is an important step toward solving that "most difficult" and "most interesting" problem.

2 Have the successive changes of life on earth been "from a lower to a higher degree of organization?" In other words, is there evidence of progressive change in species through geological time? Lyell went to great lengths to argue against this, at least in the earliest editions of the *Principles*. Wallace has a nuanced and in modern terms fairly accurate perspective: there has been "a general, but not a detailed progression." In other words, there is a general trend toward greater organic complexity over time, but this is not absolute. There are examples of what he terms "retrogression," decline in perceived complexity in some groups. However, this too accords with his "law." He gives examples of groups (Mollusca and Radiata) that existed "of the very earliest periods" and which were apparently "more highly organized than the great mass of those now existing." In any case "progression" and "apparent retrogression" both harmonize with his hypothesis, he maintains. He explains how in the next paragraph.

1   Returning to the tree analogy, here Wallace explains how the descendants in a lineage that branched from a "lower organized species" might survive later-arising and more "highly" organized descendants of the main lineage. The side-branch lineage descendants might never become as "highly organized" as those extinct forms of the main lineage, and if not the net effect is "apparent retrogression"—like "when some monarch of the forest loses a limb [and is] replaced by a feeble and sickly substitute." So while there is overall progress, perhaps, it is interrupted progress in some lineages. This insight is also found in the Species Notebook, not using the term "retrogression" but in essence addressing the issue. Lyell, he says, states that "some of the more ancient Saurians approximated more nearly in their organisation to the types of living Mammalia than do any of our existing reptiles"—in effect, they seem to have regressed, and this Lyell thinks is evidence against progressive change. Wallace replies in his notebook: "Not so if low organized mammalia branched out of <u>low</u> reptiles, fishes. All that is required for the progression is that <u>some</u> reptiles should appear before Mammalia & birds. . . . In the same manner reptiles should not appear <u>before</u> <u>fishes</u> but it matters not how soon after them" (Species Notebook, p. 37; emphasis Wallace's).

and the progression from Fishes to Reptiles and Mammalia, and also from the lower mammals to the higher, is indisputable. On the other hand, it is said that the Mollusca and Radiata of the very earliest periods were more highly organized than the great mass of those now existing, and that the very first fishes that have been discovered are by no means the lowest organized of the class. Now it is believed the present hypothesis will harmonize with all these facts, and in a great measure serve to explain them; for though it may appear to some readers essentially a theory of progression, it is in reality only one of gradual change. It is, however, by no means difficult to show that a real progression in the scale of organization is perfectly consistent with all the appearances, and even with apparent retrogression, should such occur.

Returning to the analogy of a branching tree, as the best mode of representing the natural arrangement of species and their successive creation, let us suppose that at an early geological epoch any group (say a class of the Mollusca) has attained to a great richness of species and a high organization. Now let this great branch of allied species, by geological mutations, be completely or partially destroyed. Subsequently a new branch springs from the same trunk, that is to say, new species are successively created, having for their antitypes the same lower organized species which had served as the antitypes for the former group, but which have survived the modified conditions which destroyed it. This new group being subject to these altered conditions, has modifications of structure and organization given to it, and becomes the representative group of the former one in another geological formation. It may, however, happen, that though later in time, the new series of species may never attain to so high a degree of organization as those preceding it, but in its turn become extinct, and give place to yet another modification from the same root, which may be of higher or lower organization, more or less numerous in species, and more or less varied in form and structure than either of those which preceded it. Again, each of these groups may not have become totally extinct, but may have left a few species, the modified prototypes of which have existed in each succeeding period, a faint memorial of their former grandeur and luxuriance. Thus every case of apparent retrogression may be in reality a progress, though an interrupted one: when some monarch of the forest loses a limb, it may be replaced by a feeble and sickly substitute. The foregoing remarks appear to apply to the case of the Mollusca, which, at a very early period, had reached a high organization and agreat development of forms and species in the Testaceous Cephalopoda. In each succeeding age modified species and genera replaced the former ones which had become extinct, and as we approach the present æra but few and small representatives of the group remain, while the Gasteropods and Bivalves have acquired an immense preponderance. In the long series of changes the earth has undergone, the process of peopling it with organic beings has been continually going on, and whenever any of the higher groups have become nearly or quite extinct, the lower forms which have better resisted the modified physical conditions have served as the antitypes on which to found the new races. In this manner alone, it is believed, can the representative groups at successive periods, and the risings and fallings in the scale of organization, be in every case explained.

The hypothesis of polarity, recently put forward by Professor Edward Forbes* to account for the abundance of generic forms at a very early period and at present, while in the intermediate epochs there is a gradual diminution and impoverishment, till the minimum occurred at the confines of the Palæozoic and Secondary epochs, appears to us quite unnecessary, as the facts may be readily accounted for on the principles already laid down. Between the Palæozoic and Neozoic periods of Professor Forbes, there is scarcely a species in common, and the greater part of the genera and families also disappear to be replaced by new ones. It is almost universally admitted that such a change in the organic world must have occupied a vast period of time. Of this interval we have no record; probably because the whole area of the early formations now exposed to our researches was elevated at the end of the Palæozoic period, and remained so through the interval required for the organic changes which resulted in the fauna and flora of the Secondary period. The records of this interval are buried beneath the ocean which covers three-fourths of the globe. Now it appears highly probable that a long period of quiescence or stability in the physical conditions of a district would be most favourable to the existence of organic life in the greatest abundance, both as regards individuals and also as to variety of species and generic groups, just as we now find that the places best adapted to the rapid growth and increase of individuals also contain the greatest profusion of species and the greatest variety of forms,—the tropics in comparison with the temperate and arctic regions. On the other hand, it seems no less probable that a change in the

---

* Since the above was written, the author has heard with sincere regret of the death of this eminent naturalist, from whom so much important work was expected. His remarks on the present paper,—a subject on which no man was more competent to decide,—were looked for with the greatest interest. Who shall supply his place?

1 Wallace's point reveals an understanding of the concept of common descent, but more than this it reveals his commitment to natural law. He is not absolutist about the idea of progressive change—both "risings" and "fallings" in the "scale of organization" are possible. In a departure from most other transmutationists, few as there were, he did not see evolutionary change as inherently and inexorably progressive the way some (Lamarck and Chambers, for example) envisioned the process, driven by some ill-defined principle from within or by a divine hand without. He recognized the messiness of nature, or the importance of unpredictable events shaping the history of life as when slow and steady geological and climatic changes result in a change of circumstances that proves unfavorable to the more "highly organized" species in his scenario, while the "lesser" organized species ultimately prevail.

2 With that argument, here and through the next few paragraphs Wallace criticizes Edward Forbes's polarity theory of the history of life at length, as well as offering a lucid explanation for why the fossil record is imperfect (an argument also found in the Species Notebook, pp. 92–97).

3 The premature death of Edward Forbes was remarked on earlier. Wallace's comment that "[Forbes's] remarks on the present paper,—a subject on which no man was more competent to decide,—were looked for with the greatest interest" was sincere. When he asks, "Who shall supply his place?" we see a lament that with the one person he might have counted on to engage with him over the meaning of his "law" now deceased, who would speak for Forbes and take issue with Wallace? In fact, Wallace was disappointed that no one seemed to fill the place of Forbes, and the paper was met with only silence until Darwin revealed to him that Lyell and Edward Blyth commented favorably upon it.

1   Wallace is positing a natural process to explain the oscillation in species numbers over geological time, as a counter to Forbes's "polarity." His idea is to link net increase in species numbers with periods of geological quiescence, and net decrease with periods of geological activity. As an example he identifies the "Coal formation"—Carboniferous Period—as a period of increase, whereas in "the formation immediately succeeding this . . . the poverty of forms of life is most apparent." That would be the Permian Period, the termination of which is defined by the greatest of all mass extinction events.

2   The "Coal formation" is now called the Carboniferous Period (divided into the Pennsylvanian and Mississippian in North America), so-named in 1822 for its abundance of plant-derived coal and shale (and petroleum, a later discovery). These carbon-rich formations that provide the fossil fuels of today were laid down between 360 and 300 million years ago, by modern reckoning, under tropical swampy conditions. Wallace notes the apparent decline in species numbers following what he believes to be geological upheavals of this period. He posits that "creations" of new species exceeds extinctions in periods of geological quiescence, while extinctions outpace new "creations" during periods of geological upheaval. The decline in species following the Carboniferous is a reference to the striking changes that led Roderick Murchison—Wallace's champion at the Royal Geographical Society—to propose in 1841 a new geological period he dubbed the Permian, after the old Principality of Permia in northern Russia where he conducted extensive geological investigations. The Permian Period itself does show a small drop in species abundance relative to the Carboniferous, but the terminus of the Permian is defined by the greatest of all geological upheavals, the largest mass extinction in earth history. Where Wallace is going with this is to suggest that the ebb and flow of species over time stems from natural geological processes, with no need for metaphysical models like that of Edward Forbes: note his comment that "We thus have a clue to the increase of the forms of life during certain periods, and their decrease during others, without recourse to any causes but those we know to have existed, and to effects fairly deducible from them."

physical conditions of a district, even small in amount if rapid, or even gradual if to a great amount, would be highly unfavourable to the existence of individuals, might cause the extinction of many species, and would probably be equally unfavourable to the creation of new ones. In this too we may find an analogy with the present state of our earth, for it has been shown to be the violent extremes and rapid changes of physical conditions, rather than the actual mean state in the temperate and frigid zones, which renders them less prolific than the tropical regions, as exemplified by the great distance beyond the tropics to which tropical forms penetrate when the climate is equable, and also by the richness in species and forms of tropical mountain regions which principally differ from the temperate zone in the uniformity of their climate. However this may be, it seems a fair assumption that during a period of geological repose the new species which we know to have been created would have appeared, that the creations would then exceed in number the extinctions, and therefore the number of species would increase. In a period of geological activity, on the other hand, it seems probable that the extinctions might exceed the creations, and the number of species consequently diminish. That such effects did take place in connexion with the causes to which we have imputed them, is shown in the case of the Coal formation, the faults and contortions of which show a period of great activity and violent convulsions, and it is in the formation immediately succeeding this that the poverty of forms of life is most apparent. We have then only to suppose a long period of somewhat similar action during the vast unknown interval at the termination of the Palæozoic period, and then a decreasing violence or rapidity through the Secondary period, to allow for the gradual repopulation of the earth with varied forms, and the whole of the facts are explained. We thus have a clue to the increase of the forms of life during certain periods, and their decrease during others, without recourse to any causes but those we know to have existed, and to effects fairly deducible from them. The precise manner in which the geological changes of the early formations were effected is so extremely obscure, that when we can explain important facts by a retardation at one time and an acceleration at another of a process which we know from its nature and from observation to have been unequal,—a cause so simple may surely be preferred to one so obscure and hypothetical as polarity.

I would also venture to suggest some reasons against the very nature of the theory of Professor Forbes. Our knowledge of the organic world during any geological epoch is necessarily very imperfect. Looking at the vast numbers of species and groups that have been discovered by geologists, this may be

194   Mr. A. R. Wallace *on the Law which has regulated*

doubted; but we should compare their numbers not merely with those that now exist upon the earth, but with a far larger amount *. We have no reason for believing that the number of species on the earth at any former period was much less than at present; at all events the aquatic portion, with which geologists have most acquaintance, was probably often as great or greater. Now we know that there have been many complete changes of species; new sets of organisms have many times been introduced in place of old ones which have become extinct, so that the total amount which have existed on the earth from the earliest geological period must have borne about the same proportion to those now living, as the whole human race who have lived and died upon the earth, to the population at the present time. Again, at each epoch, the whole earth was no doubt, as now, more or less the theatre of life, and as the successive generations of each species died, their exuviæ and preservable parts would be deposited over every portion of the then existing seas and oceans, which we have reason for supposing to have been more, rather than less, extensive than at present. In order then to understand our possible knowledge of the early world and its inhabitants, we must compare, not the area of the whole field of our geological researches with the earth's surface, but the area of the examined portion of each formation separately with the whole earth. For example, during the Silurian period all the earth was Silurian, and animals were living and dying, and depositing their remains more or less over the whole area of the globe, and they were probably (the species at least) nearly as varied in different latitudes and longitudes as at present. What proportion do the Silurian districts bear to the whole surface of the globe, land and sea (for far more extensive Silurian districts probably exist beneath the ocean than above it), and what portion of the known Silurian districts has been actually examined for fossils? Would the area of rock actually laid open to the eye be the thousandth or the ten-thousandth part of the earth's surface? Ask the same question with regard to the Oolite or the Chalk, or even to particular beds of these when they differ considerably in their fossils, and you may then get some notion of how small a portion of the whole we know.

But yet more important is the probability, nay, almost the certainty, that whole formations containing the records of vast geological periods are entirely buried beneath the ocean, and for ever beyond our reach. Most of the gaps in the geological series may thus be filled up, and vast numbers of unknown and unimaginable animals, which might help to elucidate the affinities of the numerous isolated groups which are a perpetual puzzle to

[* See on this subject a paper by Professor Agassiz in the 'Annals' for November 1854.—ED.]

1   In this insightful passage Wallace recognizes that each geological formation represents the merest sampling of life that existed at that time. This limited sampling, combined with the inaccessibility of many such formations to begin with, with "whole formations containing the records of vast geological periods" buried beneath the ocean and thus forever inaccessible, underscores the many inevitable gaps in the fossil record.

2   Louis Agassiz's paper was entitled "On the Primitive Diversity and Number of Animals in Geological Times," reprinted from *Silliman's American Journal of Science and Arts* of May 1854 in the November 1854 issue of the *Annals and Magazine of Natural History* [83 (second series):350–366].

1   This statement that what knowledge we have of the former world is "fragments of a vast whole" but which despite its incompleteness can give insight into
the whole echoes a passage in the Species Notebook: "Systems of Nature, compared to fragments of dissected Map or picture or a mosaic.—approximation of
fragments shew that all gaps have been filled up" (p. 52).

2   Wallace turns here to rudimentary organs—"another important series of
facts" for him, seen as not only in accordance with his law but indeed "necessary
deductions" from it. It is another way of showing that underlying structure reflects affinity: the minute limbs of snakelike lizards, anal hooks of boas (remnants
of hind limbs), "finger-bones" embedded within the paddle-like appendage of
manatees, and botanical examples like abortive stamens are all offered as examples. "Do [these] not teach us something of the system of Nature?" he asks. This
also echoes the Species Notebook, where on pages 97–100 Wallace discusses the
significance of morphological structure. There he asks, "Now what does all this
beautiful law mean, what does it teach us? Is it a substance or a shadow, a truth or
a fallacy?" Of course it is no shadow or fallacy—Wallace knows he is on to something profound; these anatomical oddities make no sense under special creation,
but are "the necessary results of some great natural law." However, while he sees
the transmutational import of rudimentary structures, he also misinterprets
them, as will become evident later in this paragraph on the following page.

the zoologist, may there be buried, till future revolutions may raise them in their turn above the waters, to afford materials for the study of whatever race of intelligent beings may then have succeeded us. These considerations must lead us to the con-clusion, that our knowledge of the whole series of the former inhabitants of the earth is necessarily most imperfect and frag-mentary,—as much so as our knowledge of the present organic world would be, were we forced to make our collections and observations only in spots equally limited in area and in number with those actually laid open for the collection of fossils. Now, the hypothesis of Professor Forbes is essentially one that assumes to a great extent the *completeness* of our knowledge of the *whole series* of organic beings which have existed on the earth. This appears to be a fatal objection to it, independently of all other considerations. It may be said that the same objections exist against every theory on such a subject; but this is not neces-sarily the case. The hypothesis put forward in this paper depends in no degree upon the completeness of our knowledge of the former condition of the organic world, but takes what facts we have as fragments of a vast whole, and deduces from them something of the nature and proportions of that whole which we can never know in detail. It is founded upon isolated groups of facts, recognizes their isolation, and endeavours to deduce from them the nature of the intervening portions.

Another important series of facts, quite in accordance with, and even necessary deductions from, the law now developed, are those of *rudimentary organs*. That these really do exist, and in most cases have no special function in the animal œconomy, is admitted by the first authorities in comparative anatomy. The minute limbs hidden beneath the skin in many of the snake-like lizards, the anal hooks of the boa constrictor, the complete series of jointed finger-bones in the paddle of the Manatus and whale, are a few of the most familiar instances. In botany a similar class of facts has been long recognized. Abortive stamens, rudimentary floral envelopes and undeveloped carpels, are of the most frequent occurrence. To every thoughtful naturalist the question must arise, What are these for? What have they to do with the great laws of creation? Do they not teach us something of the system of Nature? If each species has been created independently, and without any necessary relations with pre-existing species, what do these rudiments, these apparent imperfections mean? There must be a cause for them; they must be the necessary results of some great natural law. Now, if, as it has been endeavoured to be shown, the great law which has regulated the peopling of the earth with animal and vege-table life is, that every change shall be gradual; that no new

196   Dr. T. Wright *on some new species of* Hemipedina.

creature shall be formed widely differing from anything before existing ; that in this, as in everything else in Nature, there shall be gradation and harmony,—then these rudimentary organs are necessary, and are an essential part of the system of Nature. Ere the higher Vertebrata were formed, for instance, many steps were required, and many organs had to undergo modifications from the rudimental condition in which only they had as yet existed.   We still see remaining an antitypal sketch of a wing adapted for flight in the scaly flapper of the penguin, and limbs first concealed beneath the skin, and then weakly protruding from it, were the necessary gradations before others should be formed fully adapted for locomotion.   Many more of these modifications should we behold, and more complete series of them, had we a view of all the forms which have ceased to live. The great gaps that exist between fishes, reptiles, birds and mammals would then, no doubt, be softened down by intermediate groups, and the whole organic world would be seen to be an unbroken and harmonious system.

It has now been shown, though most briefly and imperfectly, how the law that *" Every species has come into existence coincident both in time and space with a pre-existing closely allied species,"* connects together and renders intelligible a vast number of independent and hitherto unexplained facts.   The natural system of arrangement of organic beings, their geographical distribution, their geological sequence, the phænomena of representative and substituted groups in all their modifications, and the most singular peculiarities of anatomical structure, are all explained and illustrated by it, in perfect accordance with the vast mass of facts which the researches of modern naturalists have brought together, and, it is believed, not materially opposed to any of them.   It also claims a superiority over previous hypotheses, on the ground that it not merely explains, but necessitates what exists.   Granted the law, and many of the most important facts in Nature could not have been otherwise, but are almost as necessary deductions from it, as are the elliptic orbits of the planets from the law of gravitation.

Sarawak, Borneo, Feb. 1855.

XIX.—*On some new Species of* Hemipedina *from the Oolites.* By THOMAS WRIGHT, M.D., F.R.S.E.

SINCE the publication of our paper in the August Number of the ‘Annals and Magazine of Natural History,’ on the new genus *Hemipedina* and the Synopsis of the species included

1   Here it is evident that Wallace interprets rudimentary structures not as reduced versions of an ancestrally fully formed structure, but the reverse: as evolutionarily nascent structures in the process of becoming fully formed. The
"flapper" of a penguin and limb nubs beneath or slightly protruding from the
skin of the legless reptiles represent incipient forms: the "necessary gradations
before others should be formed fully" (my emphasis). Wallace does recognize
that such structures are "an essential part of the system of Nature," however,
and as such do speak to transformation. He further sees that recognizing such
structures for what they are—organs in a transmutational process—will help
with classification. These anomalous groups "soften down" (in his odd term) the
gaps between groups, and as intermediate forms fill the gaps, "the whole organic
world [will] be seen to be an unbroken and harmonious system."

2   In this concluding paragraph of his watershed paper, Wallace reiterates the
Sarawak Law and emphasizes that the law "connects together and renders intelligible a vast number of independent and hitherto unexplained facts." In consilience fashion, he ties together classification, geographical distribution, geological
succession, and morphology (from homology and analogy to "the most singular
peculiarities of anatomical structure," namely, rudimentary structures).

3   Wallace claims superiority over previous hypotheses (Macleay, Forbes, Lyell); note the language used in his sweeping statement, "Granted the law, and
many of the most important facts in Nature could not have been otherwise, but
are almost as necessary deductions from it, as are the elliptic orbits of the planets from the law of gravitation." The parallel with the orbiting planets and satellites is noteworthy: a law that acts with the certainty of those of physics and astronomy, the dual queens of the sciences ever since Galileo and Newton, is
certainly an exalted law, and the further parallel of these all being *natural laws*
is of critical importance. Darwin later concluded *On the Origin of Species* drawing precisely the same parallel. Darwin may have taken note of Wallace's use of
the imagery here, but he also used the planet-and-satellite imagery in Transmutation Notebook N, page 36, in 1839 or 1840 (Barrett et al. 1987, 573).

The Wallace and Darwin Papers of 1858

The Wallace and Darwin papers were read at a special meeting of the Linnean Society of London on 1 July 1858 and were published in volume 3 of the Society's *Proceedings* series the following month. The published version of the papers opens with a preface by Lyell and Hooker relating the circumstances that led to this unusual reading. Two unpublished summaries by Darwin outlining his ideas followed, after which Wallace's Ternate essay was presented (ostensibly, the three were read in chronological order). Darwin's first paper consists of an extract of his 230-page "essay" written in 1844, which followed his first brief sketch of 35 pages written in 1842 (both later edited and published by his son Francis; see F. Darwin 1909). Neither of these was intended for publication. The 1844 year of composition of Darwin's essay is noteworthy as the year of publication of *Vestiges of the Natural History of Creation* and the wholesale rejection of that work by the scientific establishment. The vehement reaction to *Vestiges* may have reinforced Darwin's resolve to hold off on revealing his theory, to continue working toward an ironclad argument. In the meantime, he sealed a copy of the essay in an envelope with a letter to his wife, Emma (no. 761 of the Darwin Correspondence Project), requesting that she publish it immediately in the event of his unexpected death—he knew its revolutionary implications. But Darwin did not put it aside before asking his botanist friend Joseph Hooker, who was to become his closest scientific confidante, to critically read parts of the essay. Darwin referred to this fact in one of his June 1858 letters to Lyell: "There is nothing in Wallace's sketch which is not written out much fuller in my sketch copied in 1844, & read by Hooker some dozen years ago. About a year ago I sent a short sketch of which I have copy . . . to Asa Gray, so that I could most truly say & prove that I take nothing from Wallace" (Darwin Correspondence Project, letter 2294). Note that Darwin also mentioned a précis of his ideas about natural selection written for his friend, the botanist Asa Gray, in 1857 (see Darwin Correspondence Project, letter 2136).

For his part, Wallace's Ternate essay was not written for publication either but certainly was drafted as an exposition with an eye toward eventual publication of his ideas on how varieties diverge from their parental form and eventually themselves become new species. Wallace's essay can be read as a reply to Lyell, taking aim at Lyell's formidable anti-transmutationist arguments in the *Principles of Geology*. In almost every paragraph we find references, often direct, to statements from the *Principles* (summarized in Table 4.1).

Table 4.1. Correspondence between concepts found in the Ternate essay and the *Principles of Geology* (fourth edition; Lyell 1835). Adapted from Costa (2013b).

Ternate essay (p.)[1]	Concepts/terms/observations	*Principles*[2] (vol.:pp)
53	Domestication and anti-transmutation argument	3:437–448
54	"Struggle for existence" phrase and concept	3:9, 59, 108–109, 140, 162
	"Wild asses of the Tartarian deserts" example	3:59
	Populations increase yet appear at equilibrium	3:108–120
55	Power of population increase	3:113–115
56	Migration necessary to birds	3:66–70
58	Antelope example: variation and fleetness	2:415
	Result of "alteration of physical conditions"	3:152
	Destructiveness of locust irruptions	3:115–116, 123
	Do varieties return to original form, or not	3:162
59	Argument contra Lyell for sustained change of varieties	3:162
	Geological time: "periods of time . . . so near to infinity . . ."	1:111, 114, 127; 3:449
60	Horses, oxen turned loose on the pampas of South America	3:134–137
61	Lamarck's hypothesis of transmutation	2:407–425, 426–448, 449–465
	Long neck of the giraffe (origin of)	2:415
62	Argument for continual divergence from parental type	2:438–439
	Succession of species through time	1:222–239; 3:155, 164–166

1. Wallace 1858a.
2. Lyell 1835.

In the following presentation, note that all of the key themes tackled by Wallace in his Ternate essay were in the *Principles:* domesticated varieties, struggle for existence and "ecological" interaction (in modern terms), extinction and succession of species, abundance and rarity of populations, gradual environmental change and its perturbing effects on species, and so on. Wallace later cited Malthus as triggering his insight into the significance of struggle, but note the strikingly similar terms in which Lyell discussed the factors (in this case extrinsic) holding down populations: at the edge of the range, against a barrier, when environment is changing, "these stragglers are ready to multiply rapidly on the slightest increase or diminution of heat, rainfall, etc. that may be favorable to them" (Lyell 1835, 3:160). Some of these themes are also reflected in the Species Notebook, such as Wallace's criticism of Lyell on domestication, balance and harmony in nature, limits of variability, and effects of environmental change. Other, more direct connections between the Ternate essay and Species Notebook include the passage on bird migration in relation to food supply (found on p. 56 of the essay and pp. 108 and 155 of the notebook) and our ignorance of cause-and-effect in nature (found on p. 57 of the essay and p. 33 of the notebook).

1   The title chosen by Lyell and Hooker is confusing, seemingly at odds with the actual content of the Wallace and Darwin papers. They are not so much about the tendency of species to simply form varieties (which they were long known to do), or the mere "perpetuation" of varieties and species. Rather, the papers proposed mechanisms by which new species are formed. Did Lyell and Hooker deliberately craft the title thinking it less incendiary than one proclaiming transmutation or common descent of species?

2   Note that Lyell and Hooker consider Darwin and Wallace to have conceived "the same very ingenious theory." So did Darwin: "I never saw a more striking coincidence," he wrote in that anguished letter to Lyell; "if Wallace had my M.S. sketch written out in 1842 he could not have made a better short abstract! Even his terms now stand as Heads of my Chapters" (Darwin Correspondence Project, letter 2285). There are striking similarities but also important differences. In his follow-up letter to Lyell sent on 25 June, Darwin did notice one important difference: "We differ only, that I was led to my views from what artificial selection has done for domestic animals" (letter 2294).

3   Lyell and Hooker are rather disingenuous here; Darwin may have "unreservedly" placed his paper in their hands, but Wallace was ignorant of these proceedings. He had asked Darwin to show his essay to Lyell but said nothing about having it publicly read before a learned society or publishing it. Darwin naturally assumed that Wallace would want it published, writing to Lyell, "Please return me the M.S. which he does not say he wishes me to publish; but I shall of course at once write & offer to send to any Journal" (letter 2285).

4   In fact, neither naturalists' contribution was intended for publication as written. As of 1858 Darwin had produced no concise essay that laid out all of his ideas about transmutation. The closest in terms of length was probably his 1842 (not 1839, as Lyell and Hooker indicate here) sketch of twenty-five or so pages, quite skeletal as compared with the nearly 200-page essay he wrote two years later (both published as *Foundations of the Origin of Species;* see F. Darwin 1909). Although he had a clean copy of the 1844 essay made by the local schoolmaster, it was not intended for publication and was far too long to present along with Wallace's essay. Hooker critiqued parts of this essay in 1847, though, and Darwin later (in 1857) summarized his ideas on natural selection for Asa Gray's critical review. A copy of the summary was sent to Gray on 5 September 1857 (letter 2136 of the Darwin Correspondence). In the case of the events of June 1858 Darwin had a triple problem. He had to come up with a concise overview of his ideas very quickly, in addition to which he needed as thorough a treatment of his ideas as possible, *and* these had to be datable in order to buttress his claim of priority. No one piece of his writing fit the bill, but the extracts he prepared for Hooker and Gray constituted two overviews that together answered these three concerns. They would have to do. The first extract he sent to Lyell was derived from chapter 2 of the 1844 essay (F. Darwin 1909), and the second was his draft of the summary he prepared for Gray in 1857.

On the Tendency of Species to form Varieties; and on the Perpetuation of Varieties and Species by Natural Means of Selection. By CHARLES DARWIN, Esq., F.R.S., F.L.S., & F.G.S., and ALFRED WALLACE, Esq. Communicated by Sir CHARLES LYELL, F.R.S., F.L.S., and J. D. HOOKER, Esq., M.D., V.P.R.S., F.L.S., &c.

[Read July 1st, 1858.]

London, June 30th, 1858.

MY DEAR SIR,—The accompanying papers, which we have the honour of communicating to the Linnean Society, and which all relate to the same subject, viz. the Laws which affect the Production of Varieties, Races, and Species, contain the results of the investigations of two indefatigable naturalists, Mr. Charles Darwin and Mr. Alfred Wallace.

These gentlemen having, independently and unknown to one another, conceived the same very ingenious theory to account for the appearance and perpetuation of varieties and of specific forms on our planet, may both fairly claim the merit of being original thinkers in this important line of inquiry; but neither of them having published his views, though Mr. Darwin has for many years past been repeatedly urged by us to do so, and both authors having now unreservedly placed their papers in our hands, we think it would best promote the interests of science that a selection from them should be laid before the Linnean Society.

Taken in the order of their dates, they consist of:—

1. Extracts from a MS. work on Species*, by Mr. Darwin, which was sketched in 1839, and copied in 1844, when the copy was read by Dr. Hooker, and its contents afterwards communicated to Sir Charles Lyell. The first Part is devoted to "The Variation of Organic Beings under Domestication and in their Natural State;" and the second chapter of that Part, from which we propose to read to the Society the extracts referred to, is headed, "On the Variation of Organic Beings in a state of Nature; on the Natural Means of Selection; on the Comparison of Domestic Races and true Species."

2. An abstract of a private letter addressed to Professor Asa Gray, of Boston, U.S., in October 1857, by Mr. Darwin, in which

* This MS. work was never intended for publication, and therefore was not written with care.—C. D. 1858.

he repeats his views, and which shows that these remained unaltered from 1839 to 1857.

3. An Essay by Mr. Wallace, entitled " On the Tendency of Varieties to depart indefinitely from the Original Type." This was written at Ternate in February 1858, for the perusal of his friend and correspondent Mr. Darwin, and sent to him with the expressed wish that it should be forwarded to Sir Charles Lyell, if Mr. Darwin thought it sufficiently novel and interesting. So highly did Mr. Darwin appreciate the value of the views therein set forth, that he proposed, in a letter to Sir Charles Lyell, to obtain Mr. Wallace's consent to allow the Essay to be published as soon as possible. Of this step we highly approved, provided Mr. Darwin did not withhold from the public, as he was strongly inclined to do (in favour of Mr. Wallace), the memoir which he had himself written on the same subject, and which, as before stated, one of us had perused in 1844, and the contents of which we had both of us been privy to for many years. On representing this to Mr. Darwin, he gave us permission to make what use we thought proper of his memoir, &c.; and in adopting our present course, of presenting it to the Linnean Society, we have explained to him that we are not solely considering the relative claims to priority of himself and his friend, but the interests of science generally; for we feel it to be desirable that views founded on a wide deduction from facts, and matured by years of reflection, should constitute at once a goal from which others may start, and that, while the scientific world is waiting for the appearance of Mr. Darwin's complete work, some of the leading results of his labours, as well as those of his able correspondent, should together be laid before the public.

We have the honour to be yours very obediently,

CHARLES LYELL.
JOS. D. HOOKER.

J. J. Bennett, Esq.,
    Secretary of the Linnean Society.

I. *Extract from an unpublished Work on Species, by* C. DARWIN, *Esq., consisting of a portion of a Chapter entitled, " On the Variation of Organic Beings in a state of Nature; on the Natural Means of Selection; on the Comparison of Domestic Races and true Species."*

De Candolle, in an eloquent passage, has declared that all nature is at war, one organism with another, or with external nature.

1 It is not clear how strongly inclined Darwin was to withhold his own ideas in favor of Wallace and permit Wallace's essay to be read and published alone. There are intimations of this, as where Darwin stated in the letter to Lyell quoted earlier, "but I shall of course at once write & offer to send to any Journal." The next year he appended a postscript to a letter to Wallace making this claim explicitly: "You cannot tell how I admire your spirit, in the manner in which you have taken all that was done about publishing our papers. I had actually written a letter to you, stating that I would *not* publish anything before you had published. I had not sent that letter to the Post, when I received one from Lyell & Hooker, *urging* me to send some M.S. to them, & allow them to act as they thought fair & honourably to both of us. & I did so" (6 April 1859; Darwin Correspondence Project, letter 2449).

This seems to have overtones of guilt, if so doubtless prompted by the knowledge that Lyell and Hooker's "urging" was not unbidden; he had all but pleaded with them to help find a way that he could publish his ideas honorably (e.g., in Darwin Correspondence letters 2294 and 2295, the first of which concludes with an admission that "this is a trumpery letter influenced by trumpery feelings").

2 This statement suggests that Lyell and Hooker were aware of the question of priority. By the standards of the time, as well as our own, putting the interests of what they considered to be "science generally" above that of the matter of strict priority is ethically questionable (see Rachels 1986).

3 Darwin opens his first extract, taken from his 1844 essay, with the struggle for existence, citing de Candolle's "eloquent passage" on nature at war. As we have seen, Lyell quoted de Candolle in his own discussion of struggle in the *Principles;* it is likely that the concept of struggle in nature was impressed on both Darwin and Wallace more by their reading of Lyell than of Malthus. In this paper Darwin mentions "struggle" nine times but does not use the phrase "struggle for existence" (only Wallace uses that phrase in the 1858 papers; see notes for p. 54). He does, however, recognize different forms of struggle: "one organism with another, or with external nature."

1   Darwin's memorable expression of "the contented face of nature" refers to
the balance and harmony we mistakenly read into the natural world. This found
its way into *On the Origin of Species* as well, in a passage on the "face of nature
bright with gladness" (Darwin 1859, 62). On the contrary, what prevails in nature is "the doctrine of Malthus applied . . . with tenfold force." Malthus leads
Darwin into a description of population growth, giving an example with birds:
an initial four pairs of breeding birds grow to more than 2,000 birds in seven
years' time—and checks on that growth, pointing out that "this increase is quite
impossible."

2   Darwin then considers "practical illustrations" of population growth potential, citing first the swarming mouse populations following the death of untold
numbers of feral horses and cattle in La Plata during the extreme drought of
1826–28. Darwin visited the region of the Rio de la Plata, which forms part of
the border between Argentina and Uruguay, in the summer of 1832 while on the
Beagle voyage. In his *Journal of Researches* he recounted seeing great deposits
of bones of many different species, and hearing "several vivid descriptions of the
effect of a great drought; and the account of this may throw some light on the
cases, where vast numbers of animals of all kinds, have been embedded together.
The period included between the years 1827 and 1830 is called the 'gran seco' or
the great drought" (Darwin 1839, 155; 1845, 132–133). Rapid population expansion of introduced species is also cited, something discussed by Lyell. As in Wallace's description of this process, Darwin follows his discussion of population
growth potential with a description of the checks that must prevent growth. In
language parallel to Wallace's, he says, "Lighten any check in the least degree,
and the geometrical powers of increase in every organism will almost instantly
increase the average number of the favoured species."

Seeing the contented face of nature, this may at first well be doubted; but reflection will inevitably prove it to be true. The war, however, is not constant, but recurrent in a slight degree at short periods, and more severely at occasional more distant periods; and hence its effects are easily overlooked. It is the doctrine of Malthus applied in most cases with tenfold force. As in every climate there are seasons, for each of its inhabitants, of greater and less abundance, so all annually breed; and the moral restraint which in some small degree checks the increase of mankind is entirely lost. Even slow-breeding mankind has doubled in twenty-five years; and if he could increase his food with greater ease, he would double in less time. But for animals without artificial means, the amount of food for each species must, *on an average*, be constant, whereas the increase of all organisms tends to be geometrical, and in a vast majority of cases at an enormous ratio. Suppose in a certain spot there are eight pairs of birds, and that *only* four pairs of them annually (including double hatches) rear only four young, and that these go on rearing their young at the same rate, then at the end of seven years (a short life, excluding violent deaths, for any bird) there will be 2048 birds, instead of the original sixteen. As this increase is quite impossible, we must conclude either that birds do not rear nearly half their young, or that the average life of a bird is, from accident, not nearly seven years. Both checks probably concur. The same kind of calculation applied to all plants and animals affords results more or less striking, but in very few instances more striking than in man.

Many practical illustrations of this rapid tendency to increase are on record, among which, during peculiar seasons, are the extraordinary numbers of certain animals; for instance, during the years 1826 to 1828, in La Plata, when from drought some millions of cattle perished, the whole country actually *swarmed* with mice. Now I think it cannot be doubted that during the breeding-season all the mice (with the exception of a few males or females in excess) ordinarily pair, and therefore that this astounding increase during three years must be attributed to a greater number than usual surviving the first year, and then breeding, and so on till the third year, when their numbers were brought down to their usual limits on the return of wet weather. Where man has introduced plants and animals into a new and favourable country, there are many accounts in how surprisingly few years the whole country has become stocked with them. This increase would necessarily stop as soon as the country was fully stocked; and yet we have every reason to believe, from what is known of wild animals, that *all* would pair in the spring. In the majority of cases it is most difficult to imagine where the checks fall—though generally, no doubt, on the seeds, eggs, and young; but when we remember how impossible, even in mankind (so much better known than any other animal), it is to infer from repeated casual observations what the average duration of life is, or to discover the different percentage of deaths to births in different countries, we ought to feel no surprise at our being unable to discover where the check falls in any animal or plant. It should always be remembered, that in most cases the checks are recurrent yearly in a small, regular degree, and in an extreme degree during unusually cold, hot, dry, or wet years, according to the constitution of the being in question. Lighten any check in the least degree, and the geometrical powers of increase in every organism will almost instantly increase the average number of the favoured species. Nature may be compared to a surface on which rest ten thousand sharp wedges touching each other and driven inwards by incessant blows. Fully to realize these views much reflection is requisite. Malthus on man should be studied; and all such cases as those of the mice in La Plata, of the cattle and horses when first turned out in South America, of the birds by our calculation, &c., should be well considered. Reflect on the enormous multiplying power *inherent and annually in action* in all animals; reflect on the countless seeds scattered by a hundred ingenious contrivances, year after year, over the whole face of the land; and yet we have every reason to suppose that the average percentage of each of the inhabitants of a country usually remains constant. Finally, let it be borne in mind that this average number of individuals (the external conditions remaining the same) in each country is kept up by recurrent struggles against other species or against external nature (as on the borders of the Arctic regions, where the cold checks life), and that ordinarily each individual of every species holds its place, either by its own struggle and capacity of acquiring nourishment in some period of its life, from the egg upwards; or by the struggle of its parents (in short-lived organisms, when the main check occurs at longer intervals) with other individuals of the *same* or *different* species.

But let the external conditions of a country alter. If in a small degree, the relative proportions of the inhabitants will in most cases simply be slightly changed; but let the number of

1   This is one of Darwin's most famous metaphors, in print for the first time (it is also found in the *Origin,* but only the first edition): "Nature may be compared to a surface on which rest ten thousand sharp wedges touching each other and driven inwards by incessant blows" (Darwin 1859, 67). He exhorts his readers to ponder this—we are so firmly in the grip of the vision of balance and harmony that "much reflection is requisite" to see beyond this; "Malthus on man should be studied," he urges, and then, "Reflect on the enormous multiplying power *inherent and annually in action* in all animals [emphasis Darwin's]; reflect on the countless seeds scattered by a hundred ingenious contrivances, year after year, over the whole face of the land." Powerful expressions of population growth, indeed. And yet, he then points out, the average number of each species seems to remain constant. It is by continual epic struggle that each species holds in place (this paragraph sees the first three of the nine instances of the word "struggle" in the essay).

2   Note that Darwin here distinguishes between struggle against the elements and struggle against other organisms, both the same and different species. It is there "biotic" interactions of predation, parasition, and especially competition that assume a central role in Darwin's vision for how selection plays out. As we shall see, this is a key element of his "principle of divergence."

3   Darwin has his readers conduct a mental experiment: "let the external conditions of a country alter," in parallel with Wallace's "now let some alteration of physical conditions occur in the district." This is followed out on the next page.

1 As the environment changes, myriad ramifying effects are felt. Darwin's thought experiment takes place in relative isolation, on an island, and he envisions continued change, forming new "stations" (niches, habitats). The environmental rug, so to speak, is pulled from beneath the current well-adapted inhabitants, rendering them less well adapted. "Any minute variation in structure, habits, or instincts" that better adapt individuals to the new conditions would confer a better chance of survival upon those individuals, and likewise the offspring of those individuals would have a better chance. Darwin takes pains to underscore the magnitude of this process—the "smallest grain in the balance" makes all the difference in the long run. Then, for the first time in the paper, he draws a parallel with the improvement of domestic breeds, citing Bakewell's improvement of cattle and Western's of sheep by an "identical principle of selection." Robert Bakewell (1725–1795) was a celebrated cattle and sheep breeder from Leicestershire. Charles Callis (Lord) Western (1767–1844) was renowned for the improvement of sheep breeds.

2 Another imaginary example is offered: a predator (a canine of some kind) and its prey (rabbits and hares, the latter harder to catch than the former). The canine ordinarily preys on the abundant rabbits, but if changes in environment cause the rabbit population to fall and the hare population to grow, the "plastic organization" of the canine would eventually alter in response to this change in its food supply. Over time, Darwin argues, the sleekest, fleetest individuals would be favored with no greater difficulty than the production of the sleek, fleet greyhound morphology by dog breeders. (His example is carefully chosen to allow him to again draw the reader's attention to the parallel with domestication.) This process is based on natural heritable variation that confers even a *modicum* of advantage to some individuals; these will be selected over enormous periods of time and ultimately yield a new variety or species. This is true of plants as well as animals, as seen near the end of this paragraph. Giving the development of ever-more-downy seeds as an example (the down playing a role in seed dispersal), Darwin does not miss an opportunity to draw the domestication analogy here too, concluding in the footnote on the following page that he sees "no more difficulty in this, than in the planter improving his varieties of the cotton plant."

inhabitants be small, as on an island, and free access to it from other countries be circumscribed, and let the change of conditions continue progressing (forming new stations), in such a case the original inhabitants must cease to be as perfectly adapted to the changed conditions as they were originally. It has been shown in a former part of this work, that such changes of external conditions would, from their acting on the reproductive system, probably cause the organization of those beings which were most affected to become, as under domestication, plastic. Now, can it be doubted, from the struggle each individual has to obtain subsistence, that any minute variation in structure, habits, or instincts, adapting that individual better to the new conditions, would tell upon its vigour and health? In the struggle it would have a better *chance* of surviving; and those of its offspring which inherited the variation, be it ever so slight, would also have a better *chance*. Yearly more are bred than can survive; the smallest grain in the balance, in the long run, must tell on which death shall fall, and which shall survive. Let this work of selection on the one hand, and death on the other, go on for a thousand generations, who will pretend to affirm that it would produce no effect, when we remember what, in a few years, Bakewell effected in cattle, and Western in sheep, by this identical principle of selection?

To give an imaginary example from changes in progress on an island:—let the organization of a canine animal which preyed chiefly on rabbits, but sometimes on hares, become slightly plastic; let these same changes cause the number of rabbits very slowly to decrease, and the number of hares to increase; the effect of this would be that the fox or dog would be driven to try to catch more hares: his organization, however, being slightly plastic, those individuals with the lightest forms, longest limbs, and best eyesight, let the difference be ever so small, would be slightly favoured, and would tend to live longer, and to survive during that time of the year when food was scarcest; they would also rear more young, which would tend to inherit these slight peculiarities. The less fleet ones would be rigidly destroyed. I can see no more reason to doubt that these causes in a thousand generations would produce a marked effect, and adapt the form of the fox or dog to the catching of hares instead of rabbits, than that greyhounds can be improved by selection and careful breeding. So would it be with plants under similar circumstances. If the number of individuals of a species with plumed seeds could be increased by greater powers of dissemination within its own area (that is, if the check to increase fell chiefly on the seeds), those seeds which were provided with ever so little more down, would in the long run be most disseminated; hence a greater number of seeds thus formed would germinate, and would tend to produce plants inheriting the slightly better-adapted down*.

Besides this natural means of selection, by which those individuals are preserved, whether in their egg, or larval, or mature state, which are best adapted to the place they fill in nature, there is a second agency at work in most unisexual animals, tending to produce the same effect, namely, the struggle of the males for the females. These struggles are generally decided by the law of battle, but in the case of birds, apparently, by the charms of their song, by their beauty or their power of courtship, as in the dancing rock-thrush of Guiana. The most vigorous and healthy males, implying perfect adaptation, must generally gain the victory in their contests. This kind of selection, however, is less rigorous than the other; it does not require the death of the less successful, but gives to them fewer descendants. The struggle falls, moreover, at a time of year when food is generally abundant, and perhaps the effect chiefly produced would be the modification of the secondary sexual characters, which are not related to the power of obtaining food, or to defence from enemies, but to fighting with or rivalling other males. The result of this struggle amongst the males may be compared in some respects to that produced by those agriculturists who pay less attention to the careful selection of all their young animals, and more to the occasional use of a choice mate.

II. *Abstract of a Letter from* C. DARWIN, Esq., *to* Prof. ASA GRAY, *Boston, U.S., dated Down, September 5th, 1857.*

1. It is wonderful what the principle of selection by man, that is the picking out of individuals with any desired quality, and breeding from them, and again picking out, can do. Even breeders have been astounded at their own results. They can act on differences inappreciable to an uneducated eye. Selection has been *methodically* followed in *Europe* for only the last half century; but it was occasionally, and even in some degree methodically, followed in the most ancient times. There must have been also a kind of unconscious selection from a remote period, namely in

* I can see no more difficulty in this, than in the planter improving his varieties of the cotton plant.—C. D. 1858.

1 Darwin points out that there is another, less rigorous selection process at work, which he dubs "sexual selection." He recognizes two forms of this kind of selection: that "generally decided by the law of battle" (in modern terms, male-male competition), and selection "by the charms of their song, by their beauty or their power of courtship" as seen in many birds (in modern terms, female choice). This form of selection is "less rigorous" because it does not involve the death of the less successful individuals, but simply fewer descendants. Darwin concludes by drawing a parallel between sexual selection and the less careful forms of artificial selection practiced by plant and animal breeders. In both cases selection is less rigorous than with natural selection in that survivorship is not at stake, merely reproductive success.

2 American botanist Asa Gray, of Harvard, became in 1857 one of the few friends that Darwin took into his confidence about his species theory. He had already revealed to Gray in July 1857 (Darwin Correspondence Project, letter 2125) his "heterodox conclusion" that species are not independently created. Finding Gray sympathetic, in early September 1857 Darwin wrote that he thought that he went "as far as almost anyone in seeing the grave difficulties against my doctrine. . . . To talk of climate or Lamarckian habit producing such adaptations to other organic beings is futile. This difficulty, I believe I have sur-mounted. As you seem interested in subject, & as it is an *immense* advantage to me to write to you & to hear ever so briefly, what you think, I will enclose . . . the briefest abstract of my notions on the means by which nature makes her species" (letter 2136; emphasis in original). Darwin had sent a copy of his brief abstract, and so had the original at hand to provide to Lyell and Hooker.

3 Darwin's abstract for Gray opens with the domestication analogy, and he proceeds to describe the process of methodical selection leading to the accumulation of variations over time. In his 20 July 1857 letter to Gray he also under-scored the insights provided by domestication:

> Either species have been independently created, or they have descended from other species, like varieties from one species. I think it can be shown to be probable that man gets his most distinct varieties by preserving such as arise best worth keeping & destroying the others,—but I [should] fill a quire if I were to go on. To be brief I *assume* that species arise like our domestic varieties with *much* extinction; & then test this hypothesis by comparison with as many general & pretty well established propositions as I can find made out,—in geograph. distribution, geological history—affinities &c &c &c. (Darwin Correspondence Project, letter 2125; emphasis in original).

1   The power of selection in nature is amplified compared to that of breeders, imagined as a powerful being. This personification of a natural process later proved problematic (Wallace was to take issue with the term "natural selection" on these grounds), but his conception of selection in nature was inspired by the agricultural breeders. The rest of this paragraph emphasizes the key elements underpinning natural selection: "slight variation" in all parts (an important lesson of his barnacle studies), and the availability of "almost unlimited time." Through geology we appreciate the vastness of time needed, during which there have been "millions on millions of generations."

2   Darwin introduces his term *Natural Selection,* capitalized and italicized, which "selects exclusively for the good" of each being. "Struggle" for life or existence in nature was a well-established concept by then, but even so he feels that the magnitude of struggle is inadequately understood or appreciated. He uses the Malthusian argument to stress that checks on growth must be incessant, mere "trifling differences" determining which individuals survive and which perish (recalling the earlier comment about all parts varying in slight ways).

3   Here is the inflection point: a country "undergoing some change" as the Lyellian world inevitably does. In Darwin's formulation: (1) this itself will cause more variation (which he believes is generated by influence of the environment on the reproductive organs); (2) many individuals perish owing to the changing conditions; and (3) the survivors are exposed to the "mutual action" of new inhabitants migrating in with the new conditions—interactions Darwin believes "far more important" than climate. That is, biotic interactions (competition, predation, parasitism, mutualism) are more significant as selective forces than mere environment.

He reiterates that over the countless generations some individuals will vary in slight, profitable ways; these individuals will have a better chance of surviving, and their profitable traits may thus be "slowly increased by the accumulative action of natural selection." Thus do new varieties (and eventually new species) slowly arise. Note too that the new variety will "either coexist with, or, more commonly, will exterminate its parent form." In the final sentence of this paragraph on the next page Darwin describes how species become "adapted to a score of contingencies," stressing that this occurs though the accumulation of "those slight variations in all parts" of structure that "are in any way useful" and "during any part" of life. Note the key words: *slight* variations, in *all* parts of structure, in *any* way useful, during *any* part of life.

the preservation of the individual animals (without any thought of their offspring) most useful to each race of man in his particular circumstances. The "roguing," as nurserymen call the destroying of varieties which depart from their type, is a kind of selection. I am convinced that intentional and occasional selection has been the main agent in the production of our domestic races; but however this may be, its great power of modification has been indisputably shown in later times. Selection acts only by the accumulation of slight or greater variations, caused by external conditions, or by the mere fact that in generation the child is not absolutely similar to its parent. Man, by this power of accumulating variations, adapts living beings to his wants—may be said to make the wool of one sheep good for carpets, of another for cloth, &c.

2. Now suppose there were a being who did not judge by mere external appearances, but who could study the whole internal organization, who was never capricious, and should go on selecting for one object during millions of generations; who will say what he might not effect? In nature we have some *slight* variation occasionally in all parts; and I think it can be shown that changed conditions of existence is the main cause of the child not exactly resembling its parents; and in nature geology shows us what changes have taken place, and are taking place. We have almost unlimited time; no one but a practical geologist can fully appreciate this. Think of the Glacial period, during the whole of which the same species at least of shells have existed; there must have been during this period millions on millions of generations.

3. I think it can be shown that there is such an unerring power at work in *Natural Selection* (the title of my book), which selects exclusively for the good of each organic being. The elder De Candolle, W. Herbert, and Lyell have written excellently on the struggle for life; but even they have not written strongly enough. Reflect that every being (even the elephant) breeds at such a rate, that in a few years, or at most a few centuries, the surface of the earth would not hold the progeny of one pair. I have found it hard constantly to bear in mind that the increase of every single species is checked during some part of its life, or during some shortly recurrent generation. Only a few of those annually born can live to propagate their kind. What a trifling difference must often determine which shall survive, and which perish!

4. Now take the case of a country undergoing some change. This will tend to cause some of its inhabitants to vary slightly—

not but that I believe most beings vary at all times enough for selection to act on them. Some of its inhabitants will be exterminated; and the remainder will be exposed to the mutual action of a different set of inhabitants, which I believe to be far more important to the life of each being than mere climate. Considering the infinitely various methods which living beings follow to obtain food by struggling with other organisms, to escape danger at various times of life, to have their eggs or seeds disseminated, &c. &c., I cannot doubt that during millions of generations individuals of a species will be occasionally born with some slight variation, profitable to some part of their economy. Such individuals will have a better chance of surviving, and of propagating their new and slightly different structure; and the modification may be slowly increased by the accumulative action of natural selection to any profitable extent. The variety thus formed will either coexist with, or, more commonly, will exterminate its parent form. An organic being, like the woodpecker or misseltoe, may thus come to be adapted to a score of contingences—natural selection accumulating those slight variations in all parts of its structure, which are in any way useful to it during any part of its life.

5. Multiform difficulties will occur to every one, with respect to this theory. Many can, I think, be satisfactorily answered. *Natura non facit saltum* answers some of the most obvious. The slowness of the change, and only a very few individuals undergoing change at any one time, answers others. The extreme imperfection of our geological records answers others.

6. Another principle, which may be called the principle of divergence, plays, I believe, an important part in the origin of species. The same spot will support more life if occupied by very diverse forms. We see this in the many generic forms in a square yard of turf, and in the plants or insects on any little uniform islet, belonging almost invariably to as many genera and families as species. We can understand the meaning of this fact amongst the higher animals, whose habits we understand. We know that it has been experimentally shown that a plot of land will yield a greater weight if sown with several species and genera of grasses, than if sown with only two or three species. Now, every organic being, by propagating so rapidly, may be said to be striving its utmost to increase in numbers. So it will be with the offspring of any species after it has become diversified into varieties, or subspecies, or true species. And it follows, I think, from the foregoing facts, that the varying offspring of each species will try

1 The outcome of the selective survival and propagation of individuals with "some slight variation, profitable to some part of their economy" is the supplanting of the parental form. In this way extinction is built into Darwin's process: new varieties diverge from and eventually outcompete and replace their parent species. Note, however, that this process is not simply destructive: it is creative, underlying the exquisite adaptations of organisms like woodpeckers and mistletoe (both examples also cited in the *Origin* in this context, on p. 60).

2 Of the many difficulties anticipated by Darwin, the principle of *Natura non facit saltum*—nature does not make leaps—explains much. By this he likely means the apparent gaps in the fossil record (also addressed by the "extreme imperfection" of the record), as well as apparently unbridgeable diversity in morphology, physiology, and instinct in the organic world. This Latin phrase affirms continuity throughout organic life, both living and extinct.

3 Darwin's "principle of divergence" is not simple divergence in the sense of getting progressively more different from the parental form. Darwin (like Wallace) grasped the distinction between within-lineage change and lineage splitting. Both realized that it is the latter, branching evolution, that gives rise to the tree of life, and that this maps onto the nested hierarchy of classification. Darwin further devised a concept of competition-driven selection that translates into what we call competitive exclusion and niche partitioning. He links "diverse forms" with the diversity of species that can live in a given area, noting that "we see this in the many generic forms in a square yard of turf"—a reference to his field botany studies initiated in 1855. As he later put it (*Origin*, p. 114): "The truth of the principle, that the greatest amount of life can be supported by great diversification of structure, is seen under many natural circumstances. . . . In an extremely small area . . . where the contest between individual and individual must be severe, we always find great diversity in its inhabitants." The key point is that inexorable population pressure inherent in all species and varieties results in ever-varying offspring trying to "seize on as many and as diverse places in the economy of nature as possible." The outcome over time is diversification, and species always thus "branch and sub-branch like the limbs of a tree from a common trunk." Darwin's tree tells a tale: burgeoning twigs and branches destroy the less vigorous, and dead and lost branches represent extinct groups.

4 The experiment mentioned here (and cited in *Origin*, p. 113) is the early nineteenth-century grass plot experiment conducted at Woburn Abbey in Bedfordshire (Hector and Hooper 2002). The point for Darwin was that a mixture of taxonomically diverse grass species produced the greatest yield, which he saw as indicative of the benefit of divergence—an "ecological division of labor" to parallel the "physiological division of labor."

1   Darwin concluded his preÝcis with a comment to Gray that his sketch is "most imperfect," yet he managed to say a great deal. Note that this and the previous extract are similarly structured, opening with the domestication analogy, then building a case for selection as a natural outcome of struggle and heritable variation. An important difference is that sexual selection is introduced in the previous extract, and the "Principle of Divergence" in this one.

2   The Ternate essay. Wallace's proposed law of change has three key components: variation is abundant; there is an incessant struggle for existence resulting from population pressure and constantly (though slowly) changing geological and environmental conditions; as a result certain variants or varieties will persist over others, over time having the net effect of altering the makeup of the population relative to the parental or ancestral form. Insofar as this is a phenomenon that "always occurs if certain conditions be present," we have met the OED definition given previously for a "law" in a scientific or philosophical context. This paper is not merely an extension of the Sarawak Law paper, however. It is also a conceptual departure, as reflected by the major shift in emphasis from transmutation generally to mechanism, a process that is all about variation and the struggle for existence. "Variation" and cognates (vary, varies, variety, variable, variant, etc.) appear fifty-nine times in the Species Notebook and fifty-two times in the Ternate essay, but only eight times in the Sarawak Law paper (which is about 30 percent longer than the Ternate essay). The following notes are adapted from Costa (2013b); all references to the *Principles* are to the fourth (1835) edition, which Wallace had with him in the field.

3   Wallace acknowledges that one of the strongest arguments made for the permanence of species is the limited nature of change of domestic varieties. The tendency of such varieties to "return" to the parental type was assumed to apply to natural as well as domestic varieties. Wallace is invoking Lyell's anti-transmutationism here: this argument was advanced forcefully by Lyell in the *Principles* (3:437–448).

4   Realizing the abundance of natural varieties (and, presumably, the many degrees of difference from parental forms seen) is necessary to conquer the erroneous belief in species stability. The difficulty of distinguishing varieties from species echoes Blyth's 1835 paper on classifying varieties (Species Notebook, p. 62), as well as his own paper on "permanent and geographical varieties" (Wallace 1858c), published in January 1858 right about when Wallace wrote the Ternate paper. There are cases, albeit rare, where one slightly (though heritably) varying "race" is known to be derived from another, helping distinguish parent species from variety. Such cases undermine the "permanent invariability" of species, however, a problem that may be skirted if there are limits to variability. Wallace argues in the following paragraph against such limits, a view he says is based on a false analogy.

(only few will succeed) to seize on as many and as diverse places in the economy of nature as possible. Each new variety or species, when formed, will generally take the place of, and thus exterminate its less well-fitted parent. This I believe to be the origin of the classification and affinities of organic beings at all times; for organic beings always *seem* to branch and sub-branch like the limbs of a tree from a common trunk, the flourishing and diverging twigs destroying the less vigorous—the dead and lost branches rudely representing extinct genera and families.

This sketch is *most* imperfect; but in so short a space I cannot make it better. Your imagination must fill up very wide blanks.

C. DARWIN.

III. *On the Tendency of Varieties to depart indefinitely from the Original Type.* By ALFRED RUSSEL WALLACE.

One of the strongest arguments which have been adduced to prove the original and permanent distinctness of species is, that *varieties* produced in a state of domesticity are more or less unstable, and often have a tendency, if left to themselves, to return to the normal form of the parent species; and this instability is considered to be a distinctive peculiarity of all varieties, even of those occurring among wild animals in a state of nature, and to constitute a provision for preserving unchanged the originally created distinct species.

In the absence or scarcity of facts and observations as to *varieties* occurring among wild animals, this argument has had great weight with naturalists, and has led to a very general and somewhat prejudiced belief in the stability of species. Equally general, however, is the belief in what are called "permanent or true varieties,"—races of animals which continually propagate their like, but which differ so slightly (although constantly) from some other race, that the one is considered to be a *variety* of the other. Which is the *variety* and which the original *species*, there is generally no means of determining, except in those rare cases in which the one race has been known to produce an offspring unlike itself and resembling the other. This, however, would seem quite incompatible with the "permanent invariability of species," but the difficulty is overcome by assuming that such varieties have strict limits, and can never again vary further from the original type, although they may return to it, which, from the analogy of the domesticated animals, is considered to be highly probable, if not certainly proved.

It will be observed that this argument rests entirely on the assumption, that *varieties* occurring in a state of nature are in all respects analogous to or even identical with those of domestic animals, and are governed by the same laws as regards their permanence or further variation. But it is the object of the present paper to show that this assumption is altogether false, that there is a general principle in nature which will cause many *varieties* to survive the parent species, and to give rise to successive variations departing further and further from the original type, and which also produces, in domesticated animals, the tendency of varieties to return to the parent form.

The life of wild animals is a struggle for existence. The full exertion of all their faculties and all their energies is required to preserve their own existence and provide for that of their infant offspring. The possibility of procuring food during the least favourable seasons, and of escaping the attacks of their most dangerous enemies, are the primary conditions which determine the existence both of individuals and of entire species. These conditions will also determine the population of a species; and by a careful consideration of all the circumstances we may be enabled to comprehend, and in some degree to explain, what at first sight appears so inexplicable—the excessive abundance of some species, while others closely allied to them are very rare.

The general proportion that must obtain between certain groups of animals is readily seen. Large animals cannot be so abundant as small ones; the carnivora must be less numerous than the herbivora; eagles and lions can never be so plentiful as pigeons and antelopes; the wild asses of the Tartarian deserts cannot equal in numbers the horses of the more luxuriant prairies and pampas of America. The greater or less fecundity of an animal is often considered to be one of the chief causes of its abundance or scarcity; but a consideration of the facts will show us that it really has little or nothing to do with the matter. Even the least prolific of animals would increase rapidly if unchecked, whereas it is evident that the animal population of the globe must be stationary, or perhaps, through the influence of man, decreasing. Fluctuations there may be; but permanent increase, except in restricted localities, is almost impossible. For example, our own observation must convince us that birds do not go on increasing every year in a geometrical ratio, as they would do, were there not

1 The phrase "struggle for existence" was popularized by Thomas Robert Malthus, who used it in chapter 3 of the *Essay on Population* (1798): "and when they fell in with any tribes like their own, the contest was a struggle for existence, and they fought with a desperate courage, inspired by the rejection that death was the punishment of defeat and life the prize of victory." The idea of struggle between species is an extension of Malthus's vision of struggle between peoples. Wallace later maintained that in his formulation of natural selection he recalled Malthus, whose treatise he had read some dozen years previously. But Lyell's discussion of struggle in the *Principles*, surely fresher in Wallace's mind, could have sparked the recollection. Lyell did not quote or even cite Malthus, but he described struggle in at least four places and uses the term in three of them. In the first instance Lyell describes how "in the universal struggle for existence, the right of the strongest eventually prevails; and the strength and durability of a race depends mainly on its prolificness [*sic*]" (3:9). He next quoted de Candolle: "all the plants of a given country," says de Candolle, in his usual spirited style, "are at war one with another . . . the first which establish themselves by chance in a particular spot, tend, by the mere occupancy of space, to exclude other species . . . the more prolific gradually make themselves masters of the ground, which species multiplying more slowly would otherwise fill" (3:108–109). This is vividly described in chapter 9 (3:140), where every species must "maintain its ground by a successful struggle against the encroachments of other plants and animals." In chapter 10, finally (3:162–163), Lyell drew a parallel between human conquests and the struggles of species in nature: "A faint image of the certain doom of a species less fitted to struggle with some new condition in a region which is previously inhabited, and where it has to contend with a more vigorous species, is presented by the extirpation of savage tribes of men by the advancing colony of some civilized nation."

Wallace's statement in this paragraph that wild animals must give the "full exertion of all their faculties and all their energies . . . to preserve their own existence and provide for that of their infant offspring" is resonant with Malthusian struggle as depicted by Lyell. Wallace concludes this paragraph pointing out that struggle is a determinant of population size, and seen in this light we gain insight into why some species are abundant and others rare.

2 The "wild asses of the Tartarian deserts" were mentioned by Lyell (3:59). Wallace here makes a key point: it is mistakenly assumed that fecundity determines population abundance or scarcity, but this "has little or nothing to do with the matter," he maintains. All populations would increase if they could, yet they appear more or less stationary. Lyell addressed this in some detail (3:108–120), arguing that the number of species on earth and their relative population sizes are maintained at an equilibrium. In the natural theology tradition, the struggle for existence was seen as a mechanism by which this equilibrium was maintained.

1   We have seen earlier that Wallace attacked Lyell's idea of balance and harmony in nature in the Species Notebook. Here he does so by first framing the Malthusian formulation (previous page): species "do not go on increasing every year in a geometrical ration, as they would do, were there not some powerful check to their natural increase." He states that in a mere fifteen years a single breeding pair of birds would have some ten million descendants, but later realized that this was an underestimate. In a personal copy of the printed essay sent by his agent Samuel Stevens, Wallace wrote in the margin "°really more than two thousand <u>millions</u>!" and had this included in later reprints of the essay (Beccaloni 2008). Whether ten or two thousand million descendants in fifteen years, neither is realized in nature; in fact we would be hard pressed to establish that the overall population changes at all in 150, let alone 15, years. The population appears to be stationary, and so "it is evident . . . that each year an immense number of birds must perish—as many in fact as are born." He expands on this point with an even more startling one, calculating that whatever the average population size of a species might be, double that number must perish each year, typically by cold, hunger, and predators.

As much Lyell as Malthus is evident in this argument. Just as Lyell mentions the struggle for existence without mentioning Malthus, so too does he discuss the power of population increase and the mutual balance of species (3:108–121). Lyell framed his discussion of struggle and population pressure in dramatic terms by citing examples of prolific insects, such as an example attributed to the French entomologist René Antoine Réaumur (1683–1757). Réaumur showed that "in five generations one aphis may be the progenitor of 5,904,900,000 descendants"—and what's more, there could be twenty generations in the space of just one year (3:114)! After recounting instances of exploding populations of locusts, aphids, flies, and caterpillars, Lyell discussed the balancing "reciprocal influence" of species, checking populations.

2   Wallace continues his argument about population growth and checks, ironically discussing the passenger pigeon *(Ectopistes migratorius).* At the time this was perhaps the most abundant bird in North America, only to be tragically driven to extinction early in the next century. Lyell did not discuss passenger pigeons in the fourth edition of the *Principles,* but by the sixth edition he quoted this very passage by Wallace on the immensity of passenger pigeon populations. Wallace's discussion focuses on the abundant and stable food supply of the passenger pigeon as underlying their huge populations. To Wallace abundance and rarity all boil down to abundance and constancy of food supply. This is what drives some birds to migrate, which Lyell discussed at length (3:66–70)—another point of intersection between the Ternate essay and Lyell's *Principles.*

TENDENCY OF SPECIES TO FORM VARIETIES.   55

some powerful check to their natural increase. Very few birds produce less than two young ones each year, while many have six, eight, or ten; four will certainly be below the average; and if we suppose that each pair produce young only four times in their life, that will also be below the average, supposing them not to die either by violence or want of food. Yet at this rate how tremendous would be the increase in a few years from a single pair! A simple calculation will show that in fifteen years each pair of birds would have increased to nearly ten millions! whereas we have no reason to believe that the number of the birds of any country increases at all in fifteen or in one hundred and fifty years. With such powers of increase the population must have reached its limits, and have become stationary, in a very few years after the origin of each species. It is evident, therefore, that each year an immense number of birds must perish—as many in fact as are born; and as on the lowest calculation the progeny are each year twice as numerous as their parents, it follows that, whatever be the average number of individuals existing in any given country, *twice that number must perish annually*,—a striking result, but one which seems at least highly probable, and is perhaps under rather than over the truth. It would therefore appear that, as far as the continuance of the species and the keeping up the average number of individuals are concerned, large broods are superfluous. On the average all above *one* become food for hawks and kites, wild cats and weasels, or perish of cold and hunger as winter comes on. This is strikingly proved by the case of particular species; for we find that their abundance in individuals bears no relation whatever to their fertility in producing offspring. Perhaps the most remarkable instance of an immense bird population is that of the passenger pigeon of the United States, which lays only one, or at most two eggs, and is said to rear generally but one young one. Why is this bird so extraordinarily abundant, while others producing two or three times as many young are much less plentiful? The explanation is not difficult. The food most congenial to this species, and on which it thrives best, is abundantly distributed over a very extensive region, offering such differences of soil and climate, that in one part or another of the area the supply never fails. The bird is capable of a very rapid and long-continued flight, so that it can pass without fatigue over the whole of the district it inhabits, and as soon as the supply of food begins to fail in one place is able to discover a fresh feeding-ground. This example strikingly shows us that the procuring a constant supply of wholesome food is almost the sole condition requisite for ensuring the rapid increase of a given species, since neither the limited fecundity, nor the unrestrained attacks of birds of prey and of man are here sufficient to check it. In no other birds are these peculiar circumstances so strikingly combined. Either their food is more liable to failure, or they have not sufficient power of wing to search for it over an extensive area, or during some season of the year it becomes very scarce, and less wholesome substitutes have to be found; and thus, though more fertile in offspring, they can never increase beyond the supply of food in the least favourable seasons. Many birds can only exist by migrating, when their food becomes scarce, to regions possessing a milder, or at least a different climate, though, as these migrating birds are seldom excessively abundant, it is evident that the countries they visit are still deficient in a constant and abundant supply of wholesome food. Those whose organization does not permit them to migrate when their food becomes periodically scarce, can never attain a large population. This is probably the reason why woodpeckers are scarce with us, while in the tropics they are among the most abundant of solitary birds. Thus the house sparrow is more abundant than the redbreast, because its food is more constant and plentiful,—seeds of grasses being preserved during the winter, and our farm-yards and stubble-fields furnishing an almost inexhaustible supply. Why, as a general rule, are aquatic, and especially sea birds, very numerous in individuals? Not because they are more prolific than others, generally the contrary; but because their food never fails, the sea-shores and river-banks daily swarming with a fresh supply of small mollusca and crustacea. Exactly the same laws will apply to mammals. Wild cats are prolific and have few enemies; why then are they never as abundant as rabbits? The only intelligible answer is, that their supply of food is more precarious. It appears evident, therefore, that so long as a country remains physically unchanged, the numbers of its animal population cannot materially increase. If one species does so, some others requiring the same kind of food must diminish in proportion. The numbers that die annually must be immense; and as the individual existence of each animal depends upon itself, those that die must be the weakest—the very young, the aged, and the diseased,—while those that prolong their existence can only be the most perfect in health and vigour—those who are best able to obtain food regularly, and avoid their numerous enemies. It is, as we commenced by remarking, "a struggle for existence," in

1 Note the emphasis on food supply, "almost the sole condition" necessary to ensure rapid population increase. This refers to the passenger pigeon, but by extension Wallace sees this as a general truth.

2 Wallace summarizes the argument thus far, pointing out that *"so long as a country remains physically unchanged"* (emphasis mine), the species populations in that country "cannot materially increase." A balance ensues, or stalemate: for each species that gains an advantage and increases, another requiring the same resource must decrease. A key point in his subsequent argument will be that countries do *not* remain physically unchanged, as Lyell eloquently argued in the *Principles* (3:142–164): inexorable, slow, and steady geological and climatic change affects food supplies and perturbs the stalemate, leading to the extinction of some species. Wallace once again states that the numbers of individuals that die annually is immense: it is "a struggle for existence," in which those that succumb tend to be the weakest and "least perfectly organized." Wallace's view of selection is here eliminative: the very young, weak, and aged succumb, leaving the healthiest and most vigorous.

1   Changing gears slightly, Wallace says that the same struggle he just described among individuals "must also occur among the several allied species of a group"—suggestive of species-level selection. One passage here seems to echo Lyell: those species best able to "defend themselves against the attacks of their enemies and the vicissitudes of the seasons" will enjoy large populations, paralleling Lyell's discussion of the effects of the "vicissitudes of climate" on species (3:160). Another passage echoes a comment made in the Species Notebook: "our ignorance will generally prevent us from accurately tracing the effects to their causes," Wallace wrote, while he commented in an unrelated context on how we uncomprehendingly look at species or nature "like children looking at a complicated machine . . . the reasons of whose construction they are ignorant, and like them we constantly impute as cause what is really effect" (Species Notebook, p. 33).

2   The italics reflect the importance of these two key points to Wallace: that animal populations tend to be (or seem) stationary (kept in check by limitations of food and other factors), and that the relative abundance of a species results from its structure ("organization") and habits as they relate to survival and procuring food.

which the weakest and least perfectly organized must always succumb.

Now it is clear that what takes place among the individuals of a species must also occur among the several allied species of a group,—viz. that those which are best adapted to obtain a regular supply of food, and to defend themselves against the attacks of their enemies and the vicissitudes of the seasons, must necessarily obtain and preserve a superiority in population; while those species which from some defect of power or organization are the least capable of counteracting the vicissitudes of food, supply, &c., must diminish in numbers, and, in extreme cases, become altogether extinct. Between these extremes the species will present various degrees of capacity for ensuring the means of preserving life; and it is thus we account for the abundance or rarity of species. Our ignorance will generally prevent us from accurately tracing the effects to their causes; but could we become perfectly acquainted with the organization and habits of the various species of animals, and could we measure the capacity of each for performing the different acts necessary to its safety and existence under all the varying circumstances by which it is surrounded, we might be able even to calculate the proportionate abundance of individuals which is the necessary result.

If now we have succeeded in establishing these two points— 1st, *that the animal population of a country is generally stationary, being kept down by a periodical deficiency of food, and other checks*; and, 2nd, *that the comparative abundance or scarcity of the individuals of the several species is entirely due to their organization and resulting habits, which, rendering it more difficult to procure a regular supply of food and to provide for their personal safety in some cases than in others, can only be balanced by a difference in the population which have to exist in a given area*—we shall be in a condition to proceed to the consideration of *varieties*, to which the preceding remarks have a direct and very important application.

Most or perhaps all the variations from the typical form of a species must have some definite effect, however slight, on the habits or capacities of the individuals. Even a change of colour might, by rendering them more or less distinguishable, affect their safety; a greater or less development of hair might modify their habits. More important changes, such as an increase in the power or dimensions of the limbs or any of the external organs, would more or less affect their mode of procuring food or the range of country which they inhabit. It is also evident that most changes would affect, either favourably or adversely, the powers of prolonging existence. An antelope with shorter or weaker legs must necessarily suffer more from the attacks of the feline carnivora; the passenger pigeon with less powerful wings would sooner or later be affected in its powers of procuring a regular supply of food; and in both cases the result must necessarily be a diminution of the population of the modified species. If, on the other hand, any species should produce a variety having slightly increased powers of preserving existence, that variety must inevitably in time acquire a superiority in numbers. These results must follow as surely as old age, intemperance, or scarcity of food produce an increased mortality. In both cases there may be many individual exceptions; but on the average the rule will invariably be found to hold good. All varieties will therefore fall into two classes— those which under the same conditions would never reach the population of the parent species, and those which would in time obtain and keep a numerical superiority. Now, let some alteration of physical conditions occur in the district—a long period of drought, a destruction of vegetation by locusts, the irruption of some new carnivorous animal seeking " pastures new"—any change in fact tending to render existence more difficult to the species in question, and tasking its utmost powers to avoid complete extermination; it is evident that, of all the individuals composing the species, those forming the least numerous and most feebly organized variety would suffer first, and, were the pressure severe, must soon become extinct. The same causes continuing in action, the parent species would next suffer, would gradually diminish in numbers, and with a recurrence of similar unfavourable conditions might also become extinct. The superior variety would then alone remain, and on a return to favourable circumstances would rapidly increase in numbers and occupy the place of the extinct species and variety.

The *variety* would now have replaced the *species*, of which it would be a more perfectly developed and more highly organized form. It would be in all respects better adapted to secure its safety, and to prolong its individual existence and that of the race. Such a variety *could not* return to the original form; for that form is an inferior one, and could never compete with it for existence. Granted, therefore, a " tendency" to reproduce the original type of the species, still the variety must ever remain preponderant in numbers, and under adverse physical conditions *again alone survive.*

1   The example of the short-legged antelope comes from Lamarck via Lyell (2:415), who described how the antelope and gazelle acquired "light agile forms" because they were "compelled to exert themselves in running with great celerity" to escape predators. A variety with even a slight edge in survivorship will eventually achieve superiority in numbers. Under constant conditions populations of new varieties may meet or exceed that of the parental form, but as Lyell taught, conditions are never constant for long. One of the examples of perturbation that Wallace gives here—"destruction of vegetation by locusts"—comes from Lyell (3:116–117 and 123).

2   Following out the implications of environmental change, Wallace says that the individuals "forming the least numerous and most feebly organized variety" would soon become extinct if the pressure was great enough. With continued change the parent species next diminishes in numbers, perhaps going extinct, but the "superior variety" alone remains after other varieties and the parental species that gave rise to them are gone. Wallace then says that "on a return to favourable circumstances" this surviving variety would increase in numbers and "occupy the place of the extinct species and variety." The surviving variety already occupies the place of the extinct parental form, but in low numbers as a result of environmental adversity. Its numbers increase to the former level of the parental form when conditions improve, which it has now replaced.

3   Wallace argues that the variety replaces the parental form because it is more "perfectly developed" and "highly organized." Key here is that, contra Lyell regarding reversion of varieties, he offers a circumstance where reversion is impossible because the parental form is inferior and could not compete. This is Wallace's answer to Lyell's claim (*Principles* 3:162) that it is pointless to speculate about one species "converting" into another during environmental change because migration, if not extinction, results. Wallace sees species change as an alternative to migration or extinction. Moreover, the new-and-improved variety would eventually give rise to newer varieties, resulting over time in "several diverging modifications of form," any one of which might become dominant. "Divergence," in the sense of becoming increasingly different from the parental form over time, results from general laws.

1   The outcome of the selection dynamic that Wallace has made a case for is *"progression and continued divergence"*; significantly, a means of driving sustained change from varieties to new species in a purely natural or materialistic manner. In the Species Notebook (pp. 39–40) Wallace gave a similar scenario, which he opened by asking "what positive evidence have we that species only vary within certain limits?" He continued:

> Let us suppose that every variety of the Dog but one was to become extinct & that one [was gradually] spread over the whole world, subjected to every variety of climate & food, & domesticated by every variety of the human race. Have we any reason for supposing that in the course of ages a new series of varieties quite distinct from any now existing would not be developed—& then should the same process be repeated & one of these varieties farthest removed from the original, again be spread over the earth & be subjected to the same variety of conditions, does it not seem probable that again new varieties would be produced.

Some important additional insights come up in the remainder of this paragraph: (1) offspring varieties could end up less well adapted after all than their parental form, and go extinct; (2) variation is ubiquitous: "variations in unimportant parts might also occur"; and (3) Wallace understands the statistics of low probabilities and large numbers: given the vast time available as well as the vast numbers of individuals—approaching "so near to infinity"—even the slightest of beneficial effects of a variant will ultimately have an effect, and the population will alter. Wallace reflects Lyell's expression of both the immensity of geological time (1:111, 114, 127; 3:449) and of population growth potential (3:113–115).

2   Returning to domesticated varieties, Wallace argues that they hold no lessons for us in view of the great difference in the circumstances of animals in a state of nature versus a state of domestication in terms of activity, safety, and procuring food. Another of Wallace's corrections appears later in this paragraph, on the next page: he changed the statement "Half of [domesticated animals'] senses and faculties are quite useless" to "become quite useless" (Beccaloni 2008). He is reminding us of this as a lead-in to the next paragraph, where he contrasts domesticated and wild animals more explicitly, aiming to impress upon the reader how every aspect of animals in the wild must be "brought into full action for the necessities of existence. . . . It creates as it were a new animal, one of superior powers, and which will necessarily increase in numbers and outlive those inferior to it."

But this new, improved, and populous race might itself, in course of time, give rise to new varieties, exhibiting several diverging modifications of form, any of which, tending to increase the facilities for preserving existence, must, by the same general law, in their turn become predominant. Here, then, we have *progression and continued divergence* deduced from the general laws which regulate the existence of animals in a state of nature, and from the undisputed fact that varieties do frequently occur. It is not, however, contended that this result would be invariable; a change of physical conditions in the district might at times materially modify it, rendering the race which had been the most capable of supporting existence under the former conditions now the least so, and even causing the extinction of the newer and, for a time, superior race, while the old or parent species and its first inferior varieties continued to flourish. Variations in unimportant parts might also occur, having no perceptible effect on the life-preserving powers; and the varieties so furnished might run a course parallel with the parent species, either giving rise to further variations or returning to the former type. All we argue for is, that certain varieties have a tendency to maintain their existence longer than the original species, and this tendency must make itself felt; for though the doctrine of chances or averages can never be trusted to on a limited scale, yet, if applied to high numbers, the results come nearer to what theory demands, and, as we approach to an infinity of examples, become strictly accurate. Now the scale on which nature works is so vast—the numbers of individuals and periods of time with which she deals approach so near to infinity, that any cause, however slight, and however liable to be veiled and counteracted by accidental circumstances, must in the end produce its full legitimate results.

Let us now turn to domesticated animals, and inquire how varieties produced among them are affected by the principles here enunciated. The essential difference in the condition of wild and domestic animals is this,—that among the former, their well-being and very existence depend upon the full exercise and healthy condition of all their senses and physical powers, whereas, among the latter, these are only partially exercised, and in some cases are absolutely unused. A wild animal has to search, and often to labour, for every mouthful of food—to exercise sight, hearing, and smell in seeking it, and in avoiding dangers, in procuring shelter from the inclemency of the seasons, and in providing for the subsistence and safety of its offspring. There is no muscle of its body that is not called into daily and hourly activity; there is no sense or faculty that is not strengthened by continual exercise. The domestic animal, on the other hand, has food provided for it, is sheltered, and often confined, to guard it against the vicissitudes of the seasons, is carefully secured from the attacks of its natural enemies, and seldom even rears its young without human assistance. Half of its senses and faculties are quite useless; and the other half are but occasionally called into feeble exercise, while even its muscular system is only irregularly called into action.

Now when a variety of such an animal occurs, having increased power or capacity in any organ or sense, such increase is totally useless, is never called into action, and may even exist without the animal ever becoming aware of it. In the wild animal, on the contrary, all its faculties and powers being brought into full action for the necessities of existence, any increase becomes immediately available, is strengthened by exercise, and must even slightly modify the food, the habits, and the whole economy of the race. It creates as it were a new animal, one of superior powers, and which will necessarily increase in numbers and outlive those inferior to it.

Again, in the domesticated animal all variations have an equal chance of continuance; and those which would decidedly render a wild animal unable to compete with its fellows and continue its existence are no disadvantage whatever in a state of domesticity. Our quickly fattening pigs, short-legged sheep, pouter pigeons, and poodle dogs could never have come into existence in a state of nature, because the very first step towards such inferior forms would have led to the rapid extinction of the race; still less could they now exist in competition with their wild allies. The great speed but slight endurance of the race horse, the unwieldy strength of the ploughman's team, would both be useless in a state of nature. If turned wild on the pampas, such animals would probably soon become extinct, or under favourable circumstances might each lose those extreme qualities which would never be called into action, and in a few generations would revert to a common type, which must be that in which the various powers and faculties are so proportioned to each other as to be best adapted to procure food and secure safety,—that in which by the full exercise of every part of his organization the animal can alone continue to live. Domestic varieties, when turned wild, *must* return to something near the type of the original wild stock, *or become altogether extinct.*

1 Wallace argues that domestic varieties like fancy pouter pigeons and poodles could not survive in a state of nature. Neither, too, could race horses with their "great speed but slight endurance" nor draft horses with their "unwieldy strength" long survive turned loose on the pampas. This may be a reasonable supposition with regard to those breeds that are "abnormal, irregular, artificial," though Wallace would seem to part ways with Lyell on this point. In the *Principles* (3:134–137) Lyell described the explosive population growth of horses, oxen, and other domestic animals turned loose on the pampas, leading to the extirpation or displacement of native species. Lyell did not discuss "reversion" of these feral domestic varieties, though Wallace emphasizes with italics that "domestic varieties, when turned wild, *must* return to something near the type of the original wild stock, *or become altogether extinct.*" Wallace penned a qualifier in his personal copy of the paper: "that is, they will vary and the variations which render them best adapted to the wild state and therefore approximate them to wild animals will be preserved. Those that do not vary quickly enough will perish" (Beccaloni 2008). This is a more nuanced explanation of the process by which they revert.

1   Wallace stresses the gulf separating domesticated from wild species; the former are "abnormal, irregular, artificial." Here Wallace's emphatic case for what amounts to the irrelevance of domestic varieties for drawing inferences about the capacity of species to change in nature is a device to sidestep the standard Lyellian argument about the immutability of species based on the supposed limited extent of variability or change of domestic forms.

2   Wallace points out that while Lamarck's view of transmutation is easily refuted, this does not mean the question is settled; he aims to show that the principles he outlines show that Lamarck's hypothesis is unnecessary but give similar results (transmutation). This point about obviating Lamarck is also made in the Species Notebook, aimed at Lyell, who goes to great lengths to refute Lamarck in the *Principles* (2:407–465). Wallace discusses how the Lamarckian interpretation of organic change by "volition" or activity is not correct, giving the examples of the "powerful retractile talons" of birds of prey and felines, or the long neck of the giraffe.

3   Lyell does not discuss talons but does cite Lamarck's example of the giraffe's neck, using the then-current name of "camelopard" for this animal (2:415). Other examples reflect Wallace's own field experience: camouflage, or cryptic coloration of insects "so closely resembling the soil or the leaves or the trunks on which they habitually reside" that they are well hidden from enemies. Camouflage is explained easily by his principle in terms of selection (though he does not use that word) acting on the "varieties of many tints that may have occurred" through the sweep of time. Well-hidden beetles resting on leaves, trunks, and leaf midribs are mentioned on page 136 of the Species Notebook. Mimicry and camouflage was later the subject of several papers by Wallace.

TENDENCY OF SPECIES TO FORM VARIETIES.   **61**

We see, then, that no inferences as to varieties in a state of nature can be deduced from the observation of those occurring among domestic animals. The two are so much opposed to each other in every circumstance of their existence, that what applies to the one is almost sure not to apply to the other. Domestic animals are abnormal, irregular, artificial; they are subject to varieties which never occur and never can occur in a state of nature: their very existence depends altogether on human care; so far are many of them removed from that just proportion of faculties, that true balance of organization, by means of which alone an animal left to its own resources can preserve its existence and continue its race.

The hypothesis of Lamarck—that progressive changes in species have been produced by the attempts of animals to increase the development of their own organs, and thus modify their structure and habits—has been repeatedly and easily refuted by all writers on the subject of varieties and species, and it seems to have been considered that when this was done the whole question has been finally settled; but the view here developed renders such an hypothesis quite unnecessary, by showing that similar results must be produced by the action of principles constantly at work in nature. The powerful retractile talons of the falcon- and the cat-tribes have not been produced or increased by the volition of those animals; but among the different varieties which occurred in the earlier and less highly organized forms of these groups, *those always survived longest which had the greatest facilities for seizing their prey.* Neither did the giraffe acquire its long neck by desiring to reach the foliage of the more lofty shrubs, and constantly stretching its neck for the purpose, but because any varieties which occurred among its antitypes with a longer neck than usual *at once secured a fresh range of pasture over the same ground as their shorter-necked companions, and on the first scarcity of food were thereby enabled to outlive them.* Even the peculiar colours of many animals, especially insects, so closely resembling the soil or the leaves or the trunks on which they habitually reside, are explained on the same principle; for though in the course of ages varieties of many tints may have occurred, *yet those races having colours best adapted to concealment from their enemies would inevitably survive the longest.* We have also here an acting cause to account for that balance so often observed in nature,—a deficiency in one set of organs always being compensated by an increased development of some others—powerful wings accompanying weak fect, or great velocity making up for the absence of defensive weapons; for it has been shown that all varieties in which an unbalanced deficiency occurred could not long continue their existence. The action of this principle is exactly like that of the centrifugal governor of the steam engine, which checks and corrects any irregularities almost before they become evident; and in like manner no unbalanced deficiency in the animal kingdom can ever reach any conspicuous magnitude, because it would make itself felt at the very first step, by rendering existence difficult and extinction almost sure soon to follow. An origin such as is here advocated will also agree with the peculiar character of the modifications of form and structure which obtain in organized beings—the many lines of divergence from a central type, the increasing efficiency and power of a particular organ through a succession of allied species, and the remarkable persistence of unimportant parts such as colour, texture of plumage and hair, form of horns or crests, through a series of species differing considerably in more essential characters. It also furnishes us with a reason for that "more specialized structure" which Professor Owen states to be a characteristic of recent compared with extinct forms, and which would evidently be the result of the progressive modification of any organ applied to a special purpose in the animal economy.

We believe we have now shown that there is a tendency in nature to the continued progression of certain classes of *varieties* further and further from the original type—a progression to which there appears no reason to assign any definite limits—and that the same principle which produces this result in a state of nature will also explain why domestic varieties have a tendency to revert to the original type. This progression, by minute steps, in various directions, but always checked and balanced by the necessary conditions, subject to which alone existence can be preserved, may, it is believed, be followed out so as to agree with all the phenomena presented by organized beings, their extinction and succession in past ages, and all the extraordinary modifications of form, instinct, and habits which they exhibit.

Ternate, February, 1858.

1   Wallace draws an analogy between his "principle" (natural selection) and the steam engine's centrifugal governor, "which checks and corrects any irregularities almost before they become evident." Lyell also drew an analogy with steam engines, though in a different context. In volume 3, page 112 of the *Principles* Lyell wrote of the power of insect populations to rapidly grow and then abate, like the ability of a steam engine to bring the power of "many hundred" horses to bear instantly, and then just as quickly abate. Wallace's steam engine governor is a conservative force, reflecting natural selection's power of elimination or weeding. In the essay Wallace writes of this mechanism correcting any "unbalanced deficiency." He then returns to the bigger-picture result of his principle: "the many lines of divergence from a central type," the various modifications seen in a given organ or structure in a set of allied species, or in a succession of species. It explains, too, the tendency for more recent species to have more specialized structures than earlier, extinct species, citing Richard Owen. This echoes a similar remark Wallace made in the Species Notebook (p. 54), which also cites Owen in this context.

2   The final paragraph of the Ternate essay summarizes the main point of the paper: "a tendency in nature to the continued progression of certain classes of *varieties* further and further from the original type." Moreover, Wallace maintains that there is no reason to ascribe a priori limits on this progress (contra Lyell, in *Principles* 3:21), and this explains reversion of domestic varieties to the parental type (as argued by Lyell in *Principles* 2:438–439). It is worth noting Wallace's choice of words in the final sentence, in particular the key words "minute steps" and "in various directions"—change is gradualistic, and neither teleological nor uni-directional. Note, too, in regard to his hypothesis agreeing with "extinction and succession in past ages" that Lyell discusses both of these at length (the latter critically)—extinction in *Principles* 3:104–108, 155, and 164–166, and succession in 1:222–239.

3   Although the paper is signed "Ternate, February, 1858," recall that Wallace is known to have been on Gilolo during that month. His signing off "Ternate" instead of Gilolo has generated much discussion ever since. Some have thought he wished to attach the paper to Ternate because that was a well-known trading hub, while Gilolo was all but unknown, a thinly populated terra incognita. The most straightforward interpretation is simply that Wallace drafted his essay on Gilolo but wrote out a fair copy, likely ironing out wrinkles, once back at his base on Ternate (arriving back 1 March)—that locale was his mailing address. Whether his essay was mailed off on the 9 March steamer or the next one (discussed briefly in Chapter 5), he would have had just over a week to as long as three weeks to recopy and polish. The discrepancy is perhaps why Ternate in *February*—if the locale was updated, why not the date?

FIVE

A Striking Coincidence

The Wallace and Darwin Papers of 1858 Compared

IT HAS LONG been appreciated by scholars that Wallace's and Darwin's concepts of natural selection differed in important ways (e.g., Bowler 1976; Kottler 1985; Bulmer 2005), though at the time Darwin, Lyell, and Hooker believed them to be "the very same ingenious theory." In the main, the similarity that caused Darwin so much angst lies in the general formulation of natural selection. He and Wallace came to understand the deductive core mechanism at the heart of the evolutionary process: the idea of abundant, undirected, and heritable variation, tremendous population growth potential giving rise to a struggle for existence such that success in surviving and reproducing (achieved by a scant few) is nonrandom but depends on those chance variations. Both Darwin and Wallace suggested that as environment slowly changes selection pressure for individuals possessing those variants better suited to the new conditions "accumulate" these variations, resulting in a sustained change in the population. Consider the following parallels and intersections in their respective accounts in the 1858 papers (emphases in the originals):

Idea of Struggle for Existence

ARW: "The life of wild animals is a struggle for existence" (p. 54)

"The numbers that die annually must be immense; and as the individual existence of each animal depends upon itself, those that die must be the weakest. . . . It is, as we commenced by remarking, 'a struggle for existence'" (p. 56)

CD: "De Candolle . . . declared that all nature is at war" (p. 46)

"It is the doctrine of Malthus applied . . . with tenfold force" (p. 47)

"recurrent struggles against other species"; "struggle . . . with other individuals of the *same* or *different* species" (p. 48)

Apparent Balance or Harmony in Nature

ARW: "it is evident that the animal population of the globe must be stationary" (p. 54)

"the animal population of a country is generally stationary" (p. 57)

CD: "seeing the contented face of nature" (p. 47)

"the average percentage of . . . the inhabitants of a country . . . remains constant" (p. 48)

Variation and Its Effects

ARW: "Most or perhaps all the variations from the typical form of a species must have some definite effect, however slight, on the habits or capacities of the individuals." (p. 57)

"most changes would affect, either favourably or adversely, the powers of prolonging existence." (p. 58)

"variations in unimportant parts might also occur" (p. 59)

CD: "Now, can it be doubted . . . that any minute variation in structure, habits, or instincts, adapting that individual better to the new conditions, would tell upon its vigour and health?" (p. 49)

"In nature we have some *slight* variation occasionally in all parts" (p. 51)

"I cannot doubt that during millions of generations individuals of a species will be occasionally born with some slight variation, profitable to some part of their economy" (p. 52)

Population Increase Geometrical / Population Pressure

ARW: Reference to geometrical population growth (p. 54)

Population growth example with birds: "A simple calculation will show that in fifteen years each pair of birds would have increased to nearly ten millions!" (p. 55)

"Even the least prolific of animals would increase rapidly if unchecked" (p. 54)

"Yet at this rate how tremendous would be the increase in a few years from a single pair!" (p. 55)

CD: Reference to geometrical population growth (p. 47)

Population growth example with birds: "at the end of seven years . . . there will be 2048 birds" (p. 47)

"Even slow-breeding mankind has doubled in twenty-five years" (p. 47)

"Reflect on the enormous multiplying power *inherent and annually in action* in all animals" (p. 48)

"reflect on the countless seeds scattered by a hundred ingenious contrivances, year after year, over the whole face of the land" (p. 48)

"Now, every organic being, by propagating so rapidly, may be said to be striving its utmost to increase in numbers" (p. 52)

Checks on Population / Action of Selection

ARW: "whatever be the average number of individuals existing in any given country, *twice that number must perish annually*" (p. 55)

"On the average all above *one* become food for hawks and kites, wild cats and weasels, or perish of cold and hunger as winter comes on" (p. 55)

"The numbers that die annually must be immense; and as the individual existence of each animal depends upon itself, those that die must be the weakest—the very young, the aged, and the diseased,—while those that prolong their existence can only be the most perfect in health and vigour—those who are best able to obtain food regularly, and avoid their numerous enemies." (p. 56)

"the weakest and least perfectly organized must always succumb" (p. 57)

"of all the individuals composing the species, those forming the least numerous and most feebly organized variety would suffer first, and, were the pressure severe, must soon become extinct." (p. 58)

CD: "In the struggle it would have a better *chance* of surviving; and those of its offspring which inherited the variation, be it ever so slight, would also have a better *chance*. Yearly more are bred than can survive; the smallest grain in the balance, in the long run, must tell on which death shall fall, and which shall survive." (p. 49)

"I have found it hard constantly to bear in mind that the increase of every single species is checked during some part of its life" (p. 51)

"Only a few of those annually born can live to propagate their kind. What a trifling difference must often determine which shall survive, and which perish!" (p. 51)

"Selection acts only by the accumulation of slight or greater variations." (p. 51) "mutual action of . . . inhabitants [is] far more important to the life of each being than mere climate" (p. 52)

Geological / Environmental Change

ARW: "so long as a country remains physically unchanged, the numbers of its animal population cannot materially increase" (p. 56)

"now, let some alteration of physical conditions occur" (p. 58)

CD: "Let the external conditions of a country alter" (p. 48)

"let the change of conditions continue progressing" (p. 49)

Time

ARW: "the scale on which nature works is so vast—the numbers of individuals and periods of time with which she deals approach so near to infinity" (p. 59)

"But this new, improved, and populous race might itself, in course of time" (p. 59)

CD: "Let this work of selection on the one hand, and death on the other, go on for a thousand generations" (p. 48)

"We have almost unlimited time; no one but a practical geologist can fully appreciate this" (p. 51)

"Think of the Glacial period . . . there must have been during this period millions on millions of generations" (p. 51)

Tree Analogy / Branching Divergence

ARW: *"progression and continued divergence"* (p. 59)

"several diverging modifications of form" (p. 59)

"the many lines of divergence from a central type" (p. 62)

"continued progression . . . further and further from the original type." (p. 62)

CD: "organic beings always seem to branch and sub-branch like the limbs of a tree from a common trunk, the flourishing and diverging twigs destroying the less vigorous—the dead and lost branches rudely representing extinct genera and families." (p. 53)

These areas of intersection underscore the high degree of congruence between the Wallace and Darwin papers of 1858 and help us understand why Darwin, Lyell, Hooker, and others believed that Darwin and Wallace had devised the same theory. There are other, more trivial similarities—note, for example, that for imaginary scenarios of natural selection in action both offer a predator-prey example: Wallace a feline carnivore preying on antelope, and Darwin a canine carnivore preying on rabbits and hares. Both also chose to give bird examples in discussing exponential population growth. Most of the similarities highlighted here, however, are far more substantive: they reflect a fundamental congruence in the deductive steps leading them to natural selection (variation, population pressure, struggle, differential survival, and reproduction). However, these similarities must be considered together with differences for a full understanding of the respective thinking of these naturalists. It is in the differences especially—some obvious and others more nuanced—that we see the evidence of their independent formulation of the evolutionary mechanism. I emphasize the *mechanism* because by and large their understanding of the *fact* of transmutation, and the lines of evidence pointing to this fact, was nearly identical.

One point of departure in the lines of evidence was their use of domestication. We have seen that Wallace, in structuring his argument as a response to Lyell's antievolutionism, framed his paper in terms that acquiesced to the prevailing view that domestic varieties "infallibly revert" to the parental form (and therefore could be cited as evidence against transmutation, as Lyell did). This acquiescence is not tacit approval; rather, Wallace uses Lyell's domestication argument as means of framing his discovery of a mechanism showing how varieties can indeed "depart indefinitely from the original type." The Species Notebook and later writings show that Wallace understood domestication as human-mediated transmutation; he dedicated an entire chapter to the subject in his 1889 book *Darwinism*. What he objected to was what he saw as Darwin's overreliance on the domestication analogy as evidence for transmutation in nature. "It has always been considered a weakness in Darwin's work that he based his theory, primarily, on the evidence of variation in domesticated animals and cultivated plants. I have endeavoured to secure a firm foundation for the theory in the variations of organisms in a state of nature" (Wallace 1889, vi).

But the key differences in the evolutionary thinking of Darwin and Wallace as reflected in their 1858 papers are to be found in their conceptions of the action of natural selection, a term for the selection mechanism or dynamic coined by Darwin but unnamed by Wallace. It is useful to consider this in two related contexts: (1) in terms of how they envisioned natural selection to act, and (2) in their awareness (or not) of extensions or special forms of selection.

Action of Natural Selection

Wallace and Darwin related variation, population growth, and the struggle for existence at the heart of the mechanism of selection. They seem to be concordant with regard to the population growth and struggle concepts, but beyond that there are several key differences in their formulations. One of these concerns the unit of selection (does selection act on individual variants or varieties?), and the other concerns the relative importance of environmental versus interindividual selection (abiotic versus biotic factors). With regard to what exactly selection acts on, Bowler (1976, 1984) argued that Wallace was not so much concerned with individual variation per se as with varieties, and that his was a model of selection acting at the level of varieties and species (relatively weakly at that, asking how if even "less efficient" varieties can survive and coexist with the parental variety except in times of unusual stress selection could be strong enough to discriminate among slight, individual variants). The suggestion by Bowler was that by thinking in terms of a struggle among varieties, Wallace did not have a conception of true populational thinking—in contrast to Darwin, who appeared to have focused squarely on selection acting on slight individual variations (except perhaps in the context of family-level selection in social insects, but that came later). This is not an unreasonable charge; consider examples such as the following from the Ternate essay:

- "[There is a] general principle in nature which will cause many *varieties* to survive the parent species, and to give rise to successive variations departing further and further from the original type"
- "the varieties so furnished might run a course parallel with the parent species, either giving rise to further variations or returning to the former type"
- "The superior variety would then alone remain"
- "Now when a variety of such an animal occurs . . ."
- "New varieties would be produced"
- "all varieties in which an unbalanced deficiency occurred could not long continue their existence"

That said, however, there are other examples where Wallace seemed refer to individuals and their variation or variants:

- "The numbers that die annually must be immense; and as the individual existence of each animal depends upon itself . . ."
- "Most or perhaps all the variations from the typical form of a species must have some definite effect, however slight, on the habits or capacities of the individuals"
- "Variation in unimportant parts would also occur"
- "it is evident that, of all the individuals composing the species, those forming the least numerous and most feebly organized variety would suffer first"

There are other hints besides the Ternate essay's seeming emphasis on varieties that Wallace did not fully grasp, in the modern view, the essential mode of action of selection, such as in *The Malay Archipelago* where he wrote of the "failure of instinct" of wood-boring beetles which, attracted to the scent of a certain tree, are killed in large numbers while attempting to bore in as the tree exudes a sticky latex that entombs the hapless beetles. (This observation was made in western New Guinea, and is first recorded on page 8 of the verso side of the Species Notebook; Costa 2013a, 422). In *The Malay Archipelago* Wallace opined that "if, as is very probable, these trees have an attractive odor to certain species of borers, it might very likely lead to their becoming extinct; while other species, to whom the same odor was disagreeable, and who therefore avoided the dangerous trees, would survive" (Wallace 1869, 481). Wallace had not discovered natural selection by the time of the notebook entry, in all likelihood, but *The Malay Archipelago* came out in 1869, giving him a decade of reflection on natural selection. In the modern view selection acting on individual-level variation in attraction to the odor would suggest that *varying individuals of the same species* "to whom the same odor was disagreeable," not merely members of other species altogether, would be favored by selection, and in that way avoidance of this tree might evolve. Wholesale extinction of the species would thus be unlikely, and Wallace's evolutionary reasoning might in this instance be considered flawed.

All this said, I agree with Slotten (2004), following Kottler (1985), that the problem lies in the terms rather than the overarching idea of selection. Wallace's interchangeable use of "variety," "variation(s)," "variant individual," and in some cases "race" creates confusion, but it seems clear that Wallace did indeed have an appreciation of the importance of individual variation. Kottler (1985), in his excellent discussion of the subject, cited Ernst Mayr (1982) as pointing out that the term "variety" was often used in Wallace's day to describe variant individuals *or* populations (as in permanent variety, subspecies, etc.) and continued to be used in this way until the early twentieth century. Furthermore, to be fair to Wallace we must bear in mind, first, that his Ternate essay was hastily written, and second, that he very much had varieties on his mind in writing the essay. It is evident that the essay is largely couched in terms of refuting Lyell's assertion regarding the limited capacity of varieties to change. The words "variety" and "varieties" appear no fewer than thirty-eight times in the Ternate essay. "Variation" appears but six times in contrast, and "individual variation" and "variant" do not appear at all. Kottler (1985) further pointed out that Wallace himself clarified things twelve years later when the essay was reprinted in *Contributions to the Theory of Natural Selection* (Wallace 1870). In a footnote to his account of reversion of domestic varieties on page 40 of the reprint, Wallace wrote, "That is, they will vary, and the variations which tend to adapt them to the wild state, and therefore approximate them to wild animals, will be preserved. Those individuals which do not vary sufficiently will perish." His language here is unambiguous: *individuals* vary, and it is their *variations* that selection acts upon. Wallace was asked about this issue, as also reported by Kottler (1985). The noted Oxford naturalist E. B. Poulton (1856–1943), whose career spanned the post-*Origin* evolution debates of the 1870s through the rediscovery of Mendel and eventual "Modern Synthesis," greatly admired Darwin and Wallace. Reacting against the charge that Wallace had no concept of selection acting on individual variation, made by Yale paleontologist Henry Fairfield Osborn (1857–1935), Poulton went to get the scoop straight from Wallace himself. He wrote to the by-then elderly Wallace, and in his 1896 book *Charles Darwin and the Theory of Natural Selection* Poulton quoted Wallace's reply as follows: "I used the term 'varieties' because 'varieties' were alone recognized at that time, individual variability being ignored or thought of *no importance*. My 'varieties' therefore included individual variations" (Poulton 1896, 80; emphasis in the original).

On the subject of variation one other criticism has been leveled against Wallace. That criticism maintains that he lacked a conception of *heritability* of said beneficial variations, and as such cannot lay claim to having developed a complete theory of evolution by natural selection with Darwin. This thesis was advanced by Bock (2009), who claimed that Wallace had a concept of evolutionary change mediated by selective agents (physical environ-ment) but that intentionally or not he omitted an understanding that varia-tion must be heritable at least in part in order for selection to act upon it. I disagree in that I think Wallace intuitively understood this as a given and did not need to explicitly elaborate upon it in his Ternate essay. For example, in statements such as "If . . . any species should produce a variety having slightly increased powers of preserving existence, that variety must inevita-bly in time acquire a superiority in numbers" (p. 58), how could the popula-tion of individuals bearing a beneficial variant grow to obtain superiority in numbers if the trait was not heritable? Similarly, when Wallace wrote that a "new, improved, and populous race might itself, in course of time, give rise to new varieties" (p. 59), the "giving rise" can only be in terms of generation. And how else but by reproduction and at least some measure of heritability could "all variations have an equal chance of continuance" in domestic vari-eties (p. 60)? "Continuance" is transmission by reproduction.

Turning from Wallace's and Darwin's formulations of natural selection to how natural selection might operate, an important difference in emphasis between the two lies in the relative importance of selection mediated by the abiotic versus biotic environment—physical environment and resource needs versus competition. Kottler (1985) expresses this in terms of "hard" and "soft" selection (B. Wallace 1968)—hard selection referring to inflexible factors (lethality of freezing temperatures, say), while soft selection has incremental or relative fitness effects as, say, from competition. In fitness terms the distinction is between being "unfit" or "less fit," Kottler says. Under hard selection unfit individuals are eliminated from the population. Under soft selection less fit individuals might be eliminated by more fit individuals, but the fitness measure is relative to the presence of superior or inferior competitors, not some absolute environmental standard. Darwin emphasized competition as of paramount importance in his version of struggle and selection, explicitly writing that competition is "far more important to the life of each being than mere climate." Wallace's focus on hard selection, on the other hand, is evident in the Ternate essay where climate and food are repeatedly invoked as the factors that provide the test of selection (again, not using that term). This focus of Wallace's is curious in light of the fact that his reading of Malthus, whose *Essay on Population* he cited as provid-ing the key insight into selection, presented struggle in the context of com-peting groups of people. Moreover, Wallace would have been familiar with

Spencer's view of the outcome of struggle, where competition between groups is paramount for improvement. Spencer imagined that competition spurred individuals to ever-better adaptation to their environment, both physical and social. Consider this expression of the idea in Spencer's essay "A Theory of Population, Deduced from the General Law of Animal Fertility":

> But this inevitable redundancy of numbers—this constant increase of people beyond the means of subsistence—involving as it does an increasing stimulus to better the modes of producing food and other necessaries—involves also an increasing demand for skill, intelligence, and self-control—involves, therefore, a constant exercise of these, that is—involves a gradual growth of them. Every improvement is at once the product of a higher form of humanity, and demands that higher form of humanity to carry it into practice. (Spencer 1852, 498–499)

I suspect that this Spencerian view underlies Wallace's comment in the Species Notebook that progress is only made by interaction or engagement, not isolation: "It is only by communication, by the mingling of different races with their different customs, that improvements arise & then, how slowly! A race remaining isolated will ever remain stationary" (Species Notebook, p. 116). In his formulation of the relationship between population pressure and natural selection in the Ternate essay, however, Wallace seems to treat such environmental factors as climate, food, and predators and not conspecific head-to-head struggle. This "hard selection" focus underpins aspects of Wallace's thinking that have drawn criticism. For example, Bulmer (2005) critiqued Wallace's apparent view in the Ternate essay that superior and inferior varieties coexist until environmental change precipitates the extinction of the inferior one. Bulmer maintained that in Darwin's competition-centered formulation the inferior variety (parental form) is driven to extinction even in a constant environment—though to be fair to Wallace note that Darwin was not firm on this point: newly arisen superior varieties, he wrote in his 1858 extract, "will either coexist with, or, more commonly, will exterminate its parent form" (p. 52). Kottler (1985) summarized an argument advanced by A. J. Nicholson (1960) on Wallace's embrace of environmental and not competitive selection, pointing to the key passage of paragraph 8 of the Ternate essay:

> Now, let some alteration of physical conditions occur in the district—a long period of drought, a destruction of vegetation by locusts, the irruption of

some new carnivorous animal seeking "pastures new"—any change in fact tending to render existence more difficult to the species in question, and tasking its utmost powers to avoid complete extermination; it is evident that, of all the individuals composing the species, those forming the least numerous and most feebly organized variety would suffer first, and, were the pressure severe, must soon become extinct. The same causes continuing in action, the parent species would next suffer, would gradually diminish in numbers, and with a recurrence of similar unfavourable conditions might also become extinct. The superior variety would then alone remain, and on a return to favourable circumstances would rapidly increase in numbers and occupy the place of the extinct species and variety. (Wallace 1858e, 58)

Note that it is the "alteration of physical conditions" (including by biotic agency, such as the locusts or a new carnivore) that exerts the selective pressure and drives the "most feebly organized variety" to extinction. The "same causes continuing in action" might next affect the parental variety, which was presumably well adapted to the local environment before the hypothesized great change. Most problematic for me, but not mentioned by Kottler, is the concluding "return to favourable circumstances"—there may be an inconsistency here in that the new variety is supposed to be well adapted to the new environment, locusts and new carnivores and all, better so than their parent variety. A return to favorable circumstances is to restore *original* conditions, so the new variety should no longer be well adapted. (Admittedly, one might posit that as the only surviving variety this does not matter, especially because the "restored" environment, lacking the invading locusts or whatever, is not as challenging as the modified one.) In any case, Kottler (1985) also cites Wallace's discussion of the giraffe's neck length as illustrative of an emphasis on organism-environment interaction and not conspecific competitive interaction, and Wallace's exhortation to Darwin in 1866 to abandon the term "natural selection" because it personified a natural process. Wallace framed this argument in terms of "negative" or eliminative selection: "[nature] does not so much select special variations as exterminate the most unfavourable ones" (Darwin Correspondence, letter 5140). I concur with Kottler's (and Nicholson's) conclusion that in focusing on environmental selection Wallace failed to notice, or deemed less important, the role of competition. This is an important difference between Darwin and Wallace, all the more so insofar as competition-driven selection has creative as well as destructive potential in Darwin's view. In other words, there are conservative and creative roles for selection. Conservative (negative) selection eliminates unfit or less fit variants. This is creative only in the sense of leading to "departure," as Wallace put it, from the parental form or species. The other and in some ways more significant creative potential stemming from competition-mediated selection is ecological: niche partitioning. This lies at the heart of Darwin's "principle of divergence," which I turn to next.

Extensions and Special Forms of Selection

Darwin introduced the "principle of divergence" in the final paragraph of his second extract read in 1858, based on his letter to Asa Gray from September of 1857. Some of Darwin's severest critics have charged him specifically with the appropriation of this idea from Wallace. This is a problematic charge in two respects (discussed below and more fully in the next chapter). First, there is clear evidence that Darwin's principle had taken shape by early 1855. Second, Wallace did not hold any such view of divergence, built upon a concept of "ecological division of labor," to begin with. It is useful to consider just what Darwin meant by this "principle" and use that as a bridge to considering, finally, other aspects of the action of selection that we find in Darwin but not Wallace. That represents the final phase of my analysis of the intersections and departures seen in their thinking.

Principle of Divergence

The key phases or threads of Darwin's research program that came together to yield his principle of divergence are many and varied: classification as a nested hierarchy of affinities, the division of labor concept as it was imported into biology in the nineteenth century, Darwin's analysis of "aberrance" or variation in barnacles, biogeographic data in the form of botanical arithmetic beginning with Darwin's analysis of varieties in large versus small genera in an ecological context, and of course what I have termed "simple divergence"—branching as well as within-lineage evolution. What exactly is this principle? In essence it is the idea that competition-driven selection will lead to the evolution of ever-better competitors, one result of which is niche partitioning. "Diversify or be done for," as Browne (1980) put it: offspring variants (eventually varieties) become adapted to different niches, moving into them whether they are empty or not (in the latter case displacing current occupants). Over time these varieties, now in different lineage trajectories by virtue of their difference niches, further diverge from one another, but each also gives rise in turn to divergent subvarieties, then sub-subvarieties, and so on. Iterating this process over time inevitably yields a treelike pattern of ever-ramifying branches—the Tree of Life. Darwin be-came convinced that he was on the right track with this principle because he thought it had predictive value. Evidence is seen, he thought, in the dis-proportionate number of varieties found in species of large versus small gen-era, and in the taxonomic diversity of plants in small areas like his meadow survey plots. This was Darwin's "success breeds success" model, with a decid-edly ecological angle. The importance of that ecological context for Darwin cannot be overstated. He was led up that trail as early as 1855, when he wrote,

> On Theory of Descent, a *divergence* is implied & I think diversity of struc-ture supporting more life is thus implied. . . . I have been led to this by looking at a heath thickly clothed by heath, & a fertile meadow, both crowded, yet one cannot doubt more life supported in second than in first; & hence (in part) more animals are supported. This is not final cause, but mere results from struggle, (I must think out this last proposition). (DAR 205.3, 167, dated 30 January 1855; quoted in Ospovat 1981, 180–181)

These are observations fleshed out in *Natural Selection* and, later, the *Origin*. It is clear that Darwin considered his principle to be of central importance to his theory, not only writing specifically to Hooker and Gray about it but also making it the visual centerpiece of the *Origin*: the sole figure in the book illustrates the divergence process as Darwin saw it, not simply anagenetic and cladogenetic evolution. As an aside here, remember that Darwin's model is now considered to be correct in places and incorrect in others; competition is indeed thought to be a potent evolutionary force that can lead to competi-tive exclusion, niche partitioning, and diversification. However, note that Darwin had such faith in this competition-driven model that he discarded or at least minimized a role for isolation in the evolutionary process, imag-ining that most evolution through his principle of divergence must there-fore occur in sympatry, in expansive, continental areas with high species density. He correctly surmised that competition is fiercest in such areas, but nonetheless physical isolation, or allopatry, is now thought to be a crucial factor in the process of speciation (a view adopted by Wallace).

Several authors have provided excellent analyses of this and the other threads and how they came together into a coherent concept of divergence as a creative engine of evolution, rather than divergence being simply a descriptor of progressive change or a by-product of speciation generally (e.g., Limoges 1968; Browne 1980; Schweber 1980; Ospovat 1981; Kohn 1985; Kottler 1985; Tammone 1995). Brooks's charge in his 1984 book that Darwin "appropriated, without any acknowledgement, the concept of divergence as it appears in the *Origin of Species* from Wallace's 1855 paper and the [Ternate essay]" would seem baseless since neither of these landmark papers by Wallace articulate anything resembling Darwin's divergence principle. In fact, Wallace seemed to hold to a view of selection in which competition was not emphasized.

From the "physiological division of labor" concept, to the idea that diversification makes it possible for more species to coexist in a given area, to the emphasis on competition—these essential ingredients of Darwin's principle of divergence are lacking altogether not only in these two papers of Wallace's but in all of his writings of that period (including the Species Notebook). Kottler (1985) put it well when he wrote that "by 1857 at the very latest Darwin had formulated his principle of divergence as it appeared in both *Natural Selection* and the *Origin* . . . consequently nothing in Wallace's 1858 paper about divergence [and there is nothing about a divergence *principle*] could possibly have influenced Darwin." It is perhaps ironic that some authors have accused Darwin of taking the "principle of divergence" concept from Wallace, given that it is, after all, a flawed view of speciation— that is, speciation in sympatry—as opposed to Wallace's more correct view of the role of isolation. This is evident in the Sarawak Law paper, as well as in Species Notebook passages such as the one he copied from von Buch about how plants dwelling in "sheltered vallies & low grounds" in the Canary Islands show a pattern of "closely allied species confined to one valley or one island" (p. 90).

Other Modes of Selection

One final point should be made concerning the expressions of Darwin and Wallace's evolutionary ideas circa 1858. Darwin, by virtue of his long lead over Wallace in thinking about selection, had come to realize that there exist special forms of selection, applicable in certain circumstances. These include most importantly sexual selection and what we might call family-level selection. Of the two only the first is found in Darwin's 1858 extracts. We have seen earlier in this chapter Darwin's brief description of the two forms of

sexual selection (paragraph 5 of the first Darwin extract). There is no parallel at all regarding sexual selection in Wallace's thinking, and in fact he
was rather resistant to certain elements of the concept for many years. (For
reasons apparently as much sociopolitical as scientific, Wallace did an about-
face in 1890 and at last fully embraced the female-choice aspect of sexual
selection, particularly for humans; see Fichman 2004 and discussion of this
topic by Fichman in Richards 2005.)

That Darwin had a concept of family-level selection by 1858 is seen in
the chapter on "Difficulties" in the *Natural Selection* manuscript (Stauffer
1975, 365–367). The problem is posed by sterile insects in certain ant, bee,
and wasp societies. How could selection produce "caste polymorphism," the
sometimes radical morphological differences among members of the same
family, or sterility, the inability to reproduce found in many social insects?
This was a serious puzzle to Darwin, since selection was thought to work for
the exclusive good of the organism; not only would a sterile individual be
incapable of passing on its traits, but in some way sterility seemed associated with characteristics such as morphological diversification within the
same colony. "I confess that when this case first occurred to me, I thought
that it was actually fatal to my theory," Darwin wrote in *Natural Selection*.

In his solution Darwin connected morphological divergence among castes
to division of labor within families, joining the "physiological division of labor" concept of Milne-Edwards (1851) with the "ecological division of labor"
that lay at the heart of his principle of divergence (Darwin 1859, 111–126).
Division of labor, he thought, resulted from the struggle for existence and
increased overall productivity:

> The advantage in each group becoming as different as possible, may be com-
> pared to the fact that by division of labour most people can be supported in
> each country—Not only do the individuals of each group strive one against
> the other, but each group itself with all its members, some more numerous,
> some less, are struggling against all other groups, as indeed follows from
> each individual struggling." (DAR 205.3: 171, dated 30 January 1855; quoted
> in Ospovat 1981, 181)

But how could selection *favor* sterility, elimination of personal reproduction?
"How [is it] possible that communities of insects should come to possess
sterile females or neuters?" Darwin asked in *Natural Selection* (Stauffer
1975, 365). Here he invoked colony (family)-level selection: "it seems not
improbable, owing to the vast fecundity of the lower animals, that a certain

number of females . . . without any waste of time or vital force from breeding, might be of immense service to the community. If this were so . . . then natural selection would favour those communities, in which some of the individuals [had been rendered] in some slight degree less fertile than the other individuals" (Stauffer 1975, 366). In the *Origin* Darwin shortened his discussion of the subject, simply proposing that the great difficulty posed by morphological divergence and sterility "is lessened, or, as I believe, disappears, when it is remembered the selection may be applied to the family, as well as to the individual" (Darwin 1859, 237). In other words, selection may act on the ant or bee queen and her family as a whole, and if it is beneficial for the group to have sterile helpers and caste divisions, the queens will be selected to produce such. This mode of selection was elaborated into inclusive fitness or kin selection theory in the twentieth century by British biologists J. B. S. Haldane (1932) and especially W. D. Hamilton (1964a,b). The main point for us here is that Wallace lacked a conception of selection acting in this way. Family- or colony-level selection is another example of the subtlety of Darwin's understanding of the selection and evolution process.

Founders

Several conclusions emerge from this comparative analysis of Wallace and Darwin's 1858 papers, and more broadly their thinking as expressed in other writings such as their notebooks. But caveats first: judging one or the other more or less correct in the modern view based on their thinking as of 1858 is, in one respect at least, invalid. Consider that as of June of 1858 both Wallace and Darwin had had years of pondering lines of evidence for the *fact* of transmutation (Darwin twenty years and Wallace thirteen, to be more precise), but in regard to the *mechanism* behind transmutation, natural selection in all of its nuance, we have but one essay by Wallace quickly written upon his initial insight, while Darwin had two decades in which to ponder, refine, polish, and explore the concept. Moreover, many of the lines of evidence Darwin documented were not simply in support of transmutation per se, but in support of transmutation with an eye to natural selection as the mechanism of change. Wallace, in contrast, did not have much of a chance to consider how selection bore on the lines of pro-transmutation evidence he was pursuing.

The congruence in thinking of the two in terms of both approach and content is nonetheless remarkable. Among the points of intersection are the lines of evidence that each drew upon, in consilience fashion, to argue for the reality of transmutation, and in key elements of the mechanism of natural selection. The lines of evidence ranged widely: geographical distribution, classification, fossil succession, morphology, behavior, and more. Darwin also saw domestication as a microcosm of transmutational diversification in nature, while Wallace did not see domestication this way (though in the Species Notebook he cited domestic breeds as evidence of a capacity for transmutation). He would likely have elaborated on this argument in his planned book, but for the purposes of his brief treatment in the Ternate essay he chose to use domestication as a foil, setting up the argument for transmutation in nature in answer to Lyell's objection to transmutation based on the supposed limited capacity for change in domestic varieties.

With regard to their formulation of natural selection both Wallace and Darwin came to realize the key role played by population pressure and struggle, leading to differential survival and reproduction (the distinctness of these recognized most explicitly by Darwin through the special form of natural selection that he dubbed sexual selection—another point of departure). Early on, both thought that isolation must play a role in the origin of species, though by 1858 Darwin had downplayed this while Wallace was unwavering in recognizing its importance. Wallace's emphasis on environmental selection to the relative neglect of competition is another point of departure, with Darwin arguing that competition is paramount. This in turn led to the other significant difference in their conceptions of selection: both held a concept of divergent (branching) evolution, though Darwin elaborated this into a "principle" of divergence that he believed played a key role in generating the diversity of life.

It is clear that both Wallace and Darwin held both correct and incorrect views in terms of modern understanding: today we disagree with Darwin's almost exclusive emphasis on competition and speciation in sympatry, while we also disagree with Wallace's seeming exclusive emphasis on the abiotic environment as the mediator of selection. Wallace was correct to emphasize the role of isolation in speciation, and Darwin was correct in his grasp of sexual selection and family-level selection. Both held a fundamental understanding—indeed, were the first to do so—of transmutation as a branching process with natural selection playing a key role in the adaptation of organisms to a slowly changing environment. Moreover, both saw how this hypothesis tied together many apparently unrelated lines of evidence, from morphology and embryology to biogeography and paleontology. Lyell and Hooker were correct in writing, in their prefatory remarks to the 1858 paper, that "both [Wallace and Darwin] may fairly claim the merit of being original thinkers in this important line of inquiry." The two are equally deserving of laurels.

SIX

True with a Vengeance

From Delicate Arrangement to Conspiracy: A Guide

IN CHAPTER 4 I compared the key evolutionary ideas of Wallace and Darwin as reflected in their respective writings as of the late 1850s, alluding only briefly to the circumstances surrounding the receipt and publication by Darwin of Wallace's Ternate essay. In this chapter I explore this issue more deeply, offering a brief guide to the controversy over the timing of the arrival of Wallace's letter and manuscript, and the question of the use (or not) by Darwin of the ideas it contained. Previously I described Darwin's apparent surprise and dismay over the "striking coincidence" of Wallace's formulation. "If Wallace had my M.S. sketch written out in 1842 he could not have made a better short abstract! Even his terms now stand as Heads of my Chapters," he wrote, continuing: "your words have come true with a vengeance that I [should] be forestalled" (Darwin Correspondence Project, letter 2285). When the date of receipt of Wallace's letter was called into question, this inevitably led to speculation about questionable behavior on Darwin's part. Some plot thickeners have factored into the narrative, notably the fact that Darwin reworked substantial portions of his chapter on natural selection in the *Natural Selection* manuscript right about the time he was suspected of sitting on Wallace's manuscript (Brooks 1984, 232–235). As a result Darwin was then variously accused of stealing key ideas from Wallace, deceiving his friends about this fact, and manipulating the situation to ensure that his own (unpublished) writings would be presented to the world first, guaranteeing his priority. All supposed to have been possible by his position of wealth and privilege, having friends among the most powerful and influential in Britain, and with his rival conveniently out of the picture half a world away. Contributors to this narrative range from the work of Lewis McKinney (1972) to the progressively less measured allegations of Brackman (1980), Brooks (1984), and Davies (2008, 2013).

Matters have remained as murky as they are contentious owing to gaps in the record, including the following still-missing items, listed in chronological order (letter numbers are those of the Darwin Correspondence Project unless otherwise indicated):

- Wallace's first letter to Darwin, sent 10 October 1856 and received by Darwin April 1857 (known from Darwin's reply of 1 May 1857: letter 2086)
- Wallace's reply to Darwin's letter of 1 May 1857, sent 27 September 1857 (A fragment of this letter is extant; letter 2145)
- Wallace's original Ternate essay manuscript and accompanying letter to Darwin
- Lyell's initial reply letter to Darwin, sent upon receipt of Darwin's 18 June 1858 letter concerning Wallace's essay
- Letter from Joseph Hooker sent in late June 1858 to Darwin regarding Wallace's essay (mentioned in letter 2298, Darwin to Hooker 29 June 1858)
- Letter from Charles Lyell to Wallace, sent late June 1858 to Darwin and possibly forwarded to Wallace (see letter 2294)
- First letter from Joseph Hooker to Wallace post-Linnean readings, sent early July 1858. This was enclosed in Darwin's letter of 13 July; it is lost, but Wallace's reply to it is extant. (Wallace Correspondence, letter WCP1454, dated 6 October 1858)
- Darwin's first letter to Wallace post-Linnean readings, sent 13 July 1858. This enclosed the above letter from Hooker.
- Letter from Joseph Hooker to Darwin, sent early July (known from Darwin's reply of 11 August: letter 2321)
- Letter from Asa Gray to Darwin dated 27 July 1858
- Letter from Wallace to Darwin October 1858 (referred to in letters 2403 [Darwin to Hooker 23 January 1859] and 2405 [Darwin to Wallace 25 January 1859])
- Letter from Wallace to Darwin 30 November 1858 (known from Darwin's reply of 6 April 1859; letter 2449)

Add to these a possible letter from Darwin perhaps late summer or fall 1858 in which Darwin provided Wallace with the title and table of contents of his book-in-progress, *Natural Selection*. Wallace recorded the table of contents in his Species Notebook (Costa 2013a, 430–431).

It was at the golden anniversary celebration of the reading of the Darwin and Wallace papers in July 1908 that attention was first drawn to the fact that several of these important items were missing. Hooker (then ninety-one years old), who along with Wallace (then eighty-five) was one of the two surviving participants in the events of 1858, noted the gaps in the documentary record: "There are no letters from Lyell relating to [the events of June 1858], not even answers to Mr. Darwin's of the 18th, 25th, and 26th of June. . . . There are none of my letters to either Lyell or Darwin, nor other evidence of their having existed beyond the latter's acknowledgement of the receipt of some of them; and, most surprising of all, Mr. Wallace's letter and its enclosure have disappeared" (Hooker 1908–1909, 15–16; also quoted in Beddall 1988a, 54–55). Indeed, in the period 1856 (when Wallace initiated correspondence with Darwin) through 1860, only three (two are fragments) of ten letters known or inferred to have been sent to Darwin by Wallace survive, while of the eight or so letters Darwin sent to Wallace in that time, only one is missing. (Unfortunately, so are the letters Hooker and Lyell sent to Wallace.)

Kohn, in his 1981 review of Brackman (1980), suggested that the ratio of extant to lost Wallace letters in Darwin's correspondence was consistent with that of other correspondents, but Beddall (1988a) made the point that not all letters are equal, and surely the Wallace letters of this critical period would or should have been valued accordingly. Be that as it may, the gaps leave open the door to speculation and innuendo that does neither Darwin nor Wallace justice. It is possible, however, to cut through the perennial speculation by simply focusing on the specific claims made and assessing the extent to which they hold water, including a comparison of the evolutionary ideas that Darwin and Wallace held as of early summer 1858 in the context of the "conspiracy theory" of intellectual theft. I first give an overview of current thinking on the main claims concerning timing and priority, after which I discuss Wallace's and Darwin's ideas about natural selection in an effort to clarify the points of similarity and dissimilarity in 1858.

At issue with regard to the timing and priority issue have been (1) the precise date of receipt by Darwin of Wallace's letter and essay; (2) the question of whether Darwin took any of Wallace's ideas and passed them off as his own in the joint readings of 1858; and (3) the fairness of the action taken by Darwin's friends Lyell and Hooker, having private writings by Darwin presented with—even ahead of—Wallace's essay. The main reason that such questions have persisted stems in large part from a faulty documentary record. Neither Wallace's original Ternate manuscript nor its cover letter to Darwin appears to be extant, and nor are several other letters in the Wallace-Darwin correspondence from the 1856–1859 period, as just discussed. Creative historical sleuthing has, however, yielded a reasonably thorough reconstruction of events: some of the content of some of the missing letters can be inferred from surviving replies, and dates of various events can often be inferred from letters, journal entries, and other sources. The sequence of events is summarized in Figure 6.1.

Receipt of the Ternate Essay

The date of receipt of Wallace's letter and essay from Ternate is typically taken to be 18 June 1858. This is based on Darwin's anguished letter to Lyell dated only "Down, 18th." This is the letter announcing the receipt of Wallace's essay, in which Darwin states that Wallace "has to-day sent me the enclosed." The letter was indeed sent to Lyell in June, but Brooks (1984, 256) suggested that the "18th" refers to 18 May, though the letter was not sent to Lyell until the following month. A May arrival for Wallace's package at Down House is based on Brooks's analysis of the mail transport records from Southeast Asia, administered and run by the Dutch colonial government. Assuming that Wallace's letter and manuscript were posted on 9 March 1858, Brooks concludes that Darwin should have received it mid-May or so, or perhaps as late as early June. The 9 March posting date is itself based on Wallace's later statement (eleven years later) that following his discovery of the principle of natural selection he wrote out his essay in two or three evenings and sent it off "by the next post" to Darwin, with whom he had previously corresponded. In other accounts Wallace was consistent in maintaining that his manuscript was mailed off soon after its completion, which was probably on the island of Ternate. (Recall that Wallace had his insight—and wrote his essay—not on Ternate, but on the neighboring island of Gilolo (Halmahera) in February 1858, but he may have signed his essay "Ternate" because that was his base, his address, and the port of call for the mail steamers.) The steamers were on a monthly schedule at that

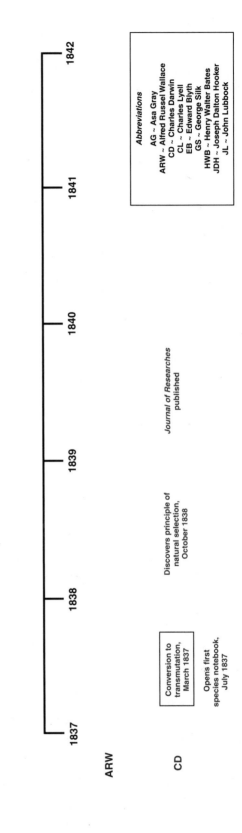

ARW

CD

Conversion to
transmutation,
March 1837

Opens first
species notebook,
July 1837

Discovers principle of
natural selection,
October 1838

Journal of Researches
published

1837   1838   1839   1840   1841   1842

Abbreviations

AG – Asa Gray
ARW – Alfred Russel Wallace
CD – Charles Darwin
CL – Charles Lyell
EB – Edward Blyth
GS – George Silk
HWB – Henry Walter Bates
JDH – Joseph Dalton Hooker
JL – John Lubbock

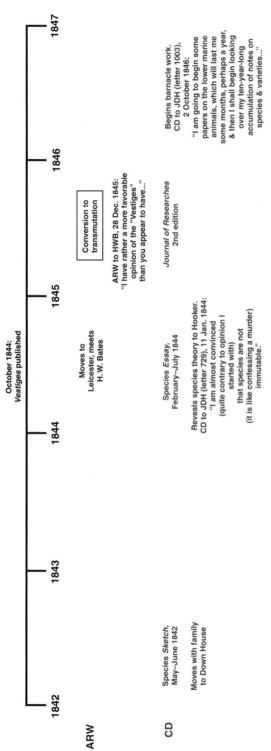

ARW

October 1844:
Vestiges published

Moves to
Leicester, meets
H. W. Bates

Conversion to
transmutation

ARW to HWB, 28 Dec. 1845:
"I have rather a more favorable
opinion of the "Vestiges"
than you appear to have…"

CD

Species *Sketch*,
May–June 1842

Moves with family
to Down House

Species Essay,
February–July 1844

Reveals species theory to Hooker.
CD to JDH (letter 729), 11 Jan. 1844:
"I am almost convinced
(quite contrary to opinion I
started with)
that species are not
(it is like confessing a murder)
immutable."

Journal of Researches
2nd edition

Begins barnacle work.
CD to JDH (letter 1003),
2 October 1846:
"I am going to begin some
papers on the lower marine
animals, which will last me
some months, perhaps a year,
& then I shall begin looking
over my ten-year-long
accumulation of notes on
species & varieties…"

1842   1843   1844   1845   1846   1847

Figure 6.1. Timeline of Wallace's and Darwin's respective paths to the discovery of evolution by natural selection, in five-year intervals from Darwin's conversion to transmutation in 1837 through Wallace's return to England in 1862. All letter citations provide Darwin Correspondence Project letter numbers.

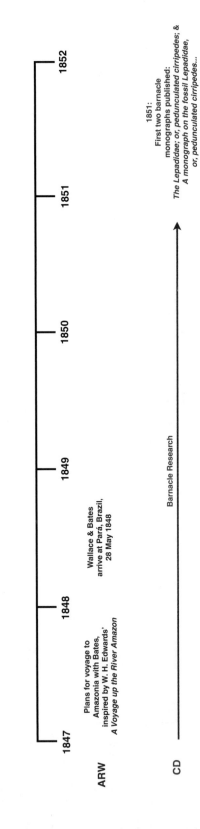

1852

1851:
First two barnacle
monographs published:
The Lepadidae; or, pedunculated cirripedes; &
*A monograph on the fossil Lepadidae,
or, pedunculated cirripedes...*

1851

1850

Barnacle Research

1849

Wallace & Bates
arrive at Pará, Brazil,
28 May 1848

1848

Plans for voyage to
Amazonia with Bates,
inspired by W. H. Edwards'
A Voyage up the River Amazon

1847

ARW

CD

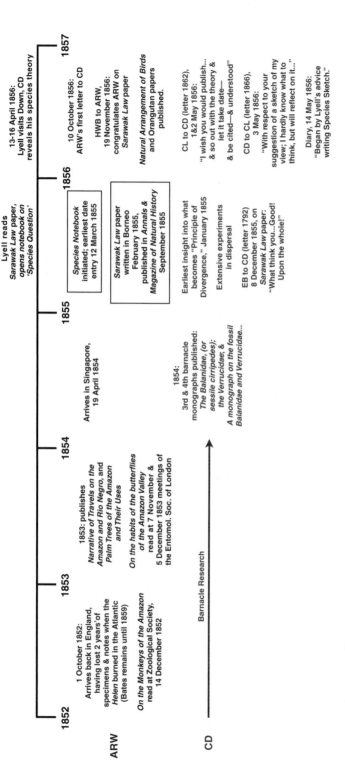

Figure 6.1. (continued)

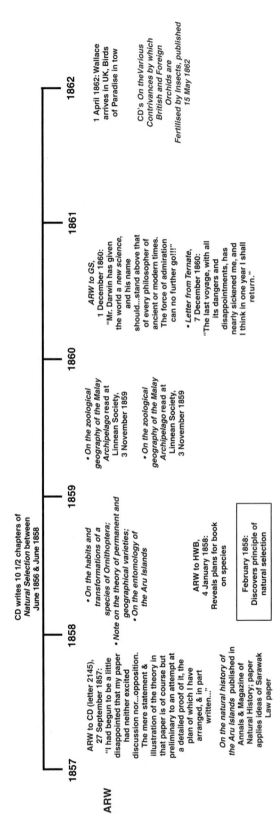

ARW

1857   1858   1859   1860   1861   1862

CD writes 10 1/2 chapters of
Natural Selection between
June 1856 & June 1858

ARW to CD (letter 2145),
27 September 1857:
"I had begun to be a little
disappointed that my paper
had neither excited
discussion nor...opposition.
The mere statement &
illustration of the theory in
that paper is of course but
preliminary to an attempt at
a detailed proof of it, the
plan of which I have
arranged, & in part
written...."

*On the natural history of
the Aru Islands* published in
Annals & Magazine of
Natural History; paper
applies ideas of Sarawak
Law paper

• *On the habits and
transformations of a
species of Ornithoptera;*
• *Note on the theory of permanent and
geographical varieties;*
• *On the entomology of
the Aru Islands*

ARW to HWB,
4 January 1858:
Reveals plans for book
on species

February 1858:
Discovers principle of
natural selection

• *On the zoological
geography of the Malay
Archipelago* read at
Linnean Society,
3 November 1859

• *On the zoological
geography of the Malay
Archipelago* read at
Linnean Society,
3 November 1859

ARW to GS,
1 December 1860:
"Mr. Darwin has given
the world a *new science,*
and his name
should...stand above that
of every philosopher of
ancient or modern times.
The force of admiration
can no further go!!!"

• *Letter from Ternate,*
7 December 1860:
"The last voyage, with all
its dangers and
disappointments, has
nearly sickened me, and
I think in one year I shall
return."

1 April 1862: Wallace
arrives in UK, Birds
of Paradise in tow

CD's *On the* *Various
Contrivances by which
British and Foreign
Orchids are
Fertilised by Insects,* published
15 May 1862

CD

On the Origin of Species 3rd edition

On the Origin of Species 2nd edition

April 1857
CD receives ARW letter of October 1856

CD to ARW (letter 2086), 1 May 1857:
"I can plainly see that we have thought much alike..."

CD to JL (letter 2123), 13-14 July 1857, thanking him for correcting CD's errors in his botanical arithmetic

CD to JDH (letter 2134), 22 August 1857: outlines species theory, including Principle of Divergence

CD to AG (letter 2136), 5 September 1857: outlines species theory, including Principle of Divergence

CD to AG (letter 2176), 29 November 1857: Thanks Gray for comments on his species theory; comments on "... what I rather vaguely call my principle of divergence..."

CD to ARW (letter 2192), 22 December 1857:
"you must not suppose that your paper has not been attended to: two very good men, Sir C. Lyell & Mr E. Blyth at Calcutta specially called my attention to it"

Ternate Essay posted to CD, March-April 1858

ARW to his mother, 1 October 1858:
"I have received letters from Mr. Darwin and Dr. Hooker, two of the most eminent naturalists in England, which have highly gratified me..."

ARW to JDH, 6 October 1858:
"I beg leave to acknowledge the receipt of your letter of July last, sent me by Mr. Darwin ..."

CD to CL (letter 2285), 18 June 1858:
"Your words have come true with a vengeance that I should be forestalled..."

30 June 1858: Lyell & Hooker communication to Linnean (letter 2299) of Wallace and Darwin papers; papers read at Linnean Soc. on 1 July.

20 July 1858
CD "began Abstract of Species book"

20 August 1858
CD-ARW papers published

ARW abandons book plan

CD to ARW (letter 2405), 25 January 1859:
"I was extremely much pleased at receiving three days ago your letter to me & that to Dr. Hooker..."

CD to ARW (letter 2449), 6 April 1859:
"You are right, that I came to conclusion that Selection was the principle of change from study of domesticated productions; & then reading Malthus I saw at once how to apply this principle..."

On the Origin of Species published, 24 November 1859

remote locale, and if Wallace's letter and essay were sent on the one de-parting Ternate on the ninth of March, it is very likely that the date of receipt at Down House would have been at least two weeks earlier than Darwin claimed. This is inferred from McKinney's (1972, 140) discovery of an extant letter Wallace posted on the 9 March steamer. That letter was ad-dressed to Frederick Bates, brother of Wallace's friend Henry Walter Bates, who was still in South America. The postmarks on the surviving letter clearly indicate that it arrived London on 3 June 1858 and was forwarded to (and stamped at) Leicester on the same day (see Figure 6.2).

If Wallace's package was received two or more weeks earlier than Darwin appeared to claim, what might the significance be? Although McKinney (1972) intimates and Brackman (1980) and Brooks (1984) explicitly charge

Figure 6.2. Postal cancellation stamps on the envelope containing a letter from Wallace to Frederick Bates, brother of his friend Henry Walter Bates, mailed 9 March 1858 (WCP367). The stamps for London and Leicester read "Ju 3 58"—June 3, 1858: the expected date of arrival of the Ternate essay at Down House assuming it was mailed at the same time as this letter. Photograph by the author, reproduced by permission of the Trustees of the Natural History Museum, London.

that Darwin pilfered ideas from Wallace and the earlier receipt date provided the time needed to do so, other historians have pointed to several
lines of evidence demonstrating that Darwin took nothing from Wallace's
manuscript—indeed, that there was nothing new to Darwin in the manuscript that could have been taken. But first it should be pointed out that the
9 March posting of Wallace's package is not settled. The fact that Wallace's
letter accompanying his essay seemed to be a *reply* to a letter from Darwin
dated 22 December 1857 has implications. In it, Wallace later said in his
autobiography (Wallace 1905, 1:363), he asked Darwin to show his essay, if
he thought it "sufficiently important," to Lyell, "who had thought so highly
of my former paper." The "former paper" is Wallace's 1855 Sarawak Law
paper, but the only way he could have known that Lyell thought highly of it
is from Darwin's letter of 22 December 1857. It was in that letter, which is
extant, that Darwin consolingly told Wallace that "two very good men" had
specially commended his 1855 paper—Lyell and Blyth (Darwin Correspondence Project, letter 2192). That Wallace's essay was intended for Lyell's
eyes is corroborated by Darwin's subsequent letter to Lyell upon arrival of
Wallace's essay, in which Darwin stated that Wallace "asked me to forward
it to you" (Darwin Correspondence Project, letter 2285). Lyell and Hooker
affirm this in their prefatory remarks to the Linnean Society papers, stating that Wallace's essay "was written at Ternate in February 1858, for the
perusal of his friend and correspondent Mr. Darwin, and sent to him with
the expressed wish that it should be forwarded to Sir Charles Lyell, if
Mr. Darwin thought it sufficiently novel and interesting" (see Chapter 4,
p. 180).

Van Wyhe and Rookmaaker (2012) pointed out that Darwin's letter must
have been received by Wallace on the very same 9 March 1858 mail steamer
on which he is supposed to have posted his essay at Ternate; the February
steamer would have been too soon. They further argued that it is unlikely
that Wallace could have both received and replied to Darwin's letter on
the same day, involving the same steamer. The likely scenario, according to
van Wyhe and Rookmaaker (2012), is that Wallace penned a letter to Darwin
upon reading the latter's letter to him, and this, together with his essay, was
posted before he departed Ternate on 25 March for a collecting trip to
New Guinea. The monthly steamer next called at Ternate in early April. In
this interpretation Wallace's "next post" was the next post *after* penning his
letter to Darwin, not the next post after penning his essay on Gilolo. In
support of their contention van Wyhe and Rookmaaker (2012) posed the

question, if Darwin received Wallace's package on 18 June 1858 as he claimed, is there any evidence of an unbroken mail route leading back to Ternate from that date, and if so when would the essay have been posted? They claimed to show such an unbroken chain of transport which involved multiple steamers, camel caravan, and train, delivering the package at Down on precisely 18 June. Moreover, this chain originates about 5 April, the very next steamer calling after Wallace's receipt of Darwin's letter on 9 March. As for time of transit, although some earlier investigators used a ten-week delivery time estimate based on a comment Wallace had made to that effect, van Wyhe and Rookmaaker (2012) showed that seventy-seven days (eleven weeks) was the typical transit time, and both the passage of Darwin's letter from Down House to Ternate (22 December 1857–9 March 1858) and their suggested passage from Ternate back to Down House (5 April–18 June 1858) are about seventy-seven days.

In reply, Davies (2012) countered that Wallace could have mailed his package to Darwin on the 9 March steamer, maintaining that it would have been a trivial matter for Wallace to pen a cover letter in reply to Darwin after receiving and reading Darwin's letter to him. More importantly, Davies argued that van Wyhe and Rookmaaker confused their mail boats, and the key boat (from Java) in their delivery scenario in fact was not contracted to transport mail collected from the islands (including Ternate). If true, this undermines van Wyhe and Rookmaaker's sequence of events. Smith (2013) further argued that Darwin's December 1857 letter to Wallace may be irrelevant in this case, insofar as Wallace's account of asking Darwin to show his essay to Lyell, "who had thought so highly of my former paper," was (1) many years later, and (2) merely a rhetorical construction. Smith contends that the comment about thinking highly of the former paper was a rhetorical device for Wallace's contemporary readers of his autobiography, not a reference to the reason he sent his essay to Darwin; if so, then the timing of receipt of Darwin's December 1857 letter is not very important.

Darwin seemed to believe that his earlier letter did have something to do with the reason that Wallace sent the Ternate essay to him with the request to show it to Lyell. In his 18 June letter to Lyell (Darwin Correspondence Project, letter 2285), Darwin wrote, "Some year or so ago, you recommended me to read a paper by Wallace in the Annals, which had interested you & as I was writing to him, I knew this would please him much, so I told him. He has to day sent me the enclosed." That he may have *believed* his letter was the impetus behind Wallace's writing to him does not mean it was, however. In any case it is evident from Wallace's extensive notes on Lyell in the Species Notebook and the clear Lyellian influence in his 1855, 1858, and other papers that he would have been keen to communicate with Lyell. Doing so through Darwin, with whom he had already been in contact, may have been an expedient way to reach Lyell (see Costa 2013b). It may be irrelevant whether Wallace's letter and manuscript to Darwin were a reply— that is, whether he asked Darwin to forward his paper to Lyell because he knew the geologist had commented favorably upon his 1855 paper, or decided to contact Lyell through Darwin because the time was ripe, without knowing of the impression made by his 1855 paper. The latter scenario simply removes Darwin's letter revealing Lyell's approbation as the impetus behind Wallace's writing to Darwin.

Regardless, however, of whether Darwin received and duly forwarded Wallace's letter and essay when he said he did, I conclude this discussion by pointing out once again that even had Darwin received Wallace's package earlier than he appeared to have claimed to Lyell, this hardly compels a conclusion of skullduggery on his part. What it more likely means is that a very human Darwin fretted, stewed, and agonized over what he should do. Considering that he appeared to believe that Wallace had developed precisely the same theory as his own, we can only imagine his devastation at realizing he had, in effect, been scooped after his many years of labor. And this on top of the stress of two stricken children at home, one of whom was to quickly succumb to his illness (baby Charles's funeral was held the very day that the Darwin and Wallace papers were read at the Linnean Society on 1 July). What is remarkable, then, and a testament to the naturalist's ultimate honesty is the fact that he did forward Wallace's package to Lyell. A lesser person might have burned it, and who would have known? But more importantly, if Darwin did fret over Wallace's essay for two weeks or even longer, is there evidence that he pilfered ideas from Wallace and incorporated them into his own writings?

Appropriation of Ideas?

Several authors have suggested that elements of Darwin's theory as presented in 1858 were taken from Wallace (e.g., Brackman 1980; Brooks 1984; Davies 2008, 2013). Two circumstances have lent credence to this claim: uncertainty over the date of receipt by Darwin of Wallace's Ternate essay (an earlier arrival creating ample opportunity for study of Wallace's manuscript), and the fact that Darwin apparently reworked a key section of his *Natural Selection* manuscript during the more than two-week interval between when he is alleged, by some, to have received the essay and when he forwarded the essay to Lyell. What are the ideas Darwin is accused of taking from Wallace? One is the imagery of the "tree of life," which likens the branching history of life to a tree. (Wallace used the tree imagery in his 1855 paper; the words "tree" and "branch" do not appear in the Ternate essay, though "divergence series" does.) This charge, first leveled by Brooks (1984, 223–227), can be dismissed as the documentary record shows clearly that Darwin had used the simile of budding and trees (even "corals") extensively over several years, beginning in the late 1830s (e.g., entries B21–27, B36–38; C145 in the Transmutation Notebooks; Barrett et al. 1987). Brooks neglected to mention this, though it was well known in 1984 when his book came out. Among other marginalia in his copy of Wallace's 1855 paper Darwin wrote "uses my simile of tree"—intimating that it was an image he long had in mind too—along with a small tree sketch much like those his notebooks. It is clear that Darwin, like Wallace, grasped early on the convenient analogy of trees with the branching and rebranching evolutionary history of life. Wallace, however, should be credited with being the first to present a non-Lamarckian evolutionary tree in a public forum. He did so descriptively in his 1855 Sarawak Law paper, and in graphical form the following year in his essay "Attempts at a Natural Arrangement of Birds" (Wallace 1856e; see p. 36).

The more important charge of intellectual theft pertains to the so-called principle of divergence, which Darwin considered to be an important part of the process of selection and organic diversification. Indeed, although Darwin had first discussed this form of divergence in 1855 and first described it as a "principle" in 1857, it is true that he expanded his treatment of the subject considerably in the spring of 1858: sixty-six new pages of material inserted into chapter 6 of his draft book *Natural Selection*. Could this have been owing to ideas found in Wallace's Ternate essay? The evidence suggests not. But first, let us review this "principle."

It is important to distinguish between divergence per se and the principle of divergence. First, though Darwin (and Wallace) did not use terms like "anagenesis" and "cladogenesis" (modern terms for within-lineage evolutionary change versus lineage splitting or branching), both naturalists understood these two forms of transmutation. As such, "divergence" can be used as a descriptor of both: descendant forms or lineages become increasingly dissimilar from their parental or ancestral species; in other words, they increasingly *diverge* from it. In regard to the concept of branching lineages Darwin and Wallace also used the tree metaphor that the branching idea naturally lends itself to—this form of divergence yields multiple lineages that have branched off at different times, giving rise to a pattern of nested sets of related species corresponding to the classification hierarchy.

Darwin's "principle of divergence" takes this a step further. As discussed in Chapter 5, it was viewed by him as a competition-driven process of diversification over evolutionary time, a process in which branching is effectively forced by competitive interaction and selection. That is, selection exerts a pressure that forces descendant lineages to diverge as a result of intense intraspecific competition for ecological resources. Over time this competition, and the selection pressure it implies, leads to niche partitioning and speciation. The end result is different species sharing common ancestry coexisting in more or less the same physical area. This does not mean their ecological needs are identical; rather, they become adapted to use resources in the same area differently, modified such that they are no longer in a dire scramble with one another.

The principle of divergence and the related concept of "divergence of character" (structure, physiology) became key elements behind the evolutionary tree-building process in Darwin's thinking. It is featured prominently in chapter 4 of the *Origin,* and the book's sole diagram (on p. 117) is designed to illustrate the process. To Darwin such a mechanism offered a better explanation for the astounding diversity of life than the chance occurrence of isolation leading to lineage splitting. He believed that it better explained, too, why the regions richest in species diversity are expansive continental regions, where extended periods of isolation would seem to be difficult to come by. Early on Darwin and Wallace were struck by the high proportion of unique species on islands, which is why the idea of speciation in isolation—what we call allopatric speciation—loomed large in their thinking. Again, as seen in Chapter 5, Wallace stuck to this view while Darwin's divergence principle provided, in his mind, a mechanism that dispensed with isolation and seemed to better explain the diversity of life on continents— speciation in sympatry. Both were partially correct in the modern view: scientists today believe as Wallace did, that geographical separation is responsible for most speciation and that speciation in sympatry is rare and evolutionarily unimportant. Darwin's focus on ecology and competition is viewed as centrally important, however, just not playing the role that Darwin assigned it in promoting sympatric speciation.

It is probably no exaggeration to say that Darwin saw his "principle of divergence" as nearly on a par with his discovery of natural selection itself. As he put it in his autobiography years later:

> I overlooked one problem of great importance; and it is astonishing to me . . . how I could have overlooked it and its solution. This problem is the tendency in organic beings descended from the same stock to diverge in character as they become modified. That they have diverged greatly is obvious from the manner in which species of all kinds can be classed under genera, genera under families, families under sub-orders, and so forth; and I can remember the very spot in the road, whilst in my carriage, when to my joy the solution occurred to me; and this was long after I had come to Down. (Barlow 1958, 120–121)

The process by which he developed his principle has been worked out in detail (e.g., see Browne 1980; Schweber 1980; Kohn 1985; Tammone 1995; and especially Ospovat 1981). It involved interweaving over two decades a remarkable diversity of ideas and observations regarding classification, biogeography, variation in different contexts, extinction, and the division of (physiological and ecological) labor concept. Here I provide an overview of the evidence that vindicates Darwin, showing that no element of his evolutionary thinking was appropriated from Wallace. Although there have been several informative treatments of this issue (e.g., Beddall 1968, 1988a; Limoges 1968; Kohn 1981; Kottler 1985; Bulmer 2005; Bowler 2009), Barbara Beddall's 1988 treatment is the most thorough. The evidence from Beddall (1988a), which I believe is definitive, may be summarized thusly:

- Darwin's earliest descriptions of a "principle of divergence" to others date to late summer 1857, in letters to Joseph Hooker (22 August) and Asa Gray (5 September), predating Darwin's reading of the Ternate essay by nearly a year. Ospovat showed that the earliest *description* of the divergence principle (though not naming it) in Darwin's notes predates the letter to Gray by two years: "On Theory of Descent, a *divergence* is implied & I think diversity of structure supporting life is thus implied. . . . I have been led to this by looking at a heath thickly clothed by heath, & a fertile meadow, both crowded, yet one cannot doubt more life supported by second than in first; & hence (in part) more animals are supported" (DAR 205.3: 167, dated 30 January 1855, emphasis Darwin's; quoted in Ospovat 1981, 180–181).

- The documentary evidence shows that what Darwin meant by his divergence principle was a competition-driven ecological division of labor, which grew out of a multiyear research project involving the problem of classification, botanical arithmetic, and the "physiological division of labor" analogy obtained from Milne-Edwards (Limoges 1968; Ospovat 1981, 170–209; Kohn 1985). There is nothing in any of Wallace's writings, and certainly not in the Sarawak Law and Ternate papers, resembling this conception of the divergence process.
- Darwin's extract read at the Linnean Society in which divergence is mentioned dated from September 1857—his letter to Gray—and not from any writings contemporaneous with Wallace's Ternate essay or the Ternate essay itself.
- Every statement found in the letter to Gray is also found in both *Natural Selection* and the *Origin.* Conversely, none of Darwin's additional material bearing on divergence and natural selection in *Natural Selection* and the *Origin* is found in Wallace's papers (e.g., the domestication analogy, sexual selection, and family-level selection to explain insect castes).

Brackman, Brooks, and Davies appear to conflate the different meanings of "divergence" when they argue that Wallace's usage of the word was the same as that in Darwin's principle. "Divergence" appears twice in the Ternate essay, once in a paragraph describing how a variety can replace a species ("this new, improved, and populous race might itself, in course of time, give rise to new varieties, exhibiting several diverging modifications of form"), and again in the penultimate paragraph on the modifications of "organized beings," which can be represented as "many lines of divergence from a central type." In both cases divergence *as branching* is described, not an ecological process of diversification through competition.

Darwin's conception of divergence developed between early 1855 and the summer of 1857, by which time it had crystallized into the centrally important "principle" he came to regard it. In that period divergence in the sense of "mere branching" was slowly modified as several more or less independent lines of research he was pursuing through much of the 1850s seemed to come together in an informative way. One such line was biogeographical, notably from about 1854 when Darwin focused on the variability of so-called aberrant genera. This soon morphed into an analysis of the variability of species in larger and smaller genera, which interested Darwin in connection with the idea that through diversification more life can be supported per unit area. Schweber (1980) and Browne (1980) documented the threads that came together to help Darwin forge a concept of an "ecological division of labor" based on Henri Milne-Edwards's "division of physiological labor" (the organization of cells and tissues into a functional whole organism). Darwin read Milne-Edwards in 1852, and it was in 1855 that he appeared to first apply the concept to the variability of species in large versus small genera. This in turn grew into a major undertaking for Darwin, calculating from dozens of botanical manuals data on average numbers of species and varieties according to the size of the taxonomic group.

Meanwhile, Lyell had taken notice of the Sarawak Law paper soon after its publication in the autumn of 1855, and he was moved to initiate what was to become a series of species notebooks of his own (Wilson 1970). The following April, visiting Darwin at Down, Lyell spoke with Darwin about Wallace's paper, and Darwin revealed his details of his theory for the first time to his friend (Lyell had long known that Darwin held "heterodox" ideas, but was not privy to the details). Soon afterward Lyell urged Darwin to publish his ideas, even a brief précis—he evidently sensed that Wallace was closer to Darwin than Darwin himself seemed to believe. Darwin was resistant to the idea of publishing a short paper, feeling that a brief treatment could not do justice to the topic, particularly given the strong sentiments against the idea of transmutation held by most of his fellow naturalists. Instead, he at last turned to a book-length treatment, bowing to Lyell's urging just two weeks later. His diary for 14 May 1856 records that he "began by Lyell's advice writing Species Sketch." Once he got going he threw himself into the project, completing four entire chapters and parts of three others over the ensuing year and two months, all the while working on related parallel projects such as his extensive program of experiments and botanical arithmetic. In the summer of 1857, however, he had a dismaying revelation concerning his calculations. His friend and neighbor John Lubbock, who assisted Darwin with many projects (Somkin 1962), pointed out an error in the way Darwin had been going about calculating his species and variety averages. Evidently Darwin saw this as a real blunder, his vexation revealed in a letter to Hooker the next day (Darwin Correspondence Project, letter 2124): "I have been making some calculations about varieties, etc., and talking yesterday with Lubbock he pointed out to me the grossest blunder which I have made in principle." Continuing that he was at a standstill until he could sort out the mess, he gave himself a solid kick in the pants: "I am the most miserable, bemuddled, stupid dog in all England, and I am ready to cry with vexation at my blindness and presumption," he wrote, signing off "Ever yours | Most Miserably | C. Darwin." To Lubbock he wrote a letter of gratitude the same day:

> You have done me the greatest possible service in helping me to clarify my brains. If I am as muzzy on all subjects as I am on proportion and chance— what a book I shall produce! . . . I am quite shocked to find how easily I am muddled, for I had before thought over the subject much, and concluded my way was fair. It is dreadfully erroneous. What a disgraceful blunder you have saved me from. I heartily thank you. (Darwin Correspondence Project, letter 2123)

Darwin concluded this letter with an impassioned "But oh! if you knew how thankful I am to you!" in the postscript. The distress he felt is obvious from the tenor of these letters, as he realized that many months of calculations from what amounted to a tall stack of botanical manuals had to be redone, beginning with retrieving all the books that had been returned to their owners when he thought he had finished with them.

But Darwin had more than he realized to thank Lubbock for. Browne (1980) suggested that his forced recalculation, which involved first dividing floras into *predetermined* larger and smaller genera (those with four or more species versus those with one to three species, for example), and then using simple proportions of species and their varieties in each category to compare the observed incidence of varieties with an expected number based on proportional relationship (Somkin 1962; Browne 1980). This is an inversion of his previous approach, and Lubbock showed him that it had predictive power: if Darwin's theory was correct, not only would species of larger genera have more varieties in absolute terms (which is to be expected by chance), but *disproportionately* more. As he put it in that letter to Lubbock:

> I have divided N. Zealand Flora as you suggested. There are 339 species in genera of 4 & upwards & 323 in genera of 3 & less. The 339 species have 51 species presenting one or more varieties—The 323 species have only 37: proportionally (as 339:323 :: 51.:48.5) they ought to have had 48 1/2 species

presenting [varieties]—So that the case goes as I want it, but not strong enough, without it be general, for me to have much confidence in. (Darwin Correspondence Project, letter 2123)

In modern terms, the new "rule-of-three" proportional device could effectively enable him to test this hypothesis. Browne (1980) posited that this realization put his related observations pertaining to the classification hierarchy, ecological division of labor, extinction, and geographical distribution in a new light. The proportion data could be interpreted as an historical signature—a result of the past long-continued action of natural selection, which he articulated as a "success breeds success" model of past, current, and future evolutionary change.

A month after his revelation from Lubbock, Darwin used the term "principle of divergence" for the first time in the late August letter to Hooker, and again in his early September letter to Gray (discussed in Chapter 5). He seemed to believe he was very much on to something, as the corrected calculations continued to support his theory. Redoing all of the calculations took many months, and while doing so he put aside the draft of his chapter on natural selection, working on other chapters while steadily working through the recalculations. He completed these by early April 1858, and recorded on 14 April that he began revising both chapters 4 and 6 in light of the new findings. This is the new material relating to divergence, which Brooks (1984) and Davies (2012) claimed amounted to intellectual theft. Whereas in the earlier draft of chapter 4, "Variation under Nature," Darwin included a discussion of divergence, in his reworked chapters he discussed large and small genera in chapter 4 while greatly expanding his discussion of divergence, now as a *principle* of divergence, in chapter 6, a change further reflected in his modifications to the draft table of contents to the book.

By coincidence Darwin had just about completed the reworking of these chapters by mid-June when he was interrupted by the receipt of Wallace's letter and essay. This is precisely the reason that "conspiracy theorists" have zeroed in on divergence—it seems rather suspicious, after all, that Darwin should rework a key element of his book just about the time that Wallace's paper should appear (and appear perhaps earlier than Darwin claimed at that). Alas for the conspiracy idea, coincidence it appears to be, as shown by the abundant documentary record revealing the steps in Darwin's trajectory, and considering that neither the Sarawak Law paper nor Ternate essay

(Chapter 5) bear any resemblance to the process that Darwin called his principle of divergence.

This does not mean that Darwin gained nothing from reading Wallace. It means that Darwin *appropriated* nothing, but it is fair to suppose that Darwin was more impressed with (and perhaps influenced by) the Sarawak Law paper than he was prepared to admit—for all of his affability, it is true that Darwin was often not nearly as diligent in acknowledging the contributions of others as he should have been. Two examples more or less bookending Darwin's career illustrate this long-standing problem: Captain Fitzroy of HMS *Beagle,* upon reading a draft of Darwin's account of the voyage, upbraided Darwin for not expressing adequate thanks or even acknowledgment to Darwin's fellow *Beagle* shipmates, who after all did much to make Darwin's researches on the voyage possible (Darwin Correspondence Project, letter 387), and forty years later Samuel Butler attacked Darwin in his book *Evolution, Old and New* (1879) for neglecting to acknowledge the influence of his predecessors, including Darwin's grandfather Erasmus Darwin (Wallace, incidentally, rallied to Darwin's defense in a review of this book, including an especially clear account of natural selection; see Wallace 1879). Darwin added a "historical sketch" to the *Origin* beginning with the third edition of 1861 to acknowledge earlier ideas on the subject, but while this was ostensibly a comprehensive treatment "of opinion on the origin of species, previously to the publication of the first edition of this work," it had some glaring omissions such as Leopold von Buch (influential to both Wallace and Darwin, as discussed in Chapters 2 and 3). Wallace had even reminded Darwin in late 1860 that von Buch must be considered among the forerunners of evolutionary thinkers, sending him a passage copied out from von Buch's flora of the Canary Islands (Darwin Correspondence Project, letter 2627), but Darwin chose not to include him in the sketch.

Wallace himself was barely mentioned in the *Origin,* in contrast to the substantial account of Wallace in Lyell's own historical sketch of the species question from the tenth edition of the *Principles* (Lyell 1868, 2:276–277). Darwin briefly mentioned Wallace's 1858 essay but not the 1855 Sarawak Law paper, yet surely realized the "evolutionary" significance of the latter, and it may even have helped gel some of his ideas. When Darwin wrote in the margin of his copy of the Sarawak Law paper, "it is all creation, why does his law hold good; he puts the facts in striking point of view," one might reasonably suppose that Wallace's remarkable paper got Darwin thinking more about the branching of species and "laws" like those of Milne-Edwards, which he called the "law of diversity" and the "law of economy" (Schweber 1980). So could Darwin have been more forthcoming or generous in acknowledging Wallace? Yes, especially if he was conscious of this effect of Wallace's paper. But this does not amount to theft. Influence and inspiration are the essence of research—if one is inspired or intrigued by an idea and that idea serves as a seed or an impetus for one's own creative pursuits, that is not theft in the least. I believe that Darwin should have been far more generous in acknowledging Wallace, both in spotlighting his younger colleague's remarkable insights in the Sarawak Law paper and in acknowledging his independent discovery of natural selection (Darwin mentioned neither the title nor the date of Wallace's paper in the *Origin*). These lapses duly noted, acknowledging sources of inspiration is a decent thing to do, but not doing so is not tantamount to intellectual theft, contrary to the claims of Davies (2008, 2013).

Was Lyell and Hooker's Action Ethical?

"He does not say he wishes me to publish, but I shall, of course, at once write and offer to send to any journal," read Darwin's letter to Lyell on 18 June 1858, "So all my originality, whatever it may amount to, will be smashed." This was followed by another letter to Lyell dated 25 June:

> I [should] be *extremely* glad **now** to publish a sketch of my general views in about a dozen pages or so. But I cannot persuade myself that I can do so honourably. Wallace says nothing about publication, & I enclose his letter.— But as I had not intended to publish any sketch, can I do so honourably because Wallace has sent me an outline of his doctrine?—I would far rather burn my whole book than that he or any man [should] think that I had behaved in a paltry spirit. Do you not think his having sent me this sketch ties my hands? I do not in least believe that he originated his views from anything which I wrote to him. (Darwin Correspondence Project, letter 2294; emphasis in original)

And yet another on the next day, 26 June:

> Forgive me for adding P.S. to make the case as strong as possible against myself. Wallace might say "you did not intend publishing an abstract of your views till you received my communication, is it fair to take advantage of my

having freely, though unasked, communicated to you my ideas, & thus prevent me forestalling you?" The advantage which I should take being that I am induced to publish from privately knowing that Wallace is in the field. It seems hard on me that I should be thus compelled to lose my priority of many years standing, but I cannot feel at all sure that this alters the justice of the case. First impressions are generally right & I at first thought it [would] be dishonourable in me now to publish. (Darwin Correspondence Project, letter 2295)

These letters elicited a swift response from Lyell and Hooker, whom Lyell deputized in the matter. Beddall (1988a) made reference to Lyell's "quick and impatient" nature, and the pressure was on and quick action was perceived necessary, perhaps because for all any of them knew Wallace had already sent off another version of his essay for publication. Lyell rallied Hooker, whom he knew was also in Darwin's confidence concerning their friend's heterodox ideas about species, and the two of them precipitately decided that selected writings by Darwin should be presented with Wallace's paper as soon as possible. At their request Darwin quickly supplied an extract of his 1844 essay, which Hooker had read and commented upon some years before, and a draft of his long September 1857 letter to Asa Gray (Darwin Correspondence Project, letter 2136), the third confidante, in which his case for the transmutation process was laid out in six numbered sections, explaining the domestication analogy, Malthusian struggle for existence, variation, natural selection, and his principle of divergence (see Chapters 4 and 5).

As it happened, a special meeting of the Linnean Society was scheduled for 1 July 1858, a rescheduling of the regular meeting from earlier in June, which was canceled owing to the recent death of former Linnean president Robert Brown. The rescheduled meeting was merely days away. Darwin's forwarded material was copied out (by Mrs. Hooker) and assembled into a format to be orally delivered the day before the meeting—in fact, it was only forwarded to the secretary of the Linnean on the evening of 30 June. Darwin's infant son in a raging fever was meanwhile declining rapidly, and it was all too much—he was glad to simply let his friends take full control and do what they thought best (knowing, to be sure, that they had his best interests in mind). Lyell and Hooker had the weight to bump another paper (by botanist George Bentham) so that the Darwin and Wallace papers could be read. Whether the order in which their papers were presented was according to the date they were written, as Lyell and Hooker maintained in their prefatory remarks, or alphabetical, as was the custom of the Linnean at the time, Darwin came out on top. Although Darwin referred more than once after the event to the fairness and impartiality of Lyell and Hooker, and was thankful that "all his originality" was not "smashed" after all, he likely had some qualms about the way his conduct would be viewed. To his relief, when Wallace at last got word of the events that transpired, he replied with the best possible response: elation, gratitude, and humility. By all accounts—including private correspondence with family and friends, where gripes otherwise carefully hidden from view would be expected to be aired—Wallace always maintained Darwin's priority and indeed lavished praise on the senior naturalist. This might be dismissed as mere Victorian class-based politeness (discussed further below); it has been suggested that Wallace may have claimed to be pleased with his consolation prize of second fiddle to the illustrious Mr. Darwin, but he had little choice but to make the best of things.

Historians who have pored over Wallace and Darwin's writings come to a different conclusion, one that I concur with. The consensus is that Wallace was genuinely happy to bow to Darwin in this case. To be instantly launched into the limelight of the world of elite British science was far more than Wallace ever expected—his paper might get Lyell's attention and a hearing, but the disappointing reception of this Sarawak Law paper was likely never far from mind. To all at once have his name alongside the likes of Darwin, Lyell, and Hooker was a coup: "This *insures me the acquaintance and assistance* of these eminent men on my return home," he wrote to his mother (WCP369, emphasis mine). If Wallace had any doubts or questions about the extent of Darwin's priority, they must have been dispelled when Darwin sent him the table of contents of his book-in-progress, *Natural Selection,* which Wallace copied into his Species Notebook (Costa 2013a, 430–431). Certainly Wallace was impressed with the *Origin,* Darwin's "abstract" of nearly 500 pages published just a year and four months after that fateful Linnean Society meeting. Wallace received his copy with Darwin's compliments in early 1860, still far from home in the Malay Archipelago. Writing to his friend George Silk of the *Origin* that September, he said, "I have read it through 5 or 6 times & each time with increasing admiration. It is the "Principia" of Natural History. It will live as long as the "Principia" of Newton. . . . Mr. Darwin has given the world a new science, and his name should, in my opinion, stand above that of every philosopher of ancient or modern times. The force of admiration can no further go!!!" (WCP373; emphases Wallace's). Similarly, he wrote to Bates soon after: "I know not how, or to whom, to express fully my admiration of <u>Darwin's book</u>. To him it would seem flattery to others self praise;—but I do honestly believe that with however much patience I had worked and experimented on the subject I could never have <u>approached</u> the completeness of his book,—its vast accumulation of evidence,—its overwhelming argument, & its admirable tone and spirit. I really feel thankful that it has not been left to me to give the theory to the world" (WCP374, emphases Wallace's).

That Wallace apparently never felt wronged is well and good, but of course it does not mean that he was *not* wronged. My conclusion is that while Darwin had priority in piecing together both a comprehensive case for the reality of transmutation and the mechanism for it, his friends did take advantage of their privileged position to see to it that his name was front and center. He had no intention to publish yet. From the perspective of our culture and theirs, Darwin, Lyell, and Hooker should have written Wallace to explain the circumstance of Darwin's earlier labors and offer to publish his essay with a work produced by Darwin for the occasion. We know today Wallace's generosity of spirit, and he likely would have wholeheartedly concurred; but they did not know his magnanimity, and decided to rush things through, bringing to mind the modern tongue-in-cheek motto "better to ask forgiveness than permission." As Rachels (1986, 22) put it, this is a "lamentable story of human weakness, in which some good men treated another good man disgracefully."

Beddall (1988a, 57) concluded that "neither Darwin nor Wallace was well served by Lyell's haste. It does not redound to Darwin's credit that his priority claim appeared to require special help, nor were Darwin's qualms properly addressed. Indeed, his express wishes were subverted, while Wallace's rights were neglected. [Lyell's] precipitous action in thus presenting unpublished works, bypassing the judgment of referees who might reject them as 'not strictly scientific,' has left some questions open, and a lingering doubt about the fairness remains." Brackman (1980) referred to the event as Lyell and Hooker's "delicate arrangement." Darwin was disingenuous in later claiming (in his autobiography) that he "cared very little whether men attributed most originality to [him] or Wallace" (Barlow 1958, 124); at the time, judging from the tone of his letters, he was dismayed that his priority would be lost even while feeling ashamed at what seemed de facto pleas for help to his friends. Add to this the tragedy of his dangerously ill infant son (who soon died), and I think we can appreciate that Darwin was truly caught up in an emotional maelstrom. To his credit, however, he certainly did not commit the ultimate act of dishonesty: he did not suppress Wallace's paper when it would have been quite easy to have done so.

Or would it have been so easy? Such an action would not only have been blatantly dishonest, it would have been risky. For all that Darwin knew, the essay he received was also sent by Wallace to other naturalists, or to a journal. Then too, if Darwin had destroyed the manuscript and then rapidly published his own paper on the subject, Wallace might have complained in a public forum, announcing in print that he had sent the same idea to Darwin, who then suddenly published the idea. Darwin had only occasional prior correspondence with Wallace, so he had no knowledge of his personality to gauge the likelihood of a reputation-damaging public response. In this scenario the best solution (i.e., the one that Darwin lost the least from) was probably the scenario that actually occurred. On the other hand, in my opinion this scenario paints too calculating and conniving a picture of Darwin. By all accounts he was quite a decent person, by the standards of his time and our own; it is more parsimonious and consistent with what we know of his personality and behavior to conclude that Darwin did the right thing because it was the right thing to do, whether or not he fretted for a day or some weeks.

What Might Have Been

Had Wallace sent his essay directly to a venue like the *Annals and Magazine of Natural History,* or the *Zoologist* (both of which had published several of his earlier papers), chances are good that it would have appeared in print in short order, eventually winning him laurels for being the first to publicly describe the mechanism of evolutionary change. What would he have called that mechanism? Although he later urged Darwin to adopt Spencer's phrase "survival of the fittest" in place of "natural selection," that will not do as a concise label for the process. Perhaps he would have dubbed the mechanism the Law of Change, the Law of Transmutation (or Transformism), or simply his Transmutation Principle. Or it is possible that the mechanism would ultimately have become known as natural selection after all? Wallace would have been the first to articulate what we now consider to be the primary mechanism of evolutionary change, but it is possible that the publication of his Ternate essay would not have prevented the appearance of a book by Darwin on the subject. Wallace was still in the Malay Archipelago with no plans to return soon. He might have been inspired to follow up his essay with another putting the arguments of the Sarawak Law paper and the Ternate essay together into a more comprehensive framework, but it is likely that a book-length exposition like *On the Organic Law of Change* (Costa 2013a) would have had to wait until his return to England, as he commented to Darwin in a letter dating to September 1857: "The mere statement & illustration of the theory in [the Sarawak Law] paper is of course but preliminary to an attempt at a detailed proof of it, the plan of which I have arranged, & in part written, but which of course requires much [research in British] libraries & collections" (WCP4080).

In the event, we know that Wallace shelved his plans for a book in the wake of the revelations of 1858. Perhaps this is not surprising; Wallace almost seemed to suggest he would defer to Darwin in a letter written to Bates in January 1858:

> I have been much gratified by a letter from Darwin, in wh[ich] he says that he agrees with "almost every word" of my [1855] paper. He is now preparing for publication his great work on species varieties [*sic*], for wh[ich] he has been collecting information 20 years. *He may save me the trouble of writing the 2nd part of my Hypothesis, by proving that there is no difference in nature between the origin of species & [varieties]* or he may give me trouble by arriving at another conclusion, but at all events his facts will be given for me to work upon. Your collections & my own will furnish most valuable materials to illustrate & prove the universal applicability of the hypothesis. (WCP366, emphasis added)

The paper Wallace refers to is his 1855 Sarawak Law paper. What is the "2nd part" of his hypothesis? January 1858 predates his discovery of natural selection, but he gives a clue that he means proof that there is essentially no difference between species and varieties. That is fundamentally a pro-transmutation argument, where species are seen as distinct and well-delineated varieties (as would be expected if new species stemmed from varieties). While that is the gist of the Ternate essay, which was written the following month, it is also at the heart of several pro-transmutation arguments given in the Species Notebook. Wallace may, then, have meant that Darwin may save him the trouble of writing his planned pro-transmutation book—in that sense, precisely what happened.

The radical new way to view species that became known as the Darwinian Revolution grew out of such a book, but perhaps we would be discussing "Wallaceism" or the "Wallacean Revolution" today had Wallace published his Ternate essay alone without the "delicate arrangement" of Lyell and Hooker, even if he decided not to produce his book in deference to Darwin. After all, despite the claims of some historians (e.g., England 1997) that the publication of the Darwin and Wallace papers in 1858 was a nonevent, the papers received considerable attention (thanks to George Beccaloni for pointing this out to me): they were soon reprinted in *The Zoologist* (1858; 16: 6263–6308), which also published two reviews (Boyd 1859 and Hussey 1859), and were noted in the *The Gardeners' Chronicle and Agricultural Gazette* (No. 40, 2 October 1858). The papers were also commented upon and criticized by Samuel Haughton, president of the Geological Society of Dublin, in his annual address given on 9 February 1859 (reported in the *Journal of the Geological Society of Dublin*, 8:151–152), and Richard Owen in his presidential address to the British Association at its 1859 meeting in Leeds (printed in the *Reports of the British Association for the Advancement of Science* [1858], xci–xciii). The papers created a stir, but perhaps their revolutionary content would have been too easily dismissed—a book-length treatment interweaving all available evidence for natural selection and transmutation, and just as importantly confronting the obvious problems such as the spotty fossil record, would have been necessary to truly plant the evolutionary flag.

A scooped (and prostrated) Darwin would probably have picked himself up within a month or two and forged on ahead with his "big species book," or some abstracted version like that we know as *On the Origin of Species*, considering that he had but a few chapters to go and had already put so much effort into gathering supporting evidence for evolution generally and a mechanism for change. A comment in one of his June 1858 letters to Lyell suggests he had not given up the idea of publishing his book even while caught up in the emotional maelstrom accompanying the arrival of Wallace's manuscript: following the oft-quoted passage "so all my originality, whatever it may amount to, will be smashed," Darwin's next sentence reads, "Though my Book, if it will ever have any value, will not be deteriorated; as all the labour consists in the application of the theory" (Darwin Correspondence Project, letter 2285). This indicates that, even if he was scooped and priority went to Wallace, Darwin still thought in terms of going ahead with his book, already quite far along in the writing. Moreover, his comment that "all the labour consists in the application of the theory" speaks to the point just made, that a book-length treatment marshaling the many lines of evidence for transmutation and natural selection would have been necessary to gain widespread acceptance of the doctrine.

Wallace's discovery acknowledged, Darwin might have gone on to declare that he too had developed such an idea and took it further than Wallace in a host of ways. He would soon have perceived the differences between his view of evolution and Wallace's—the domestication analogy, sexual selection, family-level selection—which I imagine he would have highlighted. Would Wallace have followed through on his plans for his own book once he learned of Darwin's book in the works? We cannot know, of course, but the readiness with which Wallace took his hat off to Darwin in recognition of the senior naturalist's work on the subject, combined with the difficulty of producing a book-length treatment while still in the field with no plans to return home soon, suggests that he may well have suspended his book plans. Wallace never deviated from the view that he expressed to Bates soon after reading the *Origin,* of being thankful that it had not been left to him to give the theory to the world. Fifty years later he put it this way in his acceptance speech on the occasion of being awarded the first Darwin-Wallace Medal by the Linnean Society: "*I* was then (as often since) the 'young man in a hurry'; *he,* the painstaking and patient student, seeking ever the full demonstration of the truth that he had discovered, rather than to achieve immediate personal fame" (Wallace 1908–1909, 7; see also Gardiner et al. 2008).

A scrupulously fair-minded person, Wallace seemed to place a great deal of weight on priority in the absolute sense—not the first to publish, but the first discoverer and developer of an idea. This is likely behind his perennial deference to Darwin, as seen in the acceptance speech at the Linnean Society:

> But, what is often forgotten by the press and the public, is, that the idea occurred to Darwin in October 1838, nearly twenty years earlier than to myself (in February 1858); and that during the whole of that twenty years he had been laboriously collecting evidence from the vast mass of literature of Biology, of Horticulture, and of Agriculture; as well as himself carrying out ingenious experiments and original observations, the extent of which is indicated by the range of subjects discussed in his 'Origin of Species,' and especially in that wonderful store-house of knowledge—his 'Animals and Plants under Domestication,' almost the whole materials for which works had been collected, and to a large extent systematised, during that twenty years. So far back as 1844, at a time when I had hardly thought of any serious study of nature, Darwin had written an outline of his views. (Wallace 1908–1909, 6)

We place great weight on the first to publish, and we are conditioned to think in terms of scientific rivalries—the Leibniz–Newton calculus controversy, Le Verrier and Adams and the discovery of Neptune, or Watson and Crick's race with Pauling to describe the structure of DNA come to mind—so Wallace's rather different conception of priority seems odd to us, and unfair. Did things like class and status factor into Wallace's view of the matter? Difficult to say, though it should be noted that Wallace never shied away from engaging anyone, regardless of social status, when he believed he was in the right. Or, to take a different view of the role of class and status, could Wallace's deference be seen as calculated? Recall again the comment he made to his mother, writing her shortly after hearing from Darwin and Hooker about the joint readings at the Linnean Society: "This *insures me the acquaintance and assistance of* these eminent men on my return home" (WCP369, emphasis added). Perhaps, as an outsider, being deferential, cooperative, even generous was a surefire way to instant acceptance by the scientific elite of Britain. And generous Wallace was, as Darwin repeatedly noted: "Permit me to say how heartily I admire the spirit in which [Wallace's letters to Hooker and Lyell] are written," he wrote to Wallace in January 1858, and "You cannot tell how I admire your spirit, in the manner in which you have taken all that was done about publishing our papers" in the postscript to an April 1859 letter (Darwin Correspondence Project, letters 2405 and 2449). Darwin was relieved, but Wallace, though indeed now on a road to fame smoothed by the esteem of the leading lights of British science, thus commenced his perennial role as Watson to Darwin's Holmes.

My speculations are not offered as any kind of justification for Lyell and Hooker's "delicate arrangement" in 1858 or to take away from Wallace, the brilliant "young man in hurry"; —on the contrary, I believe that the right thing for Lyell and Hooker to have done would have been to write to Wallace and suggest a joint presentation of papers. To us, knowing what we do today about Wallace's personality and his views toward the whole priority question, his response to such a letter from Lyell and Hooker (and the question of whether he would have eventually proceeded with his planned book) seems fairly predictable. Who knows, we may well still be speaking of a "Darwinian revolution" today even had Wallace's paper appeared first. Regardless, with our own standards of scientific priority and fairness in mind, the scientific and broader intellectual community needs to do better by Wallace. One hundred and fifty-plus years of Darwin's name alone as synonymous with evolution by natural selection, in all of its manifestations

(Darwinian revolution, social Darwinism, neo-Darwinism, Darwinian medicine, etc.) cannot be undone. But scientists, journalists, educators, and others could and should be diligent in henceforth referring to the Darwin-Wallace or even Wallace-Darwin theory of evolution by natural selection. The two labored along remarkably similar lines for years, and each had his epiphany along the way. These "indefatigable naturalists" were indisputably *together* our first guides to the evolution revolution.

Alfred Russel Wallace at age thirty-nine. This photograph was taken in Singapore, February 1862, shortly before Wallace's departure from the Malay Archipelago. Copyright A. R. Wallace Memorial Fund and G. W. Beccaloni, provided courtesy of G. W. Beccaloni.

Coda

The Force of Admiration

WALLACE'S PRAISE of Darwin's achievement on his reading of *On the Origin of Species* was sincere: in letters to family and friends he applauded the elder naturalist's efforts—Darwin's labors of twenty years during which he fleshed out the theory, following out the nuances and intricacies of evidence for transmutation and the action of natural selection. Wallace concluded one lengthy encomium for Darwin declaring: "the force of admiration can no further go!!!" But while Wallace lauded Darwin and deferred to his priority, he never shied away from disagreeing with Darwin over aspects of the evolutionary process. For several matters of evolutionary interpretation Wallace was Darwin's "man to apply to in a difficulty" (Darwin Correspondence, letter 5420). On other matters they sometimes argued to a draw (as over Darwin's invocation of sexual selection to explain sexual dimorphism in coloration) and sometimes disagreed profoundly (as over human evolution), but their relationship was always warm. In the wake of the events of 1858–1859 Wallace may have abandoned the book arguing for transmutation revealed in the Species Notebook, yet the notebook and other early writings reveal a deeply insightful Wallace who had clearly achieved much the same vision as Darwin, a grand vision of the "tree of life" linking all species in time and space, a tree burgeoning through natural selection. But more than this, he conceived a model of that tree of life that in important respects was actually closer to the modern view than that of Darwin, from his firm rejection of Lamarckian inheritance to his insistence on the necessity of physical barriers, allopatry, in speciation.

It was with good reason that Wallace applauded Darwin, but Wallace too deserves our "force of admiration," and better recognition as Darwin's equal and sometime-foil, whose depth and breadth of insights into transmutation and other achievements were deeply admired by his contemporaries, including such leading lights as Lyell, Huxley, and Darwin, with all of whom Wallace developed lasting friendships. Darwin even authored a successful petition to the government to grant the financially struggling Wallace a pension in his later years. The 1855 and 1858 papers garnered a mention among the many scientific accomplishments cited by Darwin in the petition:

> During his stay in [the Malay Archipelago] he sent home many scientific papers for publication, two of which were highly remarkable, viz that "On the law which has regulated the introduction of new Species," & that "On the tendency of varieties to depart indefinitely from the original type." This latter paper includes the view, which is now commonly called Natural Selection. . . . Everyone will, I believe, admit that Mr Wallace's works have added to our knowledge of an important & difficult subject namely Geographical Distribution. His essays on the colouring of animals show the extraordinary originality of his mind, & have been the parent of numerous essays by other naturalists. Many men will think that his memoir on "on the tendency of varieties to depart indefinitely from the original type &c" is of greater value even than that of his other works. (Colp 1992)

Wallace in Orbit

The titles of some biographies of Wallace invoke his outsider status, and do so explicitly in reference to Darwin: *Darwin's Moon* (Williams-Ellis 1966), *In Darwin's Shadow* (Shermer 2002), and *The Heretic in Darwin's Court* (Slotten 2004) come to mind. Yet this is not a new phenomenon; Wallace has had a supporting role in the drama of the discovery of (and advocacy for) evolution by natural selection ever since the 1858 Linnean Society readings. Increasingly he came to be seen not only as a supporting actor but as a character actor: clever, yes, but a bit eccentric and at times a loose cannon. Through his generosity and high-minded personal sense of fairness, Wallace was in no small part responsible for casting himself in this role, but it is equally fair to say that neither did Darwin or his circle try to push Wallace into the limelight, to present him as a full partner in the new and exciting evolutionary program. This is perhaps understandable when, by the late 1860s, Wallace made clear his spiritualist ideas and reservations about the applicability of natural selection to human consciousness, but the sidelining of Wallace began long before that. Consider Wallace's treatment in *On the Origin of Species.*

Darwin made passing reference of Wallace's "excellent memoir" of 1858 in his introduction to the *Origin* (1859, 2), but this was the only mention of the Ternate essay in the book. Wallace is cited in two other places in the *Origin:* a reference to the Sarawak Law paper is found on page 355, and on page 395 there is a brief mention of Wallace's "admirable zeal and researches" in connection with the anomalous distribution of species in the Malay Archipelago (Appendix 4). What became known as "Wallace's Line" (a term coined by Huxley) was to be hailed as one of Wallace's greatest discoveries, but on this subject in the *Origin* he is again cast in more of a supporting role, simply confirming the work of one Windsor Earl, to whom more space is devoted. Perhaps worse for Wallace than the inadequate treatment of his independent formulation of natural selection, Darwin used exclusive and possessive language in describing the theory throughout the book. The phrase "my theory" occurs no fewer than fifty-seven times in the *Origin* (Appendix 5), while in principle at least he could have used the more neutral "the theory" if not the more magnanimous "our theory" (the latter would have made sense only with a full discussion of Wallace's contribution, of course). To be fair, occasionally Darwin did refer to the theory as Wallace's too, but in private letters (e.g., his 18 May 1860 letter to Wallace commenting on Patrick Matthew's claim to have discovered "our view of natural selection"; Darwin Correspondence Project, letter 2807), and Wallace was at times guilty of assigning exclusive ownership to Darwin too. In his paper on "The Ornithology of Northern Celebes," for example, he made reference to "Mr. Darwin's principle of 'natural selection'" to help explain the curious nesting behavior of megapodiid birds (Wallace 1860b, 146). He assigned ownership of the theory even more explicitly to Darwin in an 1864 letter:

As to the theory of "<u>Natural Selection</u>" itself, I shall always maintain it to be actually yours & yours only. You had worked it out in details I had never thought of, years before I had a ray of light on the subject, & my paper would never have convinced anybody or been noticed as more than an ingenious speculation, whereas [your] book has revolutionized the study of Natural History, & carried away captive the best men of the present Age. All the merit I claim is the having been the means of inducing <u>you</u> to write & publish at once. (WCP1859, emphases Wallace's)

To look at this from another angle, recall Darwin's letter to Lyell of 18 June 1858, in which he wrote, "My Book, if it will ever have any value, will not be deteriorated; as all the labour consists in the application of the theory" (Darwin Correspondence Project, letter 2285). I believe Darwin's words to be prophetic, in a way he was not thinking of at the time. Darwin meant that his "species book," if he came out with it, would show his extensive "application of the theory" to diverse fields, from domestication and hybridism, instinct and morphology, paleontology and geographical distribution. In the *Origin* he commented that although covering many seemingly disparate fields, the book was in reality "one long argument." Stepping back and taking a longer view, all of Darwin's post-*Origin* works can be taken as extensions of that argument: they constitute "one longer argument," as I put it elsewhere (Costa 2009b), in representing focused studies on applications of the theory.

Darwin's first post-*Origin* book, on the pollination of orchids, is an argument for gradual coevolution of flower and pollinator and, more importantly, a case study in how suites of structures can be variously modified by selection for the same or similar ends in different groups. He concluded the book by pointing out that this is precisely what is predicted by descent with modification by natural selection. Asa Gray saw what Darwin was up to, prompting Darwin to write, "No one else has perceived that my chief interest in my orchid book, has been that it was a 'flank movement' on the enemy . . . it bears on design, that endless question" (Darwin Correspondence Project, letter 3662). Darwin became increasingly interested in botany, and his four subsequent botanical books (on insectivorous plants in 1875, crossing and selfing in 1876, forms of flowers in dioecious plants in 1877, and movement and sense perception in climbing plants in 1880) can all be seen as applications and implications of the theory of common descent by natural selection. Carnivorous and climbing plants fascinated him for their animallike qualities of movement and apparent sense perception, underscoring a physiological link between plants and animals consistent with a common origin. His analysis of the effects of crossing and selfing relates to an interest in the evolution of sexes: insofar as outcrossing is important for health and vigor, he posited that sexes evolve as selection acts to promote outcrossing. In the zoological realm, the 1868 *Variation of Plants and Animals under Domestication* represents an expansion of the first chapter of the *Origin* into a two-volume treatise making the case for the power of artificial selection. (This work also unveiled his ultimately unsuccessful theory of heredity, "pangenesis," with its strikingly Lamarckian elements.) A few years later he came out with *Descent of Man* (1871) and *On the Expression of the Emotions in Man and Animals* (1872), which together constitute his statements on human evolution and the affinity between humans and other animals (the former also elaborating on his theory of sexual selection). Even Darwin's final work, *The Formation of Vegetable Mould, Through the Action of Worms* (1881), resonates with important elements of his evolutionary vision. His case for the ability of humble earthworms to gradually shape the landscape through slow but incessant activity is a uniformitarian argument for the long-term effects of small, ordinary processes.

We have seen that Wallace, too, came out with highly acclaimed scientific works in the post-*Origin* decades, from *Contributions to the Theory of Natural Selection* and *Geographical Distribution of Animals* to *Island Life, Tropical Nature,* and *Darwinism,* on top of his immensely successful travel memoir *The Malay Archipelago.* But Wallace's scientific interests were broad, and his social interests were broader still; as Charles Smith nicely expressed in one biography:

> His insistence on argument from fact led him to scientific revelations regarding, among other things: the nature of biogeographic regions and dynamics; the nature of human racial differentiation; protective coloration, mimetic resemblance and polymorphism in animals and plants; the process of speciation; glacial motion and the causes of the Ice Age; the measurement of the age of the earth; the permanence of the continental masses; the mouthgesture theory of the origin of language; and the constraints on the existence of life on other planets in the solar system. In the social arena, he coupled a heartfelt concern for the basic rights of all individuals with a vision of the necessity for cooperative social organization to fashion a variety of startling suggestions on land reform, international trade, practical ethics, legislative reform, the future role of women in society, urban and rural planning, museum design, the use of statistics in epidemiology, the use of a paper money standard, labour issues and the standardization of consumer-oriented product information. (Smith 2002, 1158–1159)

Spiritualism, land nationalization, antivaccination, socialism—Wallace's interests in matters of philosophy and social justice were laudable, of course, but his pursuit of such wide-ranging causes also made it difficult to associate him with any one thing. If as Darwin said "all the labour consists in the application of the theory," Darwin's labors were perhaps perceived in this way in contrast to Wallace: single-minded in pursuing applications of the theory, while Wallace was less so.

As if these were not enough to dim Wallace's star, simple "star power" on Darwin's part also contributed to the Darwinian juggernaut. The Darwin name was famous, beginning with Charles's grandfather Erasmus, physician of renown and best-selling poet. Samuel Taylor Coleridge coined the term "Darwinizing" in reference to the wild evolutionary speculations of that "most inventive of philosophical men" (King-Hele 1963, 4, 86). It is but a short step from "Darwinizing" to "Darwinism," and in the early nineteenth century that term, too, was used in reference to Erasmus; for example, a review of *The Poetical Works of Percy Bysshe Shelley* (1839), edited by Mary Shelley and published by Edward Moxton, derisively referred to the "popular vortex of Darwinism" (*British and Foreign Review* [1840] 10: 105) (derisively because by then Erasmus's ornate poetic style was out of favor). In the post-*Origin* years, however, the meaning of "Darwinism" and "Darwinian" shifted in reference from Erasmus to Charles, by then fairly famous in his own right in the wake of the success of his scientific books and especially his *Journal of Researches* (later much reprinted as *Voyage of the Beagle*). Name recognition carries much weight, as any political hopeful or marketing department knows. A steady stream of works were published for or against the evolutionary thesis after 1859, a great many of them bearing "Darwinism" in the title—even that one by Wallace. Wallace's *Darwinism,* published in 1889, was neither the first nor last: at least twenty such books were published in London alone in the thirty years between 1869 and 1900 (Table 1).

Name recognition, combined with the frequent presentation of the theory as Darwin's alone (including by Darwin himself), must go a long way toward explaining the rapidity with which "Darwinism" and the "Darwinian theory" became synonymous with the theory of common descent by natural selection in the minds of most in the scientific community, as well as society at large. Wallace himself was swept by this Darwinian current, even beyond entitling his well-regarded 1889 book *Darwinism:* this book and another four of his major scientific works of the post-*Origin* years, plus his autobiography, are replete with citations of Darwin (Table 2).

Is it any wonder that in short order the idea of descent with modification by natural selection came to be associated almost exclusively with

Coda Table 1. A sample of twenty books with "Darwinism" in the title, published in London between 1869 and 1900.

Author	Title	Published
Morris, Francis Orpen	*Difficulties of Darwinism*	1869
Schleicher, August	*Darwinism Tested by the Science of Language*	1869
Laing, Sidney Herbert	*Darwinism Refuted*	1871
Stebbing, Thomas Roscoe Rede	*Essays on Darwinism*	1871
Saint Clair, George	*Darwinism and Design; Or, Creation by Evolution*	1873
Hodge, Charles	*What is Darwinism?*	1874
Schmidt, Eduard Oskar	*The Doctrine of Descent and Darwinism*	1875
MacLaren, James	*A Critical Examination of Some of the Principal Arguments For and Against Darwinism*	1876
Bateman, Sir Frederic	*Darwinism Tested by Language*	1877
Fiske, John	*Darwinism and Other Essays*	1879
Lankester, Sir Edwin Ray	*Degeneration: A Chapter in Darwinism*	1880
Walduck, Henry	*Darwinism Refuted Out of Darwin's Book*	1885
Schurman, Jacob Gould	*The Ethical Import of Darwinism*	1888
Ritchie, David George	*Darwinism and Politics*	1889
Wallace, Alfred Russel	*Darwinism: An Exposition of the Theory of Natural Selection with Some of Its Applications*	1889
Morris, Francis Orpen	*The Demands of Darwinism on Credulity*	1890
Pocock, William Willmer	*Darwinism a Fallacy*	1891
Haycraft, John Berry	*Darwinism and Race Progress*	1895
Alexander, P. Y.	*Darwin & Darwinism Pure and Mixed: A Criticism with Some Suggestions*	1899
Alexander, P. Y.	*More Loose Links in the Darwinian Armour*	1900

Coda Table 2. Occurrences of the words "Darwin," "Darwinian," and "Darwinism" in five post-*Origin* scientific books by Wallace and in his autobiography.

	Darwin	Darwinian	Darwinism
Contrib. Theory of Nat. Selection (1870)	45	0	0
Geographical Dist. of Animals, v. 1 (1876)	22	0	0
Geographical Dist. of Animals, v. 2 (1876)	12	0	0
Tropical Nature (1878)	62	0	0
Island Life (1880)	22	0	0
Darwinism° (1889)	174	23	8
My Life, v. 1 (1905)	51	0	2 (refers to 1889 book)
My Life, v. 2 (1905)	138	10	23 (mostly 1889 book references & lecture titles)

° "The Darwinian Theory" is referred to many times, and the phrase heads a section of chapter 1.

Darwin, especially outside the immediate circle who knew and admired Wallace?

Umbral Eclipse

In the sample of titles of late nineteenth-century books on Darwinism in Table 1, note that there are as many or more titles hostile to the Darwinian view as favorable to it. Titles with terms like "difficulties," "refuted," "fallacy," "tested," "loose links," and "demands on credulity" sum up the beleaguered state of the science in the latter part of the century. Many of these attacks were made on religious grounds, but others were scientific arguments. Most naturalists had come to accept the idea of descent with modification, but the opinion of naturalists on the role of natural selection in the origin of new species went from lukewarm to outright rejection in the decades after the *Origin* (see Provine 1971, chapter 1, and extensive review in Bowler 1983). A thorough discussion of the fascinating period in which Darwinism went into eclipse, as Peter Bowler expressed it in the title of his 1983 book, is beyond the scope of this discussion, but aspects of this period are worthy of notice here insofar as they played a role in the eventual deeper eclipse of Wallace. Here the eclipse analogy might more aptly be lunar than solar: the shallow and temporary penumbral eclipse of Darwin as compared to the deeper and longer umbral eclipse of Wallace.

In successive editions of the *Origin* Darwin seemed to get more and more Lamarckian, with use and disuse of organs and environmental induction of variations invoked to speed up both variability and the rate of evolutionary change (Costa 2009a, 491–495). Wallace emerged as the staunch defender of strict selectionism. Some observers at the time considered Wallace's unwavering rejection of Lamarckism in favor of the primacy of selection acting on always-abundant variation to be more correct than Darwin. English novelist, theistic evolutionist, and severe Darwin critic Samuel Butler (1835–1902) was one; he considered "Darwinism" shot through with Lamarckism, and he coined the term "neo-Darwinism" in 1880 to represent the strict selectionism of Wallace in contrast with the misled "Darwinians." To Butler, Wallace was the first neo-Darwinian; quoting Wallace's rejection of Lamarck in the Ternate essay, he noted:

> This is absolutely the neo-Darwinian doctrine, and a denial of the mainly fortuitous character of the variations in animal and vegetable forms cuts at its root. That Mr. Wallace, after years of reflection, still adhered to this view, is proved by his heading a reprint of the paragraph just quoted [from the Ternate essay] with the words "Lamarck's hypothesis very different from that now advanced"; nor do any of his more recent works show that he has modified his opinion. It should be noted that Mr. Wallace does not call his work "Contributions to the Theory of Evolution," but to that of "Natural Selection." (Butler 1880, 282–283)

After Wallace's book *Darwinism* came out Butler published a paper entitled "The Deadlock in Darwinism" in which he proclaimed Wallace "the most authoritative exponent of latter-day evolution . . . whose work, entitled 'Darwinism,' though it should have been entitled 'Wallaceism,' is still so far Darwinistic that it develops the teaching of Mr. Darwin in the direction given to it by Mr. Darwin himself . . . and not in that of Lamarck" (Butler 1908, 236). Further on he commented that Wallace had "persevered along the path of Wallaceism just as Mr. Darwin with greater sagacity was ever on the retreat from Darwinism" (p. 257). Canadian-born English physiologist George Romanes (1848–1894), close associate of Darwin, agreed; he also used the term "Wallaceism" to describe Wallace's selection-driven view of evolution versus the Lamarckism invoked by Darwinians; however, Romanes used the term critically, because he believed Wallace to be incorrect. Romanes played a central role in the ensuing debates over selection and the origin of species. The issue was how infertility between incipient species could arise by selection alone, since any crossing before mutual sterility is reached swamps differences and homogenizes the varieties. This fundamental problem at the heart of Darwin's "principle of divergence" was first articulated in 1867 by a Scottish engineer, Fleeming Jenkin. Darwin recognized the seriousness of the criticism raised by Jenkin, and his solution was Lamarckian in part: perhaps similar variations are induced in nascent varieties in response to similar environmental conditions. Some degree of isolation would help; by being restricted to a limited area groups of individuals would vary in the same way owing to their exposure to a common environment, and would interbreed only with each other, preventing the loss of any new favorable trait arising in the group through the swamping effects of blending.

Romanes believed that this served to undermine a role for natural selection in the origin of species; in that case Darwin's theory, he said, "is not, strictly speaking, a theory of the origin of *species:* it is a theory of the origin—or rather of the cumulative development—of *adaptations*" (Romanes, 1886, 345). Aiming to improve or build upon Darwin, Romanes developed a theory he dubbed "physiological selection" to explain the evolution of reproductive barriers to prevent varieties from hybridizing with parental species in the absence of physical isolation. Physiological selection led, in Romanes's view, to a sort of nongeographical isolation-in-place. This paper elicited a fiercely critical response from many naturalists, including Wallace, and marked the start of an eight-year argument between Wallace and Romanes that only ended with the latter's premature death in 1894. At issue were matters like the utility or inutility of specific characters, the swamping effects of intercrossing, and how interspecific sterility was achieved in the formation of new species. As Wallace described it, these were "the 'three great obstructions in the road of natural selection,' which Mr. Romanes believes to be insuperable by natural selection alone" (Wallace 1886, 303). Bound up with these issues was the question of the role of isolation in speciation, with Romanes increasingly arguing that isolation is a necessary part of the speciation process and Wallace insisting that, while important, isolation per se is not the sole requirement for the divergence of varieties into new species. George

(1964, 75–86), Lesch (1975), Bowler (1983, 32–33; 2004, 52–54; 2005), and Schwartz (1995) are excellent guides to this debate; for my purposes here I simply note that, as in most great debates, resolution is found somewhere in the middle ground. In this case, isolation (allopatry) came to be accepted as a key factor in speciation most of the time, but it came to be acknowledged too that in many cases it is all but impossible to disentangle isolation per se from environmental difference, however slight and subtle. Attendant on physical separation, then, is the likelihood that environmentally mediated selection must differ too, if only subtly, and can play a role in the subsequent divergence trajectory of separated subpopulations. Moreover, Romanes's "physiological selection" was recast in modified form in terms of reinforcement, and later theoretical work showed that both pre- and postzygotic mating barriers can be shaped by selection (reviewed by Otte and Endler 1989; Coyne and Orr 2004).

But this is the modern view, which only began to gel following the "Modern Synthesis" of the 1940s, a label that stems from Julian Huxley's immensely successful book *Evolution: The Modern Synthesis* (first published in 1942 and much reprinted), in which Darwinian selection (shed of its Lamarckism) was wed with new insights from genetics, systematics, and paleontology. "The death of Darwinism has been proclaimed not only from the pulpit, but from the biological laboratory; but, as in the case of Mark Twain, the reports seem to have been greatly exaggerated, since to-day Darwinism is very much alive," Huxley declared (1943, 22). A grandson of T. H. Huxley, Julian had inherited his grandfather's mantle as a great communicator of science, and his book was applauded by scientists and general readers alike. A review in *Nature* said that Huxley succeeded in "placing before the nonspecialist scientific reader concise understandable accounts of recent important researches which were often difficult to follow in the original publications" (quoted in Clark 1968, 281). The controversies that followed the rediscovery of Mendel's work in 1900 and the founding of the new discipline of genetics— arguments over the nature of mutation, the importance of continuous versus discontinuous variation, and the role of selection in effecting species change—were put to rest by developments in empirical and theoretical population genetics. Field biologists and systematists confirmed abundant natural variation in species; paleontologists defined the relative meaning of "gradualism" and linked microevolutionary processes of selection, genetic drift, and migration to macroevolutionary patterns that develop over geological time. The vision that emerged was one of gradual evolution by natural selection—the rebirth of neo-Darwinism that remains strong today (Provine 1971).

The first neo-Darwinian, however, was forgotten in the excitement. In its nearly 600 pages, Huxley's watershed book ushering in the Modern Synthesis mentioned Wallace's name precisely once, compared to some twenty citations of Darwin. Poignantly, upon the emergence of Darwinism from eclipse Wallace was left in shadow. To put this into more quantitative perspective, I undertook a survey of the indexes of twenty-five books on evolution published between 1904 and 1966, including seminal works by the founders of the Modern Synthesis (Huxley, Simpson, Haldane, Fisher, Morgan, Simpson, et al.; see Table 3). Among these influential books, Darwin merited an average of 13.12 page entries (range: 0–48) to Wallace's average of 1.44 entries (range: 0–8). Wallace had become even more of a footnote than he was in the pre-Synthesis years.

First Guides

Wallace has always had his champions, of course, though I have found it lamentable that some of those seem only to see the possibility of Wallace's star shining if Darwin's could be dimmed. I disagree; the "delicate arrangement" is what it is, the confluence of factors that have led to the preeminence of the Darwin name, penetrating even deeply into popular culture (one of very few scientists to have done so); these are what they are. Whatever we may think about all of these circumstances, there is no denying the scintillating brilliance of Darwin. But the recent Wallace centennial year has provided a golden opportunity to celebrate a scintillating Wallace, and happily there is now much renewed attention on his too-often-underrated creativity and contributions. In my estimation, though, too often plaudits for Wallace are given simply for his codiscovery of natural selection and his work in biogeography. Those unfamiliar with Wallace, the casual reader or student who gets only snippets about discovering natural selection, something about a "line" somewhere far away, and suggestions of a scandal end up with a grotesquely skewed picture of the man and the history. It becomes too easy to dismiss that discovery of natural selection as luck, a one-off coup (after all, was he not mainly collecting birds and bugs amid some tropical islands?), and the rest as sour grapes or fodder for conspiracists. It was my goal in this centennial year to introduce to a broader audience a different Wallace: the eager, young, driven Wallace hot on the evolutionary trail—one whose energy, optimism, and creativity seem boundless as he crisscrossed the two-thousand-mile-wide Malay Archipelago with its seeming infinitude of islands great and small, teeming with exotic species and peoples.

Wallace's most important field notebook of the period—his Species Notebook—is, to me, the best antidote for the malady of Wallace nay-saying, dismissal, or minimizing. It is also the finest document with which to tell Wallace's story of discovery in those pre-*Origin* years. It reveals Wallace in "consilience mode," assembling evidence and arguments for his planned pro-transmutation book: morphology, geographical distribution, embryology, fossils, habit versus structure, variation and new varieties, scenarios for gradual species change, skewering the arguments for design, balance and harmony in nature, and refuting claims that variation and change are limited. The depth and breadth of Wallace's "evolutionary" investigations in that period—the myriad subjects discussed in his notebook and in the myriad papers written in bungalows and jungle huts—these are astonishing, but astonish all the more in the context of Wallace's social background and lack of formal scientific training, wealth, or connections as he set out to travel and pursue the mystery of the origin of species. To have succeeded in that audacious pursuit given the formidable odds really does constitute an epic achievement. Surely Wallace's own laudatory words for Darwin, that the *force of admiration* for the senior naturalist could no further go, are equally appropriately applied to Wallace himself. He and Darwin were our first guides to a revolutionary new understanding of our world, and ourselves. Their mutual grand vision is captured, in echoes of one another, in their respective closing words from the *Origin* and the Sarawak Law paper:

> There is grandeur in this view of life, with its several powers, having been originally breathed into a few forms or into one; and that, whilst this planet has gone cycling on according to the fixed law of gravity, from so simple a beginning endless forms most beautiful and most wonderful have been, and are being, evolved.

> It has now been shown . . . how the law that *"Every species has come into existence coincident both in time and space with a pre-existing closely allied species,"* connects together and renders intelligible a vast number of independent and hitherto unexplained facts. . . . Granted the law, and many of the most important facts in Nature could not have been otherwise, but are almost as necessary deductions from it, as are the elliptic orbits of the planets from the law of gravitation. (Emphasis in original)

Coda Table 3. Number of index entries for Darwin and Wallace in a selection of twenty-five works in evolution and genetics published between 1904 and 1966.

Author	Title	Edition Consulted	Length (pp.) (excl. bibliog. & index)	Darwin entries	Wallace entries
Weissmann, A.	*The Evolution Theory*	1904	396	18	2
DeVries, H.	*Species and Varieties, Their Origin by Mutation*	1905	826	32	5
Bateson, W.	*Problems of Genetics*	1913	250	7	0
Lull, R. S.	*Organic Evolution*	1917	691	20	2
Ford, E. B.	*Mendelism and Evolution*	1931	102	4	0
Shull, A. F.	*Evolution*	1936	287	6	8
Waddington, C. H.	*An Introduction to Modern Genetics*	1939	408	0	0
Huxley, J.	*Evolution: The Modern Synthesis*	1943	578	20	1
Schmalhausen, I. I.	*Factors of Evolution*	1949	284	11	0
Lindsay, A. W.	*Principles of Organic Evolution*	1952	365	6	2
Simpson, G. G.	*The Meaning of Evolution*	1952	348	8	0
Huxley, J.	*Evolution in Action*	1953	176	10	0
Simpson, G. G.	*Life of the Past: An Introduction to Paleontology*	1953	185	4	1
Darlington, C. D.	*The Evolution of Genetic Systems*, 2nd ed.	1958	240	8	0

Fisher, R. A.	The Genetical Theory of Natural Selection, 2nd ed.	1958	284	24	8
Bonner, J. T.	The Ideas of Biology	1962	161	11	1
Waddington, C. H.	The Nature of Life	1962 1926	125	14	0
Morgan, T. H.	The Theory of the Gene, Revised ed.	(1932) 1932	321	6	0
Haldane, J. B. S.	The Causes of Evolution (reprint)	(1990) 1937	204	30	1
Dobzhansky, T.	Genetics and the Origin of Species (reprint)	(1982) 1944	321	10	0
Simpson, G. G.	Tempo and Mode in Evolution (reprint)	(1984) 1954	218	1	0
Cain, A. J.	Animal Species and Their Evolution (reprint)	(1960) 1955	183	5	1
Dobzhansky, T.	Evolution, Genetics, and Man	(1959) 1963	378	48	3
Mayr, E.	Populations, Species, and Evolution	(1970) 1966	409	16	1
Williams, G. C.	Adaptation and Natural Selection (reprint)	(1974)	273	9	0
		25 works	Average:	13.12	1.44
			Range:	0–48	0–8

"Granted the law." Another echo of Darwin, who long before had speculated in a notebook on what would ensue when we "once grant" the very same law: "Once grant that species [of] one genus may pass into each other . . . & whole fabric totters & falls." "Look abroad," he continued, in a breathless stream-of-consciousness burst: "Study gradation study unity of type study geographical distribution study relation of fossil with recent. the fabric falls!" (C Notebook; Barrett et al. 1987, 76–77)—the very same "vast number of independent and hitherto unexplained facts" that Wallace saw "connect[ed] together and render[ed] intelligible" in his own remarkable notebook.

There have always been those in the scholarly community who have admired Wallace for his insights and achievements; it is my hope that the present work, together with *On the Organic Law of Change*, will inspire an even wider appreciation of Wallace's personal odyssey of evolutionary discovery and the insights he achieved, and honor him as we do his illustrious colleague as a star of the first magnitude.

Appendixes

Bibliography

Acknowledgments

Notes on the Text and Illustrations

Index

Appendix 1

The thirty-nine authors cited by Wallace in the Species Notebook.* Twenty-one of these (54 percent) were also cited by Darwin (Transmutation Notebooks [Barrett et al. 1987] or *Natural Selection* [Stauffer 1975]), in eight (21 percent) of which Wallace and Darwin referred to the same work (see Appendix 2). In most cases Wallace and Darwin make different uses of these authors and works; in only two instances did they cite the same information from authors.

*Pagination follows Wallace's in the Species Notebook; see Costa (2013a).

Table A.1. Works Cited by Wallace in Species Notebook.

| | Author | Work | Location in Species Notebook (Costa 2013a) | Darwin cited[a]: | | |
				Same author?	Same work?	Same info?
Recto	Agassiz, L.	Lake Superior	140, 145–147	Yes	No	
	Bentham, J.	Book of Fallacies	102	No		
	Bleeker, P.	? [ichthyological publications]	27	No		
	Blyth, E.	London Magazine of Natural History	62	Yes	Yes	No
	Boswell, J.	Life of Johnson	56	Yes	Yes	
	Buch, L. von	Description physiques des îsles Canaries	90, 141	Yes	Yes	Yes
	Combe, G.	Constitution of Man	155	No		
	Dampier, W.	A New Voyage Round the World	152	Yes	Yes	No
	Darwin, C.	Journal of Researches	60	—	—	
	Dufour, P. [La Croix, P.]	History of Prostitution	91	No		
	Durmont-d'Urville, J. M.	Annales des sciences naturelles	105	Yes	Yes	Yes
	Erichson, W. F.	Several possible	177	No		
	Fullom, S. W.	Marvels of Science and Other Testimonies to Holy Writ	155	No		
	Huc, E. R.	L'Empire Chinois	57	Yes	Yes	No
	Johnston, G.	Botany of the Eastern Borders	119	No		
	Knight, C.	Cyclopedia of Natural History	12, 31, 32	No		
	Knighton, W.	Tropical Sketches	8	No		
	Lesson, R.-P.	Voyage autour du monde: Sur la corvelle La Coquille	b, 105–106, 129	Yes	Yes	No
	Lindley, J	Introduction to Botany	59	Yes	No	
	Lyell, C.	Principles of Geology	27, 34–53, 142–143, 149–150	Yes	Yes	Some

Milne-Edwards, H.	Manual of Zoology	111	Yes	?	
Müller, S.	Annalen der Erdkunde	110	Yes	No	
Owen, R.	On the Anthropoid Apes, and Their Relations to Man	63	Yes	No	
Owen, R.	Proc. of the Geol. Soc. London 16 May 1855	54	Yes	No	
Pope, A.	Essay on Man	143, 177	No		
Ramsay, A. C.	Reports of the British Association for 1854	34	Yes	No	
Renan, E.	Essai de Morale et de Critique	154–155	No		
Scott, W.	The Pirate	107	Yes	No	
Siebold, P. F. B., & Melville de Carnbee, P.	Le Moniteur des Indes-Orientales et Occidentales	110	No		
Somerville, M.	Physical Geography	7	Yes	No	
Spencer, H.	First Principles of a New System of Philosophy	178	No		
Strickland, H., & A. G. Melville	The Dodo and Its Kindred	55	Yes	No	
Thomson, J.	Archives Entomologiques . . .	130, 133	No		
White, A.	Catalogue of Coleopterous Insects	133	No		
White, G.	Natural History of Selborne	108	Yes	Yes	No
Zollinger, H.	Die Land- und Süsswasser-Mollusken von Java	110	No	No	
Verso					
Comte, A.	Philosophie Chimique et Biologique	c	Yes	No	
Erichson, W. F.	Faune Ent. d'Allemagne	c	Yes	No	
Abbott, J.	?	32	No	No	

39 citations

*Cited in Transmutation Notebooks and/or Natural Selection manuscript.

Appendix 2

The eight publications from Appendix 1 cited by both Wallace in the Species Notebook* and Darwin in the Transmutation Notebooks (Barrett et al. 1987) and/or the *Natural Selection* manuscript (Stauffer 1975). Note that this list includes only works cited in the Species Notebook, and as such does not include other works that both naturalists are known or thought to have read (e.g., Humboldt, Chambers, and Malthus). Wallace and Darwin gleaned the same general information from just three of the authors that they read in common in the period of the Species Notebook (von Buch, Lyell, and D'Urville), though in other cases they gleaned the same information from different authors (see Chapter 3).

Blyth, Edward. 1835. An attempt to classify the "varieties" of animals . . .
Magazine of Natural History 8: 40–53.
    WALLACE: Species Notebook: p. 62
    DARWIN: Transmutation Notebooks: C70
                 Natural Selection: p. 324

Buch, Léopold von. 1836. *Description physique des îsles Canaries.* Paris: G. Levrault.
    WALLACE: Species Notebook: pp. 90, 141
    DARWIN: Transmutation Notebooks: RN137, 150; A39, 40, 42; B156–159, 164; D69
                 Natural Selection: p. 523

*Pagination follows Wallace's in the Species Notebook; see Costa (2013a).

Dampier, William. 1698–1703. *A New Voyage Round the World*. London: James Knapton.
 WALLACE: Species Notebook: p. 152
 DARWIN: Transmutation Notebooks: RN8–10, 15; C266; E182
              Natural Selection: none

Durmont-d'Urville, Jules-Sébastian-César. 1825. De la distribution des fougères sur la surface du globe terrestre. *Annales des sciences naturelles* 6: 51–73.
 WALLACE: Species Notebook: p. 105
 DARWIN: Transmutation Notebooks: C16
              Natural Selection: none

Huc, Évariste Régis. 1854. *L'Empire Chinois*. Paris: Librairie de Gaume Frères. (ARW, 1854 French edition (?); CD, 1855 English edition)
 WALLACE: Species Notebook: p. 57
 DARWIN: Transmutation Notebooks: C16
              Natural Selection: p. 60

Lesson, René-Primevère. 1826–1830. *Voyage autour du monde: Sur la corvelle La Coquille*. Paris: A. Bertrand.
 WALLACE: Species Notebook: pp. 105–106
 DARWIN: Transmutation Notebooks: RN62, 101, 102; B31, 54, 220, 234, 249; C16–29, 276; E42; ZEd5, 7, 11, 13
              Natural Selection: none

Lyell, Charles. *Principles of Geology*. London: John Murray. (ARW, 4th edition in Sp. Notebook; CD, multiple editions)
 WALLACE: Species Notebook: pp. 27, 34–53, 142–143, 149–150
 DARWIN: Transmutation Notebooks: 84 citations in RN, A, B, C, D, E, GR, M (**)
              Natural Selection: 157, 173, 177, 180, 187, 202, 207, 208, 219, 223, 247, 370, 535, 546, 547, 561, 562, 583

White, G. *The Natural History and Antiquities of Selborne* (1st pub. 1789; many editions)
 WALLACE: Species Notebook: p. 108
 DARWIN: Transmutation Notebooks: C248, 254, 275
              Natural Selection: 179, 186, 258, 489, 503, 504, 522, 572

**RN44, 52, 57, 60, 61, 63, 65, 67, 68, 70, 79, 82–84, 88, 100, 115; A7, 11, 79, 85, 95, 116, 121; B6, 10–13, 23, 59, 63, 69, 81–82, 87, 91, 96, 115, 116, 153, 155–157, 170, 172, 200, 201, 202, 249; C39, 53, 137, 106, 168, 270; D21, 39, 60, 104, 134; E4, 26, 35, 38, 65, 105, 109, 167; GR109; M128.

Appendix 3

Correspondence of key evolutionary topics, themes, and observations given by Wallace in the Species Notebook° with similar writings by Darwin including entries from the Transmutation Notebooks (Barrett et al. 1987), *Natural Selection* manuscript (Stauffer 1975), and *On the Origin of Species* (Darwin 1859). S = similar topical treatments or conclusions, D = dissimilar treatments or conclusions; in most cases Wallace and Darwin are congruent in their view or treatment of topics.

° Pagination follows Wallace's in the Species Notebook; see Costa (2013a).

Table A.3. Correspondence between key evolutionary topics between Wallace (Species Notebook) and Darwin (Transmutation Notebooks, Natural Selection manuscript, and *On the Origin of Species*).

	Wallace		Darwin
Geology and Paleontology			
Geological change; uplift, subsidence	pp. 34, 50, 92–97, 119, 153	S	RN77; A8, 85, 113, 118 *Origin* pp. 296, 300
Intermediate nature of common ancestors	pp. 37, 54 (Owen)	S	C110, 201, 257; E88, 119–120 *Natural Selection* pp. 262, 264, 384 *Origin* pp. 280; Owen cited p. 329
Gaps in fossil record	pp. 92–97	S	C216 *Natural Selection* pp. 262–264 *Origin* pp. 279–302
Geological succession (Continuous and gradual change)	pp. 37–38, 45–49, 54, 92–97, 145	S	E126–127 *Natural Selection* p. 284 *Origin* pp. 312–317
Progressive development, parallelism	pp. 36–38, 145	S	B117, E60 *Origin* p. 338; cites Agassiz p. 449
Iceberg theory / Erratics	pp. 142–143	D	RN99, 114; A20, 110; E37–38 *Natural Selection* pp. 545–547, 561 *Origin* pp. 363, 373, 393, 399
Didelphys of the Oolite	p. 37	S	B87–88, 219; D62; E128

Morphology and Affinity			
Morphological affinities, analogies, transitions	pp. 76–83, 97–100	S	B162, E91 *Origin* pp. 434–439 *Origin* pp. 179–186; 191–193 (Trans. habits, organs) *Origin* pp. 329–336 (Affinities of extinct forms, etc.) *Natural Selection* pp. 262–264; 384–386
Eyes of cave/burrowing animals	pp. 60–61	S	*Natural Selection* pp. 295–296 *Origin* pp. 137–139 ("Abortive organs"—C215–216; D59)
Embryology informs classification	p. 144–145	S	*Origin* pp. 338, 449
Branching, tree, divergence	pp. 37, 38, 44, 54	S	B23–27, 36–38; C145 *Natural Selection* pp. 249–250 *Origin* pp. 130, 317, 432
Geographical Distribution			
Distribution, relationships	pp. 7, 8, 27, 54, 110, 145–147, 151, 177	S	B261, 275; C80; D23 *Origin* chs. 11, 12
Islands / island biota, etc. St. Helena, Galápagos, Canary Islands, Cape Verde Islands	pp. 46–49, 90, 151	S	B98, 100, 102–104, 156–158, 278; C54, 209; E90–91, 100, 104 *Natural Selection* pp. 256–257, 561 *Origin* pp. 364, 398–399

(Continued)

Table A.3. (continued)

	Wallace		Darwin
Tenerife—species interaction, isolation	p. 90	S	B158, C184
Climate change affecting distribution	pp. 50–52, 145, 147	S	*Natural Selection* pp. 535–540, 564–565 *Origin* pp. 365–382
Instinct and Habit			
Habits and structure not in accordance; Transitional habits, form	pp. 32–33, 53	S	C81, 105 *Natural Selection* pp. 341–350 *Origin* pp. 179–186
Bees' cells	pp. 168, 173–177	S/D	N70, 77, 82 *Natural Selection* pp. 512–516 *Origin* pp. 224–234
Instinct; "proceed by degrees" (Graduated series)	pp. 149, 166	S	C159, 196 *Natural Selection* pp. 382, 477, 480 *Origin* pp. 209–210
Bird migration	p. 155	S	C159; N77–N78 *Natural Selection* pp. 490–495 *Origin* pp. 212–214
Ants and aphids	pp. 137–140	D	*Natural Selection* pp. 382, 480 *Origin* ch. 7, pp. 210–211

Birds' nests (Variation in nest-building behavior)	pp. 112–119, 166–167	D	C189; N68–70; *Natural Selection* pp. 498–505; *Origin* p. 211
Human-Primate Relationship and Human Variation			
"Man"—reason vs. instinct	pp. 166, 168–172	S/D	N69, 70, N77–N81; Extensive entries in M notebook; Queries on Expression (1867)
Human-primate relationship	pp. 64, 91	S	C204; D61; E68, 89; M128, 129
Orangutans / behavioral obs.	pp. 20–27	S	C79; M85, 107, 127, 129, 138–140; N13, 94
Human races	pp. 63–66, 100, 104–106, 134	S	C174, 204, 217, 234; Extensive entries in M notebook; Varieties of human race (Darwin et al. 1841)
Arguments and Observations for Transmutation			
Beautiful facts of morphology . . .	pp. 98–99	S	"cosmogonists of old"; *Origin* p. 167
Caution attributing adaptive function (Ostrich bone air cells)	p. 112	S	Mammalian skull sutures; *Origin* p. 197; *Natural Selection* p. 377
Resistance to acceptance of transmutation (Like prejudiced reception of earlier ideas)	pp. 34, 39	S	*Origin* p. 282

(continued)

Table A.3. (continued)

	Wallace		Darwin
Domestication, supposed limits of variation in domestic breeds	pp. 39, 41–45, 149–150	S/D	Many entries in B, C, D, M, N, Q&E Notebooks *Origin* pp. 30–38 (Microcosm of natural selection)
New varieties arising—examples	p. 59	S	QE 15v *Natural Selection* p. 127
Varieties: origin and fate	pp. 45, 57–59; 62, 90, 150	S/D	B7, 11, 17, 209; D23 *Natural Selection* pp. 226–234, 238 *Origin* pp. 51–52, 57 (Varieties as incipient sp.); 111–126 (Divergence of character)
Criticism of claims of "design"	pp. 12, 31, 32	S	C106; E23–24; MAC167 *Natural Selection* pp. 380–382 *Origin* pp. 202–203
Struggle for existence; Lack of harmony and balance	pp. 49–50, 146–147	S	D134–135; E114 *Natural Selection* pp. 175, 186 *Origin* "glad face of nature . . ." p. 62
Primroses and cowslips	p. 42	S/D	C194; E16, 113, 141; QE1v, 5 *Natural Selection* pp. 128–133 *Origin* pp. 49–50

Sources: Wallace: pages given from Species Notebook (LINSOC MS 180; Costa 2013a)

Darwin: Transmutation Notebooks (Barrett et al. 1987), *Natural Selection* manuscript (Stauffer 1975), *On the Origin of Species* (Darwin 1859)

Appendix 4

Wallace's name is mentioned four times in *On the Origin of Species* (Darwin 1859): twice in the introduction in reference to Wallace's independent discovery of evolution by natural selection (inducing Darwin to publish), once in chapter 11 in reference to the Sarawak Law, and once in chapter 12 in reference to Wallace's biogeographical work in the Malay Archipelago.

Origin pp. 1–2

My work is now nearly finished; but as it will take me two or three more years to complete it, and as my health is far from strong, I have been urged to publish this Abstract. I have more especially been induced to do this, as Mr. Wallace, who is now studying the natural history of the Malay archipelago, has arrived at almost exactly the same general conclusions that I have on the origin of species. Last year he sent to me a memoir on this subject, with a request that I would forward it to Sir Charles Lyell, who sent it to the Linnean Society, and it is published in the third volume of the Journal of that Society. Sir C. Lyell and Dr. Hooker, who both knew of my work—the latter having read my sketch of 1844—honoured me by thinking it is advisable to publish, with Mr. Wallace's excellent memoir, some brief extracts from my manuscripts.

Origin p. 355

This view of the relation of species in one region to those in another, does not differ much (by substituting the word variety for species) from that

lately advanced in an ingenious paper by Mr. Wallace, in which he concludes, that "every species has come into existence coincident both in space
and time with a pre-existing closely allied species." And I now know from
correspondence, that this coincidence he attributes to generation with
modification.

Origin p. 395

Mr. Windsor Earl has made some striking observations on this head in regard to the great Malay Archipelago, which is traversed near Celebes by a
space of deep ocean; and this space separates two widely distinct mammalian faunas. On either side the islands are situated on moderately deep
submarine banks, and they are inhabited by closely allied or identical
quadrupeds. No doubt some few anomalies occur in this great archipelago,
and there is much difficulty in forming a judgment in some cases owing to
the probable naturalisation of certain mammals through man's agency; but
we shall soon have much light thrown on the natural history of this archipelago by the admirable zeal and researches of Mr. Wallace.

Appendix 5

The phrase "my theory" appears fifty-seven times in *On the Origin of Species* (Darwin 1859), reflecting Darwin's possessive view of the idea of descent with modification by natural selection, to the exclusion of Wallace's share in the discovery. The more neutral phrase "the theory" appears thirteen times in reference to natural selection or descent with modification (pp. 5, 188, 194, 206, 235, 237, 243, 245, 281, 302, 317, 320, 322); "our theory" appears zero times.

p. 56   for if this had been so, it would have been fatal to my theory

p. 111 *Divergence of Character.*—The principle, which I have designated by this term, is of high importance on my theory

p. 154 as in the case of the wing of the bat, it must have existed, according to my theory

p. 161 As all the species of the same genus are supposed, on my theory

p. 162 yet we ought, on my theory

p. 171 the greater number are only apparent, and those that are real are not, I think, fatal to my theory.

p. 173 By my theory these allied species have descended from a common parent

p. 176 two varieties are supposed on my theory to be converted and perfected into two distinct species

p. 179 Lastly, looking not to any one time, but to all time, if my theory be true

p. 184 we might expect, on my theory, that such individuals would occasionally have given rise to new species

p. 189   my theory would absolutely break down.

          according to my theory, there has been much extinction.

p. 190   This doctrine, if true, would be absolutely fatal to my theory.

p. 201   it would annihilate my theory.

p. 203   We have in this chapter discussed some of the difficulties and
          objections which may be urged against my theory.

p. 206   if we include all those of past times, it must by my theory be strictly true.

          On my theory, unity of type is explained by unity of descent.

p. 210   Again as in the case of corporeal structure, and conformably with my
          theory

p. 230   I could show that they are conformable with my theory

p. 239   when I do not admit that such wonderful and well-established facts at
          once annihilate my theory.

p. 242   this is by far the most serious special difficulty, which my theory has
          encountered.

p. 243   I do not pretend that the facts given in this chapter strengthen in any
          great degree my theory.

p. 246   the fertility of their mongrel offspring, is, on my theory

p. 280   this, perhaps, is the most obvious and gravest objection which can be
          urged against my theory.

          In the first place it should always be borne in mind what sort of
          intermediate forms must, on my theory, have formerly existed.

p. 281   It is just possible by my theory, that one of two living forms might
          have descended

p. 296   all the fine intermediate gradations which must on my theory have
          existed between them

p. 297   we do find the kind of evidence of change which on my theory we
          ought to find.

p. 301   transitional forms, which on my theory assuredly have connected all
          the past and present species

p. 302   pressed so hardly on my theory.

p. 305   it would be an insuperable difficulty on my theory,

p. 306   and it cannot on my theory be supposed,

p. 307   Consequently, if my theory be true

          which on my theory no doubt were somewhere accumulated before
          the Silurian epoch

p. 310   will undoubtedly at once reject my theory.

p. 311   These several facts accord well with my theory.

p. 316   and the rule strictly accords with my theory.

p. 317   This gradual increase in number of the species of a group is strictly
           conformable with my theory.

p. 337   more recent forms must, on my theory

p. 341   It must not be forgotten that, on my theory

p. 350   This bond, on my theory, is simply inheritance

p. 354   which on my theory have all descended from a common progenitor

p. 355   my theory will be strengthened

           the species, on my theory, must have descended from a succession of
           improved varieties

p. 381   on my theory of descent with modification

p. 385   allied species, which, on my theory, are descended

p. 390   This fact might have been expected on my theory

p. 407   With respect to the distinct species of the same genus, which on my
           theory must have spread from one parent-source

p. 410   On my theory these several relations throughout time and space are
           intelligible

p. 429   the greater must be the number of connecting forms which on my
           theory have been exterminated and utterly lost.

p. 430   they are due on my theory to inheritance in common.

p. 446   Let us take a genus of birds, descended on my theory from some one
           parent-species

p. 463   this is the most obvious and forcible of the many objections which may
           be urged against my theory.

           For certainly on my theory such

p. 465   and they have changed in the manner which my theory requires

           Such is the sum of the several chief objections and difficulties which
           may justly be urged against my theory

p. 482   Any one whose disposition leads him to attach more weight to
           unexplained difficulties than to the explanation of a certain
           number of facts will certainly reject my theory.

Bibliography

Electronic / Online Bibliographic Resources

The Alfred Russel Wallace Correspondence Project, wallaceletters.info/
Referred to as simply the Wallace Correspondence Project (WCP), this
resource is directed by George Beccaloni at the Natural History Museum
in London. This project's publicly accessible electronic archive of Wallace-
related documents is named Wallace Letters Online (WLO), www.nhm.ac
.uk/wallacelettersonline. WLO contains metadata, digital scans, and
transcripts of all known letters sent to and written by Wallace, as well as a
selection of other important Wallace-related manuscripts. Note that the
documents catalogued in WLO each have a unique identifier, known as a
WCP number, and that these are cited in this volume.

The Alfred Russel Wallace Page, people.wku.edu/charles.smith/index1.htm
Created and maintained by Charles H. Smith, Western Kentucky University, this online archive includes comprehensive Wallace bibliographic
records, many with full text. Works are uniquely designated with an "S"
number on this site; these numbers are included for Wallace works listed in
the "Literature Cited" section that follows.

Darwin Correspondence Project, www.darwinproject.ac.uk
University of Cambridge–based digital archive of the letters of Charles
Darwin; directed by Jim Secord et al. Darwin letter citations in this book
include the Darwin Correspondence Project letter number.

Darwin Online, darwin-online.org.uk/contents.html
Directed by John van Wyhe, this is the most extensive scholarly website on
Darwin, featuring complete transcriptions and scans of manuscripts,
published works, private papers, and other documents. Made possible by a
consortium of universities, museums, libraries, and other institutions.

Manuscripts

The following Wallace manuscripts located at the Linnean Society of London are cited in this work:

1. Field Notebooks

LINSOC-MS179: Field Notebook, 1854–1861.
LINSOC-MS180: Field Notebook, 1855–1859 ("Species Notebook," transcribed and annotated by Costa 2013a. Scan of the original can be viewed at: linnean-online.org/wallace_notes.html).

2. Journals

Wallace's "Malay Journals" consist of four notebooks with sequentially numbered entries:
LINSOC-MS178a: first Malay Journal, June 1856–March 1857; entries 1–68.
LINSOC-MS178b: second Malay Journal, March 1857–March 1858; entries 69–128.
LINSOC-MS178c: third Malay Journal, March 1858–August 1859; entries 129–192.
LINSOC-MS178d: fourth Malay Journal, October 1859–May 1861; entries 193–245.

Literature Cited

Agassiz, L. 1854. On the primitive diversity and number of animals in geological times. *Annals and Magazine of Natural History* 83 (2nd series): 350–366.

Agassiz, L., and J. E. Cabot. 1850. *Lake Superior: Its Physical Character, Vegetation, and Animals, Compared with Those of Other and Similar Regions.* Boston: Gould, Kendall and Lincoln.

Baer, K. E. von. 1828. *Entwickelungsgeschichte der Theire: Beobachtung und Reflexion.* Konigsberg: Borntrager.

Barlow, N. (ed.). 1958. *The Autobiography of Charles Darwin 1809–1882: With Original Omissions Restored [edited by and with appendix and notes by his grand-daughter Nora Barlow].* London: Collins.

Barrett, P. H. (ed.). 1977. *The Collected Papers of Charles Darwin,* vol. 1. Chicago: University of Chicago Press.

Barrett, P. H., P. J. Gautrey, S. Herbert, D. Kohn, and S. Smith (eds.). 1987. *Charles Darwin's Notebooks, 1836–1844.* Ithaca, N.Y.: Cornell University Press.

Bartholomew, M. 1973. Lyell and evolution: An account of Lyell's response to the prospect of an evolutionary ancestry for man. *British Journal for the History of Science* 6: 261–303.

Bartley, M. M. 1992. Darwin and domestication: Studies on inheritance. *Journal of the History of Biology* 25: 307–333.

Bates, H. W. 1863. *The Naturalist on the River Amazons.* London: John Murray.

Beccaloni, G. 2008. Wallace's annotated copy of the of the Darwin-Wallace paper on natural selection. In C. H. Smith and G. Beccaloni (eds.), *Natural Selection and Beyond: The Intellectual Legacy of Alfred Russel Wallace,* 91–101. Oxford: Oxford University Press.

Beddall, B. G. 1968. Wallace, Darwin, and the theory of natural selection: A study in the development of ideas and attitudes. *Journal of the History of Biology* 1: 261–323.

Beddall, B. G. 1972. Wallace, Darwin, and Edward Blyth: Further notes on the development of evolution theory. *Journal of the History of Biology* 5: 153–158.

Beddall, B. G. 1988a. Darwin and divergence: The Wallace connection. *Journal of the History of Biology* 21: 1–68.

Beddall, B. G. 1988b. Wallace's annotated copy of Darwin's *Origin of Species. Journal of the History of Biology* 21: 265–289.

Berry, A. (ed.). 2002. *Infinite Tropics: An Alfred Russel Wallace Anthology.* London: Verso.

Berry, A. 2008. "Ardent beetle hunters": Natural history, collecting, and the theory of evolution. In C. H. Smith and G. Beccaloni (eds.), *Natural Selection and Beyond: The Intellectual Legacy of Alfred Russel Wallace,* 47–65. Oxford: Oxford University Press.

Berry, A., and J. Browne. 2008. The other beetle hunter. *Nature* 453: 1188–1190.

Blyth, E. 1835. An attempt to classify the "varieties" of animals, with observations on the marked seasonal and other changes which naturally take place in various British species, and which do not constitute varieties. *Magazine of Natural History* 8: 40–53.

Bock, W. J. 2009. The Darwin-Wallace myth of 1858. *Proceedings of the Zoological Society* 62: 1–12.

Bowler, P. J. 1976. Alfred Russel Wallace's concepts of variation. *Journal of the History of Medicine* 31: 17–29.

Bowler, P. J. 1983. *The Eclipse of Darwinism: Anti-Darwinian Evolution Theories in the Decades Around 1900.* Baltimore, Md.: Johns Hopkins University Press.

Bowler, P. J. 1984. Wallace and Darwinism. [Review of Brooks 1984, *Just Before the Origin.*] *Science* 224: 277–278.

Bowler, P. J. 2003. *Evolution: The History of an Idea,* 3rd ed. Berkeley: University of California Press.

Bowler, P. J. 2004. The specter of Darwinism: The popular image of Darwinism in early twentieth-century Britain. In A. Lustig, R. J. Richards, and M. Ruse (eds.), *Darwinian Heresies,* 48–68. Cambridge: Cambridge University Press.

Bowler, P. J. 2005. Revisiting the eclipse of Darwinism. *Journal of the History of Biology* 38: 19–32.

Bowler, P. J. 2009. Darwin's originality. *Science* 323: 223–226.

Boyd, T. 1859. [Review of] On the Tendency of Species to Form Varieties. *Zoologist* 17: 6357–6359.

Brackman, A. C. 1980. *A Delicate Arrangement: The Strange Case of Charles Darwin and Alfred Russel Wallace.* New York: Times Books.

Bronn, H. G. 1842–1843. *Handbuch einer Geschichte der Natur.* 2 vols. Stuttgart: Schweizerbart.

Brooks, J. L. 1984. *Just Before the Origin: Alfred Russel Wallace's Theory of Evolution.* New York: Columbia University Press.

Browne, J. 1980. Darwin's botanical arithmetic and the "principle of divergence," 1854–1858. *Journal of the History of Biology* 13: 53–89.

Buch, L. von. 1825. *Physikalische Beschreibung der Kanarische Inseln.* Berlin: Druckerei der Königlichen Akademie der Wissenschaften. Translated by C. Boulanger as *Description physique des îsles Canaries, suivie d'une indication des principaux volcans du globe* (Paris: F. G. Levrault, 1836).

Bulmer, M. 2005. The theory of natural selection of Alfred Russel Wallace FRS. *Notes & Records of the Royal Society* 59: 125–136.

Burchfield, J. D. 1998. The age of the Earth and the invention of geological time. Geological Society of London, *Special Publications,* 143: 137–143.

Butler, S. 1879. *Evolution, Old and New, or, The Theories of Buffon, Dr. Erasmus Darwin, and Lamarck, as Compared with That of Mr. Charles Darwin.* London: Hardwicke and Bogue.

Butler, S. 1880. *Unconscious Memory.* London: David Bogue.

Butler, S. (R. A. Streatfeild, ed.). 1908. *Essays on Life, Art and Science.* London: A. C. Fifield.

Camardi, G. 2001. Richard Owen, morphology and evolution. *Journal of the History of Biology* 34: 481–515.

Camerini, J. R. 1993. Evolution, biogeography, and maps: An early history of Wallace's Line. *Isis* 84: 700–727.

Candolle, A. P. de. 1820. Géographie botanique. In F. G. Levrault (ed.), *Dictionnaire des Sciences Naturelles.* Strasbourg, Paris: F. G. Levrault. 18: 359–422.

Carpenter, W. B. 1841. *Principles of General and Comparative Physiology,* 2nd ed. London: John Churchill.

Chambers, R. 1844. *Vestiges of the Natural History of Creation.* London: John Churchill.

Claeys, G. 2008. Wallace and Owenism. In C. H. Smith and G. Beccaloni (eds.), *Natural Selection and Beyond: The Intellectual Legacy of Alfred Russel Wallace*, 235–262. Oxford: Oxford University Press.

Clark, R. W. 1968. *The Huxleys.* New York: McGraw-Hill.

Colp, R., Jr. 1992. "I Will Gladly Do My Best": How Charles Darwin obtained a Civil List Pension for Alfred Russel Wallace. *Isis* 83: 3–26.

Corsi, P. 1978. The importance of French transformist ideas for the second volume of Lyell's *Principles of Geology. British Journal for the History of Science* 2: 1–25.

Costa, J. T. 2009a. *The Annotated Origin: A Facsimile of Charles Darwin's* On the Origin of Species. Cambridge, Mass.: Harvard University Press.

Costa, J. T. 2009b. The Darwinian revelation: Tracing the origin and evolution of an idea. *BioScience* 59: 886–894.

Costa, J. T. 2013a. *On the Organic Law of Change: A Facsimile Edition and Annotated Transcription of Alfred Russel Wallace's 'Species Notebook' of 1855–1859.* Cambridge, Mass.: Harvard University Press.

Costa, J. T. 2013b. Engaging with Lyell: Alfred Russel Wallace's Sarawak law and Ternate papers as reactions to Charles Lyell's *Principles of Geology. Theory in Biosciences* 132: 225–237.

Coyne, J. A., and H. A. Orr. 2004. *Speciation.* Sunderland, Mass.: Sinauer Associates.

Darwin, C. R. 1839 (2nd ed. 1845). *Journal of Researches into the Natural History and Geology of the Countries Visited During the Voyage of H.M.S. Beagle Round the World.* [Voyage of the Beagle.] London: John Murray.

Darwin, C. R. 1859. *On the Origin of Species by Means of Natural Selection.* London: John Murray.

Darwin, C. R. 1867. Queries about expression. In R. B. Freeman and P. J. Gautrey (eds.), Charles Darwin's queries about expression. *Bulletin of the British Museum of Natural History* (historical series) 4 (1972): 205–219.

Darwin, C. R. 1868. *The Variation of Animals and Plants under Domestication.* 2 vols. London: John Murray.

Darwin, C. R. 1871. *The Descent of Man, and Selection in Relation to Sex.* 2 vols. London: John Murray.

Darwin, C. R. 1872. *On the Expression of the Emotions in Man and Animals.* London: John Murray.

Darwin, C. R. 1880. [Memorial of A. R. Wallace for a Civil List Pension.] CUL-DAR91.95–98, transcribed by John van Wyhe. Available at Darwin Online: darwin-online.org.uk.

Darwin, C. R. 1881. *The Formation of Vegetable Mould, Through the Action of Worms, with Observations On Their Habits.* London: John Murray.

Darwin, C. R. and A. R. Wallace. 1858. "On the tendency of species to form varieties; and on the perpetuation of varieties and species by natural means

of selection" by Charles Darwin and Alfred Wallace [communicated by Sir
Charles Lyell and Joseph D. Hooker to the LSL meeting of 1 July 1858].
Journal of the Proceedings of the Linnean Society: Zoology 3(9): 45–62.

[Darwin, C. R., J. E. Gray, T. Hodgkin, J. C. Prichard, R. Taylor, N. Wiseman,
W. Yarrell, and J. Yates.] 1841. Queries respecting the human race, to be
addressed to travellers and others. Drawn up by a Committee of the British
Association for the Advancement of Science, appointed in 1839. Report of
the British Association for the Advancement of Science, at the Glasgow
meeting, August 1840, 10: 447–458.

Darwin, F. (ed.). 1909. *The Foundations of the Origin of Species: Two Essays
Written in 1842 and 1844.* Cambridge: Cambridge University Press.

Davies, R. 2008. *The Darwin Conspiracy—Origins of a Scientific Crime.*
London: Golden Square Books.

Davies, R. 2012. How Charles Darwin received Wallace's Ternate paper 15
days earlier than he claimed: A comment on van Wyhe and Rookmaaker
(2012). *Biological Journal of the Linnean Society* 105: 472–477.

Davies, R. 2013. 1 July 1858: What Wallace knew; what Lyell thought he knew;
what both he and Hooker took on trust; and what Charles Darwin never
told them. *Biological Journal of the Linnean Society* 109: 725–736.

Edwards, W. H. 1847. *A Voyage Up the River Amazon, Including a Residence
at Pará.* London: John Murray.

England, R. 1997. Natural selection before the *Origin:* Public reactions of some
naturalists to the Darwin-Wallace papers (Thomas Boyd, Arthur Hussey,
and Henry Baker-Tristram). *Journal of the History of Biology* 30: 267–290.

Fabre, E. 1854. On the species of aegilops of the south of France, and their
transformation into cultivated wheat. Translated from the French. *Journal
of the Royal Agricultural Society of England* 15: 167–180.

Fichman, M. 2001. Science in theistic contexts: A case study of Alfred Russel
Wallace. *Osiris* 16: 227–250.

Fichman, M. 2004. *An Elusive Victorian: The Evolution of Alfred Russel
Wallace.* Chicago: University of Chicago Press.

Forbes, E. 1854. On the manifestation of polarity in the distribution of
organized beings in time. *Notices of the Proceedings of the Meetings of the
Members of the Royal Institution* 1: 428–433.

Gardiner, B., R. Milner, and M. Morris (eds.). 2008. Wallace defends Darwin's
priority—50 years on. Survival of the Fittest, pp. 45–47. *The Linnean*,
Special Issue No. 9.

George, W. 1964. *Biologist Philosopher: A Study of the Life and Writings of
Alfred Russel Wallace.* London: Abelard-Schuman.

Gray, A. 1856. Statistics of the flora of the northern United States. *American
Journal of Science and Arts* 12: 204–232.

Gross, C. 2010. Alfred Russel Wallace and the evolution of the human mind. *Neuroscientist* 16: 496–507.

Haber, F. C. 1968. Fossils and the idea of a process of time in natural history. In B. Glass, O. Temkin, and W. L. Straus, Jr. (eds.), *Forerunners of Darwin: 1745–1859*, 222–261. Paperback edition. Baltimore, Md.: Johns Hopkins University Press.

Haldane, J. B. S. 1932. *The Causes of Evolution.* London: Longmans, Green. [1990 reprint by Princeton University Press, Princeton, N.J.]

Hamilton, W. D. 1964a,b. The genetical evolution of social behaviour. I and II. *Journal of Theoretical Biology* 7: 1–16, 17–52.

Hector, A., and R. Hooper. 2002. Darwin and the first ecological experiment. *Science* 295: 639–640.

Herschel, J. F. W. 1830. *A Preliminary Discourse on the Study of Natural Philosophy.* London: Longman, Rees, Orme, Brown, and Green.

Holland, J. 1996. Diminishing circles: W. S. Macleay in Sydney, 1839–1865. *Historical Records of Australian Science* 11: 119–147.

Hooker, J. D. 1908–1909. [Address] In *The Darwin-Wallace Celebration Held on Thursday, 1st July 1908, by the Linnean Society of London.* London (Linnean Society): Longmans, Green, February 1909: 12–16.

Huc, E. R. 1854. *L'Empire Chinois: Faisant suite à l'ouvrage intitulé Souvenirs d'un voyage dans la Tartarie et le Thibet,* 2nd ed. 2 vols. Paris: Librairie de Gaume Frères.

Hussey, A. 1859. [Review of] On the Tendency of Species to Form Varieties. *Zoologist* 17: 6474–6475.

Huxley, J. 1943. *Evolution: The Modern Synthesis.* New York: Harper and Brothers.

Jenkin, F. 1867. The origin of species. *North British Review* 46: 149–171.

Jones, G. 2002. Alfred Russel Wallace, Robert Owen and the theory of natural selection. *British Journal of the History of Science* 35: 73–96.

King-Hele, D. 1963. *Erasmus Darwin.* New York: Charles Scribner's Sons.

Kohn, D. 1981. On the origin of the principle of diversity. [Review of Brackman 1980: *A Delicate Arrangement.*] *Science* 213: 1105–1108.

Kohn, D. 1985. Darwin's principle of divergence as internal dialog. In D. Kohn (ed.), *The Darwinian Heritage,* 245–257. Princeton, N.J.: Princeton University Press.

Kottler, M. J. 1974. Alfred Russel Wallace, the origin of man, and spiritualism. *Isis* 65: 144–192.

Kottler, M. J. 1985. Charles Darwin and Alfred Russel Wallace: Two decades of debate over natural selection. In D. Kohn (ed.), *The Darwinian Heritage,* 367–432. Princeton, N.J.: Princeton University Press.

Lamarck, J.-B. 1809. *Philosophie zoologique, ou exposition des considérations relatives à l'histoire naturelle des animaux* [. . .]. Paris.

Latham, R. G. 1850. *The Natural History of the Varieties of Man*. London: John van Voorst.

Lawrence, W. 1819. *Lectures on Physiology, Zoology and the Natural History of Man*. London: J. Callow (1822 reprint; W. Benbow).

Lesch, J. E. 1975. The role of isolation in evolution: George J. Romanes and John T. Gulick. *Isis* 66: 483–503.

Lesson, R.-P. 1826–1830. *Voyage autour du monde: Sur la corvette La Coquille*. Paris: A. Bertrand.

Limoges, C. 1968. Darwin, Milne-Edwards et le principe de divergence. *Actes XII*ᵉ Congres International d'Histoire des Sciences, Paris 8: 111–115.

Lindley, J. 1837. Remarks upon the botanical affinities of Orobanche. *London & Edinburgh Philosophical Magazine and Journal of Science* (3rd series) 11: 409–411.

Lindley, J. 1841. *Elements of Botany: Structural, Physiological, Systematical, and Medical. Being a Fourth Edition of the Outline of the First Principles of Botany*. London: Taylor and Walton.

Lindley, J. 1844. Editorial. *Gardeners' Chronicle and Agricultural Gazette* 33 (17 August): 555.

Lyell, C. 1831–1833. *Principles of Geology*. 3 vols. London: John Murray.

Lyell, C. 1835. *Principles of Geology*, 4th ed. 4 vols. London: John Murray.

Lyell, C. 1868. *Principles of Geology*, 10th ed. 2 vols. London: John Murray.

Macculloch, J. 1837. *Proofs and Illustrations of the Attributes of God, From the Facts and Laws of the Physical Universe; Being the Foundation of Natural and Revealed Religion*. 3 vols. London: James Duncan.

Mallet, J. 2008. Wallace and the species concept of the early Darwinians. In C. H. Smith and G. Beccaloni (eds.), *Natural Selection and Beyond: The Intellectual Legacy of Alfred Russel Wallace*, 102–113. Oxford: Oxford University Press.

Mallet, J. 2009. Alfred Russel Wallace and the Darwinian species concept: His paper on the swallowtail butterflies (Papilionidae) of 1865. *Gayana* 73(2) (supp.): 42–54.

Malthus, T. R. [with notes by J. Bonar]. 1965. *First Essay on Population, 1798*. New York: Augustus M. Kelley, Bookseller.

Marchant, J. 1916. *Alfred Russel Wallace: Letters and Reminiscences*. 2 vols. London: Cassel.

Mayr, E. 1982. *The Growth of Biological Thought*. Cambridge, Mass.: Belknap Press of Harvard University Press.

McCartney, P. J. 1976. Charles Lyell and G. B. Brocchi: A study in comparative historiography. *British Journal for the History of Science* 9: 175–189.

McKinney, H. L. 1966. Alfred Russel Wallace and the discovery of natural selection. *Journal of the History of Medicine and Allied Sciences* 21: 333–357.

McKinney, H. L. 1972. *Wallace and Natural Selection.* New Haven, Conn.:
Yale University Press.

Meehan, T. (ed.). 1879. The doctrine of the morphology. *The Gardener's
Monthly and Horticulturist* 21: 278.

Meyer, A. B. 1895. How was Wallace led to the discovery of natural selection?
Nature 52: 415.

Mill, J. S. 1840. [An article on Coleridge.] *London and Westminster Review*
March 1840: No. 65. Reprinted in *The Collected Works of John Stuart Mill,
Volume X: Essays on Ethics, Religion, and Society,* ed. John M. Robson
(Toronto: University of Toronto Press, 1985).

Milne-Edwards, H. 1851. *Introduction à la zoologie générale.* Paris: Victor
Masson.

Moore, J. 1997. Wallace's Malthusian moment: The common context revisited.
In Bernard Lightman (ed.), *Victorian Science in Context,* 290–311.
Chicago, Ill.: University of Chicago Press.

Nelson, G. 2008. The two Wallaces then and now. In B. Gardiner, R. Milner, and
M. Morris (eds.), Survival of the Fittest, pp. 25–34. *The Linnean,* Special
Issue No. 9.

Nicholson, A. J. 1960. The role of population dynamics in natural selection. In
S. Tax (ed.), *Evolution After Darwin,* 477–522. Chicago, Ill.: University of
Chicago Press.

Ospovat, D. 1976. The influence of K. E. von Baer's embryology, 1828–1859:
A reappraisal in light of Richard Owen's and William B. Carpenter's
palaeontological application of "von Baer's law." *Journal of the History
of Biology* 9: 1–28.

Ospovat, D. 1981. *The Development of Darwin's Theory: Natural History,
Natural Theology, and Natural Selection, 1838–1859.* Cambridge:
Cambridge University Press.

Otte, D., and J. Endler (eds.). 1989. *Speciation and Its Consequences.* Sunderland, Mass.: Sinauer Associates.

Paul, D. B. 1988. The selection of the "survival of the fittest." *Journal of the
History of Biology* 21: 411–424.

Peabody, W. B. O. 1840. A report on the birds of Massachusetts made to the
legislature in the session of 1838–9. *Boston Journal of Natural History* 3:
65–266.

Pope, A. 1734. *Essay on Man.* London.

Poulton, E. B. 1896. *Charles Darwin and the Theory of Natural Selection.*
London: Cassell.

Prichard, J. C. 1851. *Researches into the Physical History of Mankind,* 4th ed.
5 vols. London: Houlston and Stoneman. [First edition of 1813 published in
2 vols.]

Provine, W. B. 1971. *The Origins of Theoretical Population Genetics.* Chicago,
   Ill.: University of Chicago Press.

Rachels, J. 1986. Darwin's moral lapse. *National Forum* (Phi Kappa Phi) 66(3):
   22–24.

Rennie, J. 1839. *Natural History of Birds: Their Architecture, Habits, and
   Faculties.* New York: Harper and Brothers.

Rennie, J. 1844. *Bird-Architecture.* Rev. ed. London: Charles Knight.

Richards, E. 1987. A question of property rights: Richard Owen's evolutionism
   reassessed. *British Journal for the History of Science* 20: 129–171.

Richards, E. 2005. The whole Wallace: Mapping the multi-dimensional man.
   [Review of M. Fichman, *An Elusive Victorian: The Evolution of Alfred
   Russel Wallace.*] *Metascience* 14: 237–241.

Richards, R. A. 1997. Darwin and the inefficacy of artificial selection. *Studies
   in the History and Philosophy of Science* 28: 75–97.

Romanes, G. 1886. Physiological selection: An additional suggestion on the
   origin of species. *Journal of the Linnean Society* (Zoology) 19: 337–411.

Rudwick, M. J. S. 1970. The strategy of Lyell's *Principles of Geology. Isis* 61: 5–33.

Rudwick, M. J. S. 1972. *The Meaning of Fossils: Episodes in the History of
   Palaeontology.* London: MacDonald.

Rudwick, M. J. S. 1998. Lyell and the *Principles of Geology.* In D. J. Blundell
   and A. C. Scott (eds.), *Lyell: The Past Is the Key to the Present.* Geological
   Society Special Publication 143, pp. 3–15.

Ruse, M. 1989. *The Darwinian Paradigm: Essays on Its History, Philosophy,
   and Religious Implications.* London: Routledge.

Schwartz, J. S. 1984. Darwin, Wallace, and the *Descent of Man. Journal of the
   History of Biology* 17: 271–289.

Schwartz, J. S. 1995. George John Romanes's defense of Darwinism: The
   correspondence of Charles Darwin and his chief disciple. *Journal of the
   History of Biology* 28: 281–316.

Schweber, S. S. 1980. Darwin and the political economists: Divergence of
   character. *Journal of the History of Biology* 13: 195–289.

Secord, J. A. 1981. Nature's fancy: Charles Darwin and the breeding of pigeons.
   Isis 72: 162–186.

Secord, J. A. 2000. *Victorian Sensation: The Extraordinary Publication,
   Reception, and Secret Authorship of Vestiges of the Natural History of
   Creation.* Chicago, Ill.: University of Chicago Press.

Shermer, M. 2002. *In Darwin's Shadow: The Life and Science of Alfred Russel
   Wallace: A Biographical Study on the Psychology of History.* Oxford:
   Oxford University Press.

Sloan, P. R. 1985. Darwin's invertebrate program, 1826–1836: Preconditions
   for transformism. In D. Kohn (ed.), *The Darwinian Heritage*, 71–120.
   Princeton, N.J.: Princeton University Press.

Slotten, R. A. 2004. *The Heretic in Darwin's Court: The Life of Alfred Russel Wallace.* New York: Columbia University Press.

Smith, C. H. 1992. Alfred Russel Wallace on Spiritualism, Man, and Evolution: An Analytical Essay. Torrington, Conn.: Privately published. Available at http://people.wku.edu/charles.smith/essays/ARWPAMPH.htm.

Smith, C. H. 2002. Wallace, Alfred Russel (1823–1913). In W. J. Mander and Alan P. F. Sell (eds.), *The Dictionary of Nineteenth-Century British Philosophers,* vol. 2, 1156–1160. Bristol, U.K.: Thoemmes Press.

Smith, C. H. 2008. Wallace, spiritualism, and beyond: "Change," or "no change"? In C. H. Smith and G. Beccaloni (eds.), *Natural Selection and Beyond: The Intellectual Legacy of Alfred Russel Wallace,* 391–423. Oxford: Oxford University Press.

Smith, C. H. 2013. A further look at the 1858 Wallace-Darwin mail delivery question. *Biological Journal of the Linnean Society* 108: 715–718.

Somkin, F. 1962. The contributions of Sir John Lubbock, F.R.S. to the 'Origin of Species': Some annotations to Darwin. *Notes and Records of the Royal Society of London* 17: 183–191.

Spencer, H. 1852. A theory of population, deduced from the general law of animal fertility. *Westminster Review* 57: 468–501.

Spencer, H. 1862. *First Principles of a New System of Philosophy.* London: Williams and Norgate.

Spencer, H. 1891. *Essays: Scientific, Political & Speculative.* 3 vols. London: Williams and Norgate.

Stauffer, R. C. 1975. *Charles Darwin's Natural Selection: Being the Second Part of His Big Species Book Written from 1856 to 1858.* Cambridge: Cambridge University Press.

Stevenson, B. 2009. Samuel Stevens, naturalist (1817–1899). *Micscape Magazine,* issue 166. Available at www.microscopy-uk.org.uk/mag/indexmag.html.

St. John, S. 1879. *The Life of Sir James Brooke, Rajah of Sarawak.* Edinburgh: William Blackwood and Sons.

Strickland, H. 1841. On the true method of discovering the natural system in zoology and botany. *Annals and Magazine of Natural History* 6: 184–194.

Sulloway, F. J. 1979. Geographic isolation in Darwin's thinking: The vicissitudes of a crucial idea. *Studies in the History of Biology* 3: 23–65.

Tammone, W. 1995. Competition, the division of labor, and Darwin's principle of divergence. *Journal of the History of Biology* 28: 109–131.

Tiedemann, F. (W. Bennett, Trans.). 1826. *The Anatomy of the Foetal Brain, with a Comparative Exposition of Its Structure in Animals.* Edinburgh: John Carfrae and Son.

Van Wyhe, J. (ed.). 2009. *Charles Darwin's Shorter Publications 1823–1883.* Cambridge: Cambridge University Press.

Van Wyhe, J., and K. Rookmaaker. 2012. A new theory to explain the receipt of Wallace's Ternate essay by Darwin in 1858. *Biological Journal of the Linnean Society* 105: 249–252.

Vorzimmer, P. J. 1969. Darwin's questions about the breeding of animals (1839). *Journal of the History of Biology* 2: 269–281.

Wallace, A. R. 1852. On the monkeys of the Amazon. *Proceedings of the Zoological Society of London* 20: 107–110. [S8]

Wallace, A. R. 1853. *A Narrative of Travels on the Amazon and Rio Negro, With an Account of the Native Tribes, and Observations on the Climate, Geology, and Natural History of the Amazon Valley.* London: Reeve. [S714]

Wallace, A. R. 1854. On the habits of the butterflies of the Amazon Valley. *Transactions of the Entomological Society of London* 2 (new series), part 7: 253–264. [S13]

Wallace, A. R. 1855. On the law which has regulated the introduction of new species. *Annals and Magazine of Natural History* 16 (2nd series): 184–196. [S20]

Wallace, A. R. 1855a. Borneo [letter dated 25 May 1855, from Si Munjon Coal Works, Sarawak, Borneo]. *The Literary Gazette and Journal of the Belles Lettres, Science, and Art* no. 2023: 683b–684a. [S22]

Wallace, A. R. 1856a. On the habits of the Orang-utan of Borneo. *Annals and Magazine of Natural History* 18 (2nd series): 26–32. [S26]

Wallace, A. R. 1856b. Some account of an infant "Orang-utan." *Annals and Magazine of Natural History* 17 (2nd series): 386–390. [S23]

Wallace, A. R. 1856c. On the Orang-utan or Mias of Borneo. *Annals and Magazine of Natural History* 17 (2nd series): 471–476. [S24]

Wallace, A. R. 1856d. A new kind of baby. *Chambers's Journal* 6 (3rd series): 325–327. [S30]

Wallace, A. R. 1856e. Attempts at a natural arrangement of birds. *Annals and Magazine of Natural History* 18 (2nd series): 193–216. [S28]

Wallace, A. R. 1856f. Observations on the zoology of Borneo. *Zoologist* 14: 5113–5117. [S25]

Wallace, A. R. 1857a. [Letter concerning collecting, dated 21 August 1856, Ampanam, Lombock.] *Zoologist* 15: 5414–5416. [S31]

Wallace, A. R. 1857b. On the Great Bird of Paradise, *Paradisea apoda* Linn.; 'Burong mati' (Dead bird) of the Malays; 'Fanéhan' of the Natives of Aru. *Annals and Magazine of Natural History* 20 (2nd series): 411–416. [S37]

Wallace, A. R. 1857c. On the natural history of the Aru Islands. *Annals and Magazine of Natural History* 20 (supp.): 473–485. [S38]

Wallace, A. R. 1858a. On the Arru Islands [communicated to the RGS meeting of 22 February 1858]. *Proceedings of the Royal Geographical Society of London* 2(3): 163–170. [S41]

Wallace, A. R. 1858b. On the entomology of the Aru Islands. *Zoologist* 16: 5889–5894. [S40]

Wallace, A. R. 1858c. Note on the theory of permanent and geographical varieties. *Zoologist* 16: 5887–5888. [S39]

Wallace, A. R. 1858d. On the habits and transformations of a species of *Ornithoptera*, allied to *O. priamus*, inhabiting the Aru Islands, near New Guinea. *Transactions of the Entomological Society of London* 4 (new series), part 7: 272–273. [S36]

Wallace, A. R. 1858e. On the tendency of varieties to depart indefinitely from the original type. *Proceedings of the Linnean Society* 3: 53–62. [S43]

Wallace, A. R. 1859. Correction of an important error affecting the classification of the Psittacidae. *Annals and Magazine of Natural History* 3 (3rd series): 147–148 (February 1859: no. 14, 3rd series). [S46]

Wallace, A. R. 1860a. On the zoological geography of the Malay Archipelago [communicated at the Linnean Society meeting of 3 November 1859]. *Journal of the Proceedings of the Linnean Society: Zoology* 4: 172–184. [S53]

Wallace, A. R. 1860b. The ornithology of Northern Celebes. *Ibis* 2: 140–147. [S57]

Wallace, A. R. 1863a. On the physical geography of the Malay Archipelago. *Proceedings of the Royal Geographical Society* 7: 217–234. [S78]

Wallace, A. R. 1863b. Remarks on the Rev. S. Haughton's paper on the bee's cell, and on the *Origin of Species*. *Annals and Magazine of Natural History* 12 (3rd series): 303–309. [S83]

Wallace, A. R. 1864. The origin of human races and the antiquity of man deduced from the theory of "Natural Selection." *Journal of the Anthropological Society of London* 2: clviii–clxx. [S93]

Wallace, A. R. 1865a. On the varieties of man in the Malay Archipelago. *Transactions of the Ethnological Society of London* 3 (new series): 196–215. [S82]

Wallace, A. R. 1865b. On the phenomena of variation and geographical distribution as illustrated by the Papilionidae of the Malayan Region [a paper read at the Linnean Society meeting of 17 March 1864]. *Transactions of the Linnean Society of London* 25, part I: 1–71. [S96]

Wallace, A. R. 1867. The philosophy of birds' nests. *Intellectual Observer* 11(6): 413–420. [S136]

Wallace, A. R. 1869. *The Malay Archipelago; The Land of the Orang-utan and the Bird of Paradise*. New York: Harper and Brothers. [S715]

Wallace, A. R. 1870. *Contributions to the Theory of Natural Selection*. London: Macmillan. [S716]

Wallace, A. R. 1873. Perception and instinct in the lower animals. *Nature* 8: 65–66. [S227]

Wallace, A. R. 1875. *On Miracles and Modern Spiritualism: Three Essays*. London: John Burns. [S717]

Wallace, A. R. 1876. *The Geographical Distribution of Animals; With A Study of the Relations of Living and Extinct Faunas as Elucidating the Past Changes of the Earth's Surface.* 2 vols. London: Macmillan. [S718]

Wallace, A. R. 1878. *Tropical Nature and Other Essays.* London: Macmillan. [S719]

Wallace, A. R. 1879. [Review of S. Butler, *Evolution Old and New.*] *Nature* 20: 141–144. [S311]

Wallace, A. R. 1880. *Island Life: Or, The Phenomena and Causes of Insular Faunas and Floras, Including a Revision and Attempted Solution of the Problem of Geological Climates.* London: Macmillan. [S721]

Wallace, A. R. 1886. Romanes *versus* Darwin: An episode in the history of the evolution theory. *Fortnightly Review* 40 (new series): 300–316. [S389]

Wallace, A. R. 1889. *Darwinism: An Exposition of the Theory of Natural Selection, with Some of Its Applications.* London: Macmillan. [S724]

Wallace, A. R. 1891a. [Letter to the Editor.] *Nature* 44: 518–519. [S440]

Wallace, A. R. 1891b. *Natural Selection and Tropical Nature: Essays on Descriptive and Theoretical Biology.* London: Macmillan. [S725]

Wallace, A. R. 1898. *The Wonderful Century: Its Successes and Its Failures.* London: Swan Sonnenschein. [S726]

Wallace, A. R. 1905. *My Life: A Record of Events and Opinions.* 2 vols. London: Chapman and Hall. [S729]

Wallace, A. R. 1908–1909. [Address on receiving the Darwin-Wallace Medal on 1 July 1908.] In *The Darwin-Wallace Celebration Held on Thursday, 1st July 1908, by the Linnean Society of London.* London (Linnean Society): Longmans, Green, February 1909: 5–11. [S656]

Wallace, A. R. 1909. [Dr. A. R. Wallace and Woman Suffrage.] *The Times* (London) no. 38880: 10d (11 February 1909). [S671]

Wallace, A. R., and H. W. Bates. 1849. Journey to explore the province of Pará. [Extract of letter dated 23 October 1848 from Wallace and Bates to S. Stevens.] *Annals and Magazine of Natural History* 3 (2nd series): 74–75. [S3]

Wallace, B. 1968. *Topics in Population Genetics.* New York: W. W. Norton.

Weissenborn, W. 1838. On the transformation of oats into rye. *Magazine of Natural History* 2: 670–672.

Whewell, W. 1837. *History of the Inductive Sciences, from the Earliest to the Present Times.* 3 vols. London: John W. Parker.

Whewell, W. 1839. Address to the Geological Society, delivered at the anniversary, on the 15th of February, 1839. *Proceedings of the Geological Society of London* 3 (1838–1842): 61–98.

Whewell, W. 1840. *The Philosophy of the Inductive Sciences, Founded Upon Their History.* 2 vols. London: John W. Parker.

Whewell, W. 1853. *Of the Plurality of Worlds.* London: John W. Parker.

Williams-Ellis, A. 1966. *Darwin's Moon: A Biography of Alfred Russel Wallace.* London: Blackie.

Wilson, L. G. (ed.). 1970. *Sir Charles Lyell's Scientific Journals on the Species Question.* New Haven, Conn.: Yale University Press.

Wood, R. J. 1973. Robert Bakewell (1725–1795) Pioneer animal breeder and his influence on Charles Darwin. *Folia Mendeliana* 8: 231–242.

Wood, S. 1995. The first use of the terms "Homology" and "Analogy" in the writings of Richard Owen. *Archives of Natural History* 22: 255–259.

Acknowledgments

This book has its genesis in my recent homage to Wallace, the annotated transcription of Wallace's Species Notebook published in 2013 by Harvard under the title *On the Organic Law of Change.* That presentation of the Species Notebook was to have included notebook content analysis and an in-depth examination of Wallace's thinking in relation to that of his colleague Darwin. It soon became clear, however, that the annotated notebook itself was going to be a hefty volume, and the planned analysis would have made it beyond hefty—too unwieldy for readers, and distracting from its most important feature: the notebook itself. I am grateful to my editor Michael Fisher at Harvard University Press for the happy solution of producing this companion volume; thank you, Michael, and thanks too to Susan Boehmer, Lauren Esdaile, Anne Zarrella, and Christine Thorsteinsson at HUP for your ideas, assistance, and infinite patience. I thank Keith Kuhn for his work on the book design, and I am especially appreciative of Kimberly Giambattisto of Westchester Publishing Services for her tremendous assistance with copyediting and proofing. The Linnean Society of London and the A. R. Wallace Literary Estate made the project possible by kindly granting permission to publish and analyze the Species Notebook, and I owe Linnean librarians Lynda Brooks and Elaine Charwat a debt of gratitude for their considerable assistance. I am indebted to Dana Fisher of the Ernt Mayr Library, Harvard University, for kindly scanning the Wallace 1855 and Darwin-Wallace 1858 papers for HUP. The manuscript benefited greatly from the comments and criticisms of Janet Browne and George Beccaloni; I hope I have done their criticisms justice, particularly in places where we differ in opinion or interpretation. I have had the great pleasure and benefit of discussing Wallaceana and Darwinana with George Beccaloni and Andrew Berry over the years, and I am especially grateful to George for being so generous with his time in responding to my myriad questions and ideas, and reading and rereading draft chapters. George's work with the Wallace Correspondence Project, Wallace Memorial Fund, and Wallace Collection at London's Natural History Museum, as well as his informative Wallace-related blog posts and articles, all proved immensely helpful to me and constitute distinguished service to the scholarly community and to Wallace's memory. I also thank Gillian Bentley, Matthias Glaubrecht, Randal Keynes, Bruce Kogut, Ahren Lester, Jim Moore, Jack Werren, and John van Wyhe for helpful information or sharing ideas over the course of this project. Very special thanks go to my wife Leslie Costa, indefatigable transcriber of a significant portion of the Species Notebook, helpful but no-nonsense editor, and mother of our spirited boys. Neither this book nor *On the Organic Law of Change* could have been completed in a timely manner without Leslie's painstaking efforts. I wrote this book, finally, as a 2012–2013 Fellow of the *Wissenschaftskolleg zu Berlin*. Sincere thanks to the *Wissenschaftskolleg* and the leadership of Western Carolina University and the University of North Carolina General Administration for making my leave from Highlands Biological Station possible, to my fellow Wiko Fellows and other friends who collectively and individually contributed so much to that memorable year for me and my family, and to the Wiko's fabulous librarians Sonja Grund, Anja Brockmann, Marianne Buck, Kirsten Graupner, and Thomas Reimer for their ever-cheerful assistance in procuring all manner of obscure literature.

Notes on the Text and Illustrations

This book makes extensive reference to Alfred Russel Wallace's Species Notebook of 1855–1859, Linnean Society of London manuscript no. 180; page citations refer to Wallace's notebook pagination. Readers wishing to refer to the Species Notebook can access it digitally on the Linnean Society website (linnean-online. org/54022/) or consult the full transcription with annotations published, with facsimile pages of the notebook, by Harvard University Press (Costa 2013a).

The copyright on the content of the Species Notebook and other literary works by Alfred Russel Wallace that were unpublished at the time of his death belongs to the A. R. Wallace Literary Estate, including the Species Notebook and many letters. The works quoted here are licensed under Creative Commons Attribution-NonCommercial-ShareAlike 3.0 Unported. To view a copy of this license, visit creativecommons.org/licenses/by-nc-sa/3.0/legal-code.

The story of Wallace's Species Notebook appearing in the introduction is adapted from an article by the author published in *Evolve* magazine (2013, 17: 30–33), a publication of the Natural History Museum, London.

All quotations of Wallace letters come from the Wallace Correspondence Project / Wallace Letters Online database of the Natural History Museum, London: Beccaloni, G. W. (ed.). 2012. Wallace Letters Online, www.nhm.ac .uk/wallacelettersonline.

These letters are cited throughout by Wallace Correspondence Project (WCP) number.

All quotations of Darwin's letters come from the Darwin Correspondence Project database of the University of Cambridge (www.darwinproject.ac.uk/), cited throughout by letter number.

Illustration Credits

The map of Wallace's travels and Figures 2.1 and 2.2 are reproduced from Alfred Russel Wallace, *The Malay Archipelago*, 4th ed. (London: Macmillan, 1872), courtesy of the Staatsbibliothek zu Berlin.

The images of the Species Notebook (Linnean Society of London MS 180) are provided courtesy of the Linnean Society and the A. R. Wallace Literary Estate.

The portrait of Alfred Russel Wallace was generously provided by G. W. Beccaloni. Copyright A. R. Wallace Memorial Fund and G. W. Beccaloni.

Figure 1.1 is taken from Robert Chambers, *Vestiges of the Natural History of Creation* (London: John Churchill, 1844), courtesy of the Staatsbibliothek zu Berlin.

Figure 1.2 is taken from Charles Darwin, *The Descent of Man, and Selection in Relation to Sex,* volume 2 (London: John Murray), courtesy of the Museum of Comparative Zoology's Ernst Mays Library (Harvard University) and Kathy Horton.

Figure 1.3 is taken from Alfred Russel Wallace, "Attempts at a Natural Arrangement of Birds" (*Annals and Magazine of Natural History,* 1856, 18 [2nd series]: 193–216), courtesy of the Natural History Museum, London, and Biodiversity Heritage Library (www.biodiversitylibrary.org/).

The facsimile page scans of the Wallace 1855 and Darwin and Wallace 1858 papers were provided courtesy of the Museum of Comparative Zoology, Harvard University.

Index

Abbott, John, 285
Actualism. *See* Uniformity/
    uniformitarianism
Aegelops ovata, 95, 139
Africa: hornbills, structure vs. habit,
    37, 68, 69, 121; locus of human origins,
    45; species affinity with Canary
    Islands, 74, 75, 76, 115; species
    change, 98
Affinity (homology, relationship), 37–38,
    66, 77–81, 90, 105, 108, 110–113, 115,
    154, 170, 173, 291; and classification,
    66, 77–78, 80, 110–111, 173, 225; island
    with mainland species, 29, 118. *See also*
    Analogy; Homology
Agassiz, Louis, 72, 80, 98, 102, 109, 110,
    111, 114, 133, 164, 284, 290
Alfuros, Alfures, Alfurus, 55, 89
Allen, Charles, 27, 32
Amazon River: as geographical barrier,
    23, 25–26, 145; Wallace *Narrative
    of Travels* on, 22, 23, 36, 45, 64, 129,
    239
Amazonia, 3, 9, 15, 21, 22, 24, 25, 26, 44,
    62, 143, 144; palms of, xi, 24; plans for
    expedition to, 9, 20–21, 25, 238; tribal
    languages of, 45. *See also* Bates, Henry
    Walter
Amboyna (Ambon), 43, 58, 60
Ammonites, 103
Ampanam, Lombock, 38

Anagenesis, 54, 146, 147, 153, 226, 246.
    See also Cladogenesis
Analogy (morphological), 37–38, 66, 68,
    78, 173. *See also* Affinity; Homology;
    Structure
Ants, 56, 85, 98; and aphids, 56, 292
Aru Islands, 6, 38, 40, 41, 42, 43, 44, 45,
    46, 47, 49, 51, 75, 88, 116, 240; "fossil"
    river channels of, 40–41. *See also*
    Dobbo; Wanumbai; Wokan
Aspalax, 113
Australia, xii, 19, 38, 39, 40, 42, 116

Bacan. *See* Batchian
Baer, Karl Ernst von, 101, 111
Balance (in nature), criticism of.
    See Design; Harmony; Natural theology
Bali, xii, 38, 39, 40, 45, 89, 157; (Baly), 38.
    See also Lombock
Bamboo, 48
Banda, 58, 60
Barnacles, Darwin's research on, 143,
    190, 225, 237–239
Barriers (geographical), 23, 25, 145, 146,
    157, 177, 265, 274. *See also* Riverine
    Barrier Hypothesis
Batchian (Bacan), 56, 89
Bates, Frederick, 242; Henry Walter, x,
    xii, 9, 15, 17, 18, 20, 21, 41, 44, 45, 47,
    48, 55, 60, 144, 145, 221, 236 242, 257,
    259, 261

Bats, 8, 67, 78, 102

Bees: putative instincts of, 66, 84, 85, 105, 118, 119, 120, 123, 143, 292; cell construction by, 84, 85, 105, 118–120, 123. See also *Melipona domestica*

Beetles, x, 81, 107, 210, 220, 285

Belemnites, 103

Bentham, George, 255; Jeremy, 284

Birds: air cells of bones, 81, 130, 293; "Attempts at a Natural Arrangement" of, 34–38, 54, 239, 246; biogeographic relationships, xii, 29, 39, 40, 57, 158; evolutionary relationships, 78, 79–81, 94, 162; flightless, 68; migration, 176, 177, 198, 292; nest architecture, 66, 82, 83, 85, 86, 87, 122, 123, 124, 125, 293; population growth examples with, 182, 215, 216; rearing in isolation, untried experiment with, 85; structure and habit, 68, 81, 121, 122, 132. See also Birds of paradise; Cockatoo; Cock-of-the-rock; Hornbills; Hummingbirds; Kingfishers; Maleo bird; Megapodiidae; Oriole; Pigeons; Parrots; Umbrella-bird

Birds of paradise, 8, 40, 48, 55, 60, 240. See also *Paradisea*

Bleeker, Pieter, 113, 284

Blyth, Edward, 31, 42, 49, 100, 140, 165, 194, 236, 241, 243, 284, 287

Borneo, xii, 27, 29, 31, 32, 34, 39, 42, 43, 90, 116, 144, 149, 157, 239

Boswell, James, 284

Botanical arithmetic, 16, 113–114, 116, 241, 249, 250–251. See also Principle of divergence

Bouru (Buru), 60

Branching, and evolutionary divergence, 3, 9, 78, 94, 102, 105, 106, 109, 111–112, 130, 141, 146, 153–154, 193, 213, 217, 225, 230, 246–247, 249, 254, 291. See also Anagenesis; Cladogenesis

Brocchi, Giovanni Battista, 92

Brooke, Sir James, 27, 29, 32, 43, 149

Bucerotidae, 69

Buch, Christian Leopold von, 74, 75, 76, 107, 113, 118, 140, 157, 227, 253, 284, 287

Buffon, Comte de, 75, 92–93, 131

Buru. See Bouru

Butler, Samuel, 253, 273

Butterflies, xiv, 24, 29, 30, 43, 44, 75, 157, 239, 240

Camouflage, and mimicry, x, 53, 210

Canary Islands, 74, 75, 76, 113, 117, 118, 140, 157, 227, 253, 287, 291

Candolle, Alphonse de, 133; Augustin Pyramus de, 16, 116, 132, 133, 181, 197, 215

Carboniferous period, 166

Carpenter, William, 19

Cave animals, adaptations of, 113, 291

Celebes (Sulawesi), 38, 39, 56, 57, 58, 59, 60, 88, 89, 90, 267

Cephalopterus, 22–23

Ceram (Seram), 58, 60, 89

Cetaceans, 78

Chambers, Robert, 17–19, 165, 287. See also *Vestiges of the Natural History of Creation*

Chinese: emperor, 100; Imperial rice variety, 108

Cineraria, 74, 75

Cladogenesis, 54, 146, 147, 153, 226, 246. See also Anagenesis

Classification, 35, 66, 77–78, 80, 110–111, 146, 153, 154, 173, 193, 230, 247, 248, 249, 252; and embryology, 80, 291; Natural system of, 78, 146, 153, 154; of varieties, 100; Quinarian system of, 154

Climate: and selection, 189, 190, 202, 217, 222, 223; does not predict species relationships, 41–42, 96–97, 114–115; effects of gradual change of, 98–99, 114–115, 146, 149, 202, 217, 292; Lyellian change in, 3, 98–99, 104, 146

Coal: geological formations of, 166; mines in Simunjon, Borneo, 13, 32

Cock-of-the-rock, 22, 23

Cockatoo, 38, 158

Coleoptera, Coleopterous, x, 81, 285. See also Beetles

Coleridge, Samuel Taylor, 123, 270

Collecting, collections, x, xi, xii, xiv, 3, 7, 11, 13, 20, 21, 22, 23, 25, 26, 27, 29, 31, 32, 38, 40, 48, 49, 55, 58, 60, 61, 68, 69,

75, 77, 243, 276; bearing on species origins, x, 3, 7, 15, 18, 20, 22, 26, 27, 48; importance of locality information, 21, 26; loss of Amazonian, xi

Combe, George, 284

Common descent, 79, 80, 165, 178, 268, 270

Competition, 189, 190, 193, 222–224, 226, 227, 230; and population pressure, 51, 87, 105, 107, 134, 193, 194, 215, 216, 218, 223, 230; competitive exclusion, 72, 87, 89, 118, 193, 226, 247, 249; with introduced species, 72, 133, 182. *See also* Malthus, Thomas Robert; Struggle for existence

Comte, Auguste, 285

Consilience (of Inductions): defined, 12, 65–66; in Wallace's and Darwin's method, 30, 103, 104, 142, 173, 230, 277

Cotingidae. *See* Cock-of-the-rock

*Ctenomys,*112. *See also* Tucotuco

Cuvier, Georges, 80, 84, 99, 112

Cyclopedia of Natural History, Knight's, 67, 121, 284

Dampier, William, 108, 284, 288

Darwin, Charles: books by, 268–269; consilience method of, 30, 103, 104, 142, 173, 230, 277; discovery of natural selection, 236; extracts by, read at Linnean Society, 175, 180–195, 227–228; letter to Emma on species theory, 175; letters with Wallace, 239–241; observations of children by, 127; petition for pension award for Wallace, 63, 266; praised by Wallace, 12, 60, 256–257, 265; putative appropriation of Wallace's ideas by, 245–254; receipt of Ternate manuscript by, 235–245; Transmutation Notebooks, 76, 104, 107, 108, 112, 114, 116, 117, 119, 123, 128, 134, 135, 138, 140, 246, 285, 287–288, 290–294; view of natural selection (compared with Wallace's), 144, 214–218, 219–227, 230. *See also* Delicate arrangement; Family-level selection; Natural

selection; *On the Origin of Species*; Principle of divergence; Sexual selection

Darwin, Emma, 175

Darwin, Erasmus, 17, 253, 270

Darwinism: eclipse of, 272–273, 275, 276; in book titles, 271. *See also* Neo-Darwinism

Darwinism (Wallace), xvii, 6, 13, 61, 87, 218, 269, 272

Darwinizing, 270

Delicate arrangement, 5, 257, 260, 262

Design: critique of, in Species Notebook, 8, 66, 67–73, 81, 97, 98, 119, 121, 129, 130, 131, 132, 141, 277, 294; Underlying argument against, in Darwin's orchid book, 268. *See also* Harmony; Natural theology

Development or change: progressive, 9, 18, 29, 79, 93–94, 111, 161, 162, 165, 227, 290; in *Vestiges*, 18–19; Lyell's arguments against, 9, 94, 111, 162. *See also* Embryology; Progressive development, theory of; Succession, geological; Transmutation

Didelphis, Didelphys, 94, 111, 112, 290

Disharmony, 116

Distribution: geographical, 3, 9, 18, 19, 63; and the Sarawak Law, 30, 153, 157, 173; Darwin's interest in, 108, 113–118, 189, 252, 268, 280, 287–288; geological changes affecting, 150; in the Species Notebook, 66, 73–77; Lyell on, 20, 149; of the Malay Archipelago, 39, 40–42, 57–58; Wallace's interest in, 15, 16, 21, 22, 24, 25, 28, 32, 39, 40–42, 57–58, 73, 89, 96, 113–118, 266, 267, 277, 287–288, 291–292; *The Geographical Distribution of Animals,* xiv, 14, 61, 269

Divergence: in isolation, 117, 139–140, 141, 157, 158, 247, 274–275; of character, 247, 294, 297; of lineages, 37, 54, 136, 146, 137, 176, 198, 205–206, 213, 217, 229, 246, 249, 274, 291; Principle of, 134, 141, 193–194, 225–227, 228, 230, 241, 246, 247, 249, 252–253, 255, 274. *See also* Branching; Ecological division of labor; Trees

Dobbo, Aru Islands, 40, 41
Dogs, 10, 11, 95, 123, 136, 137, 186, 198; transmutation of breeds, 10–11, 95, 137
Domestication/domestic varieties: and transmutation, 3, 8, 50, 52, 53, 84, 95–96, 135–137, 176–177, 189, 206, 218, 230, 249, 294; Darwin's use of as analogy, 52, 189, 230, 241, 249, 255, 261, 268; in *Principles of Geology*, 176–177, 194, 213, 230; reversion of, 135–136, 194, 206, 210, 213, 218, 221; variation in, 130, 135–137, 294; Wallace's use of, as evidence for transmutation, 10, 95–96, 136–137, 198, 209, 294; Wallace on, in Ternate essay, 84, 176, 177, 197–198, 206, 209, 210, 213, 218, 221
Dorey, New Guinea, 55
Drift, erratic blocks and, 109, 110, 290
Dufour, Pierre, 284
Duivenboden, Mr., 48, 55
Dyaks, 32, 90, 126

Earl, Windsor, 267
Ecological division of labor, 54, 193, 225, 228, 248, 249, 252. *See also* Divergence; Physiological division of labor; Woburn Abbey
Ectopistes migratorius, 198, 201
Embryology: and classification, 110–111, 291; and "law of parallelism," 101, 103; as evidence of transmutation, 3, 8, 18–19, 66, 103, 141, 231; in *Vestiges*, 18–19. *See also* Baer, Karl Ernst von
Endemism, 16, 117; and islands, 58, 74–75, 96–97, 113, 116–117, 158; island antiquity and, 96–97, 117, 157, 158. *See also* Divergence; Islands; Speciation; Transmutation
Erichson, W. F., 284, 285
Ethics: ethical sense, 62, 66, 67, 70–71, 269; and reading of Darwin-Wallace papers, 181, 254–258. *See also* Delicate arrangement; Morality; Religion; Spiritualism
Ethnology: ethnographic observations, 47, 49, 90, 143; Darwin's interest in, 128–129; Wallace's interest in, 44–46,

49, 52, 89, 90, 129. *See also* Human races; Morality
Evolution. *See* Transmutation
Experience, and learning, 66, 81–87, 118–125, 127, 141. *See also* Instinct
Extinction, 10, 31, 35, 37, 42, 43, 51, 55, 58, 93, 97, 98, 101, 128, 147, 161, 166, 189, 193, 198, 201, 205, 213, 220, 223, 224, 248, 252, 298
Eyes, reduced in darkness-loving animals, 113, 291

Fabre, Esprit, 95–96
"Face of nature" metaphor; 133, 182, 215, 294. *See also* Wedges metaphor
Family-level selection, 219, 227, 228, 230, 249, 261
Fish: biogeographic relationships, 158; branching order of, 9, 94, 162; embryological development, 18–19
Flora: biogeographical relationships, 42, 57, 76, 116; Darwin's botanical arithmetic, 250–251; of Canary Islands, 74, 113, 118, 253; eastern North America and eastern Asia compared, 76, 114–115; fossil, 114; naturalized, 133–134
Flores, 89
Forbes, Edward, 27, 28, 30, 153, 165, 166, 173
Fossils: fossil record, xii, 3, 8, 9, 114, 141, 150, 153, 154; affinity with extant species, 29, 102; and progressive development, 9, 18, 29, 79–80, 93–94, 111, 230; flora, 114; gaps in record, 35, 102, 109, 112, 165, 169, 193, 260, 290; generalized structure of earlier, 9–10, 79, 94, 101–102, 111, 114; intermediacy of common ancestors, 109; of Europe, 114; of the Carboniferous Period, 166; of the Permian Period, 166; reality of, 80, 130–131. *See also* Ammonites; Belemnites; *Didelphis*; *Ichthyosaurus*; Labyrinthidonts; *Rhynchosaurus*
Fullom, Stephen Watson, 284

Galápagos Islands, 10, 29, 73, 74, 113, 116, 117, 157, 291
Galela, 89, 90

Geological periods: Carboniferous, 166; Permian, 166; Silurian, 298; Tertiary, 34, 72

Geology. *See* Fossils; Geological periods; Gradualism; Succession; Time; Uniformity/uniformitarianism; Uplift; Volcanic islands

George, Henry, 135

Gilolo (Halmahera), 11, 45, 46, 49, 50, 55, 56, 72, 87, 88, 89, 90, 118, 213, 235, 243; grasslands of, 55, 72, 87, 89, 118

Gould, John, 29

Gradualism: gradual change (geological, climatological, transmutational), 3, 11, 37, 43, 61, 83, 92, 93, 96, 98, 99, 102, 106, 109, 113, 114, 115, 135, 136, 139, 146–147, 149, 161, 177, 197, 198, 213, 224, 268, 269, 275, 277, 290. *See also* Lyell, Charles; Progressive development; Succession, geological; Uniformity/uniformitarianism; Uplift

Grant, Robert Edmond, 17

Grasses: of Gilolo and llanos, 55, 72, 87, 89, 118

Gray, Asa, 114, 175, 178, 189, 194, 225, 226, 233, 236, 241, 248, 249, 252, 255, 268

Guiana, 22, 26

Habit: and instinct, 58–59, 81–87, 108, 118–125; and structure, 36–37, 68, 72, 98, 105, 115, 121, 125, 142, 202, 277, 291, 292. *See also* Instinct

Halmahera. *See* Gilolo

Harmony (in nature): criticism of in *Species Notebook*, 8, 66, 72, 98, 105, 132–135, 141, 177, 277, 294; Darwin's criticism of, 182, 185; Wallace and Darwin's views compared, 215. *See also* Design; Natural theology

Heliconia butterflies, 24

Henslow, John Stevens, 96

Herbert, William, 138

Herschel, John, 65, 66

Homology, 38, 66, 78, 80, 105, 130, 173. *See also* Affinity; Analogy

Hooker, Joseph Dalton, xiii, xiv, xvii, 11, 56, 57, 62, 139, 144, 175, 178, 181, 189, 214, 217, 226, 231, 233–234, 235, 236, 237, 241, 243, 248, 250, 252, 254–257, 260, 262; Mrs. [Frances Harriet Hooker], 255. *See also* Delicate arrangement

Hornbills, 37, 68, 69, 121

Horses, 34, 72, 176, 182, 209, 213

Huc, Evariste Régis, 108, 139, 284, 288

Human evolution: primate relationships, 47, 90, 108, 125–129, 269, 293; taillike structures as evidence for, 90, 125; Wallace's reservations about, xiv, xvi, 1, 61–62, 71, 86, 129

Human races: Alfurus (Alfuros), 55, 89; Dyaks, 32, 90; Malays, 46, 88, 89, 127; Malayan-Papuan relationship, 45–46, 49, 52, 55–56, 88–90; observations of, 44, 45, 56, 66, 87–90, 107, 127, 288; variation and origin, 44–45, 66, 87–91. *See also* Instinct, in humans

Humboldt, Alexander von, 16, 21, 28, 77, 107, 113, 114, 116, 287

Hummingbirds, xii, 158

Huxley, Julian, 275, 278; Thomas Henry, 58, 62, 84, 86, 266, 267, 275, 276

Icebergs, 110, 143; iceberg theory, fatal argument concerning, 110

Ichthyosaurus, 102

India: hornbills, structure vs. habit, 37, 68, 69, 121

Indians, purported navigational instinct, 85, 86

Infants: suckling instinct, 85, 86, 120; death of Darwin's son, 12, 255, 257; behavioral observations, 127

Instinct: and experience/learning, 66, 68, 81–87, 120, 122, 123, 124–25; and habit, 58–59, 118–123, 125, 292; and reason, 83–84, 87, 99, 293; Darwin on, 123–125; experimental test for, 85–86, 120, 121; "failure of" in boring beetles, 220; in humans, 85–87, 122–123, 127, 129; of bees in cell-building, 84, 105, 118–120; of birds in nest-building, 83, 122–123, 124; variation in, 82, 84, 105, 123, 124, 186, 215, 292

Introduced species, and competition, 72, 182

Islands and Archipelagos. *See* Aru Islands; Bali; Batchian; Canary Islands; Celebes; Flores; Galápagos Islands; Gilolo; Jamaica; Java; Ké Islands; Lombock; Madeira; New Guinea; Saint Helena; Sicily; Timor

Islands: antiquity and endemism, 58, 96–97, 117, 157, 158; endemism, 58, 74–75, 96–97, 113, 116–117, 158; isolation and speciation, 96–97, 116–117, 140, 145, 157, 227; relationship of species to those of nearest mainland, 10, 29, 96–97, 116, 117, 158

Isolation: and speciation: 96–97, 116–117, 140, 145, 157, 227, 247; divergence resulting from, 117, 139–140, 141, 157, 158, 247, 274–275; in Blyth, 140; in von Buch, 74, 75, 76, 140, 157, 227. *See also* Divergence; Endemism; Islands; Speciation; Transmutation

Jamaica, 114
Java, 38, 39, 60, 89, 114, 157, 244, 285
Jenkin, Fleeming, 274
Johnston, George, 284
Journal of Researches / Voyage of the Beagle (Darwin), 10, 16, 107, 112, 113, 182, 236, 237, 270, 284; Galápagos observations from, 73, 96, 117
Jumaat, death of, 11, 55
Justice, social. *See* Socialism

Kai Islands. *See* Ké Islands
Kangaroo, xii
Ké Islands, 40, 45, 46, 75, 89
Khang-Hi, Emperor, 100. *See also* Rice, imperial
Kin selection. *See* Family-level selection
Kingfishers, 35, 68
Knight, Charles, 67, 284; See also *Cyclopedia of Natural History,* Knight's
Knighton, William, 284
Kuching, Sarawak, 27, 32

Labyrinthidonts, 102
La Croix, Paul (Pierre Dufour), 284

Lake Superior (Agassiz), 102, 284; Wallace criticisms of, 111, 133; accounts of invasive plants in, 133
Lake Superior, erratic blocks around, 110
Lamarck, Jean-Baptiste, 17, 19, 29, 52, 94, 96, 111, 113, 165, 176, 205; Lamarckism, 94, 210, 246, 265, 269, 273, 274, 275
Latham, Robert Gordon, 45–46, 88
Law, scientific, defined, 149, 194
Law of succession, 101. *See also* Succession, geological; Progressive development
Lawrence, William, 45, 128, 129
Learning. *See* Experience
Lepidoptera, 24, 81. *See also* Butterflies; *Ornithoptera*
Lesson, René Primevère, 89, 107–108, 284, 288
Library of natural history, Wallace's proposal for, 8
Lindley, John, 16, 38, 108, 137, 139, 284
Link, Heinrich Friedrich, 100, 108, 137
Linnaeus, 43
Linnean Society of London: Darwin-Wallace medal, 261; reading of Darwin-Wallace papers at, xii, xiv, 5, 12, 39, 144, 243, 245, 255–256, 262, 266; Wallace notebooks and journals at, 300; Wallace "Species Notebook" at, 2, 5–6, 13–14, 15, 300
Llanos, 72
Locusts, 71, 98, 205, 223, 224
Lombock (Lombok), xii, 38, 39, 75, 157
Lubbock, John, 236, 250, 251, 252
Lyell, Charles: xiii, xiv, xv, 5, 8, 9, 10, 11, 12, 16, 19, 20, 21, 28, 29, 30, 31, 42, 49, 50, 52, 56, 57, 62, 63, 66, 67, 71, 74, 83, 91, 92, 93, 94, 95, 97, 98, 99, 100, 101, 107, 109, 110, 111, 114, 115, 125, 131, 132, 133, 135, 137, 138, 144, 146, 149, 161, 162, 165, 175, 176, 177, 178, 181, 182, 189, 194, 197, 198, 201, 202, 205, 206, 209, 210, 213, 214, 217, 218, 221, 230, 231, 233, 234, 235, 236, 239, 241, 243, 244, 245, 246, 250, 253, 254, 255, 256, 257, 260, 262, 266, 268, 284, 287, 288; and the Darwin-

Wallace papers, 175–181, 254–263;
anti-transmutationism of, 8, 10, 20, 29,
83, 92, 99–100, 175, 194, 218, 230; as
object of Sarawak Law and Ternate
papers, 175–177; as Wallace's "evolutionary foil," 12, 146, 230; influence on
Wallace, 16–17; Lyellian change, 3, 11,
29, 39, 93, 104, 109, 146, 149, 157, 210,
245; notebooks on "species question,"
28, 31, 250; on lessons from domestic
varieties, 95–96, 135–139, 176–177,
194, 213, 230; on limits of variability, 9,
10, 67, 95–96, 135–139, 177, 194, 210;
Wallace's critique of in the Species
Notebook, 8–11, 66–67, 91–103, 125,
259. *See also* Delicate arrangement;
Gradualism; *Principles of Geology*

Macassar, Makassar (Ujung-Pandang), 32,
38, 39, 41, 43, 47, 49, 75, 88
Macleay, William, 154, 173
Macrocephalon (Maleo bird), 58–59
Macroura, 103
Madeira, 25, 26
Malacca, 27, 35, 145, 157
Malay Archipelago: Wallace's papers on,
41, 43, 44, 46, 57, 75, 240; Wallace's
travels in, 4, 6, 47, 61, 277. *See also*
Sarawak Law; Ternate essay
[*The*] *Malay Archipelago* (Wallace), 1, 8,
13, 44, 69, 89, 91, 129, 220, 269;
illustrations from, 4, 69, 91
Malay Journals, Wallace's, 46, 48, 49, 56,
88, 89; summarized, 300
Malays, 46, 88, 89, 127
Malaysia, xi, 26, 27, 35
Maleo bird (*Macrocephalon*), 58–59
Malthus, Thomas Robert, xii, 51, 52,
87, 107, 134–135, 177, 181, 182, 185,
197, 215, 222, 241, 287; *Essay on
Population,* 51, 107, 134, 197, 222;
Malthusian(ism), 52, 134, 190, 197,
255. *See also* Population(s); Struggle
for existence
Mammals, 7, 19, 78, 81, 94, 108, 121, 130,
132, 291. *See also* Bats; Cetaceans;
Dogs; Manatees; Marsupials; Mias;
Monkeys; Orangutan

Manatees, 170
Marsupials, 94, 111, 112; kangaroo, xii.
See also Didelphis, Didelphys
Materialism, criticism of, 71
Matthew, Patrick, 59, 267
Mauritia palms, 24
Mayr, Ernst, xiv, 221, 279
McKinney, H. Lewis, 6, 8, 16, 22, 27, 30,
46, 47, 49, 52, 87, 134, 233, 242
Megapodiidae, 38, 58–59, 267
Megapodius, 38
Melipona domestica, 119
Melville, A. G., 285
Melville de Carnbee, P., 285
Menado (northern Celebes), 56, 58
Mesmerism, 62. *See also* Phrenology;
Spiritualism
Mias, 32, 126
Migration, 20, 98, 176, 177, 275, 292
Mill, John Stuart, 122, 123
Milne-Edwards, Henri, 81, 228, 249, 250,
254, 285
Mimicry. *See* Camouflage
Modern Synthesis, 221, 275, 276, 278;
founders of, 276
Mollusks, 103, 113, 114, 158, 161, 285
Monogenism, 127–128
Monkeys, 25, 26, 239
Monsoon, 6, 40, 41
Morality: moral capacity, 44, 62, 70, 129;
of native peoples, 44. *See also* Ethics
Morphology. *See* Structure
Mouthparts, 81
Müller, Salomon, 285
Murchison, Roderick, 166
Mysol, 89

[A] *Narrative of Travels on the Amazon and
Rio Negro* (Wallace), 22, 36, 45, 64, 239
Natura non facit saltum, 193
Natural selection: accumulation of
variations, 45, 190, 216; counterpart to
artificial selection, 136–137, 187, 269;
Darwin's discovery of, 236; timing of
discovery, 105–107; Wallace and
Darwin's views of compared, 144,
214–218, 219–227, 230; Wallace's
centrifugal governor analogy, 52–53,

Natural selection *(Continued)*
213; Wallace's criticism of term, 50,
190, 224, 258; Wallace's defense of, 1,
13; Wallace's description of, 50–53;
Wallace's discovery of, ix, xii-xiii, 2, 11,
46, 49–50, 87–88, 125, 133, 147, 197;
Wallace's reservations over applicability
to humans, xiv, xv-xvi, xvii, 62, 71, 86,
267. *See also* Family-level selection;
Physiological selection; Principle of
divergence; Sexual selection
Natural Selection (Darwin manuscript),
107, 108, 109, 115, 228, 232, 246, 285,
287–288, 290–294
Natural theology, 37, 67, 68, 71, 72, 97,
121, 197. *See also* Design; Harmony
Neo-Darwinism, 263, 273, 276. *See also*
Darwinism; Wallaceism
Nests, birds', 66, 82, 83, 85, 86, 87, 122,
123, 124, 293. *See also* Birds; Instinct
New Guinea, xi, xii, xvi, 6, 11, 26, 40, 41,
42, 43, 48, 55, 56, 60, 89, 116, 220, 243.
See also Dorey, New Guinea; Waigiou,
New Guinea
New Zealand, flora, 251
Newman, Edward, 25

On the Organic Law of Change, 1, 2, 6,
12, 48, 63, 259, 280
On the Origin of Species, as "one long
argument," 268; citations of Wallace in,
267, 297–299; praise for by Wallace,
12, 60, 256–257, 265; publication of,
241; references to "my theory" in,
297–299
Oolite, 111, 112, 290
Orangutan, 6, 7, 13, 32, 90, 91, 105, 125,
126, 127, 141, 239, 293; canine teeth of,
32–33; hunting, 7, 13, 32, 90; Jenny, 90,
126, 127; observations of infant, 7,
90–91, 105, 125, 126, 141, 293;
variation in, 32. *See also* Mias
Oriole, Baltimore, variable nesting habits
of, 124
Orinoco, Brazil, 72, 118
Ornithoptera, 29, 43, 44, 240; divergence
in, 43–44
Osborn, Henry Fairfield, 221

Owen: Richard, 29, 38, 77, 78, 80, 84, 86,
101, 102, 109, 112, 154, 213, 260, 285,
290; Robert, 81, 82, 285

Paleontology, 29, 108, 109, 153, 231, 268,
275, 278, 290. *See also* Fossils
Palms: of the Amazon, xi, 24; *Mauritia,*
24; Sago, 6; Wallace book on, xi
Papilio, 44
Papilionids, 75
Papuans, 46, 87, 88, 89, 127
Paradisea: P. apoda, 40; *P. papuana,* 40;
P. rubra, 60. *See also* Birds of paradise
Parallelism, law of, 101, 103
Parrots, 36
Peabody, William B. O., 124
Permian period, 166
Phrenology, 62. *See also* Mesmerism;
Spiritualism
Physiological division of labor, 193, 227,
228, 248, 249. *See also* Ecological
division of labor; Principle of
divergence
Physiological selection, 274. *See also*
Romanes, George
Pictet, François Jules, 29
Pigeons: collected, 75; domestic varieties
of, 209; Passenger, 198, 201
Plants: fossil, 114; climbing, 268;
homologous structures of, 79–80, 130;
insectivorous, 143, 268; transmutation
of, 112, 116. *See also* Flora
Polarity theory, Edward Forbes's, 27, 30,
153, 165, 166
Polygenism, 128
Pope, Alexander, 83, 285
Population(s): appear stationary, 197, 202,
215; explosive increase of locust and
ant, 71, 72, 98, 133, 176, 205, 223–224;
method for estimating global species,
77. *See also* Competition; Malthus,
Thomas Robert, Malthusian(ism);
Struggle for existence
Poulton, Edward Bagnall, 221
Primrose, 95, 138, 294
Principle of divergence, 54, 134, 141,
193–194, 225–227, 228, 230, 241, 246,
247, 249, 252–253, 255, 274. *See also*

Branching; Divergence; Ecological
division of labor; Natural selection;
Physiological division of labor; Trees
Principles of Geology (Lyell): anti-
transmutation arguments of, 8–11, 50,
91–103; arguments critiqued in Species
Notebook, 8, 66, 91–103, 129–132, 135,
284, 288; influence in Sarawak Law
paper, 29, 146, 161; influence in Ternate
essay, 175–177, 197–213; struggle for
existence in, 52, 176, 177, 197; Wallace
cited in, 63, 253; Wallace's early
reading of, 19–20. *See also* Lyell,
Charles
Prichard, James Cowles, 45, 128
Progressive development, theory of, 9, 18,
94, 111, 290. *See also* Succession,
geological
Proteus, 113
Pyrethrum, 74, 75

Queen Victoria. *See* Victoria, Queen
Queries about Expression, Darwin's, 127,
293
Quinarian system. *See* Classification

Races, human. *See* Human races
Ramsay, Andrew Crombie, 285
Ray, John, 75
Reeve, Lovell Augustus, 158
Relationships, among species. *See* Affinity
Religion, xv, 62, 70, 71, 73, 128
Renan, Ernest, 62, 71, 129, 285
Rennie, James, 82
Reversion, 52, 95, 96, 135, 136, 205, 209,
213, 221
Rhynchosaurus, 102
Rice, imperial (new variety), 100, 108, 139
Riverine Barrier Hypothesis, 25. *See also*
Barriers
Romanes, George, 273–275
Rudimentary structures, evolutionary
interpretation of, 30, 141, 146, 153, 170,
173
Rupicola rupicola, 22, 23

Sadong River, Borneo, 13
Sago (palm), 6

Sahoe, Ternate, 55, 89, 90
Saint Helena, 76, 117, 157, 291
Saint John, Spencer, 27
Sarawak, island of, 6, 13; "Law," xii, xiii, 5,
6, 7, 9, 12, 13, 15, 27–31, 34, 35, 38, 41,
42, 43, 47–54, 63, 96, 98, 100, 104, 115,
125, 141, 143, 144–173, 194, 227,
239–240, 243, 246, 249, 250, 252–253,
254, 256, 259, 267, 277
Sarawak Law: facsimile and analysis,
144–173; paper summarized, 27–31;
stated, 7, 28, 51, 150, 277
Scott, Walter, 285
Sexual selection, 1, 63, 189, 194, 227–228,
230, 249, 261, 265, 269
Sicily, 116
Siebold, Philipp Franz Balthasar von,
285
Silk, George, 60, 236, 256
Silurian period, 298
Simunjon, Borneo: coal mines, 13, 32;
River, 31
Singapore, 6, 26, 27, 29, 32, 34, 38, 60,
239, 264; portrait taken at, 264;
Wallace's arrival at, 27, 239
Skull sutures, 130
Socialism, 269; Owenite, 81, 82, 134
Somerville, Mary, 285
Speciation: allopatric, 75, 118, 146, 226,
247, 265, 275; sympatric, 226, 227, 230,
247, 248. *See also* Islands; Isolation;
Varieties
Species: concept, biological, xiv, 63, 75,
140; doubtful, 138; fixity, belief in, 8,
10, 11, 93, 102; varieties as incipient,
141, 274, 294
Species change. *See* Gradualism;
Reversion; Transmutation
Species Notebook: and consilience,
66–67; at Linnean Society of London,
2, 6, 14, 15; critique of Lyell in, 8–12,
30, 66, 91–101; ethnographic entries in,
47; overview, 5–14; publication of, 1–2;
reference to planned pro-transmutation
book in, 9, 99; transmutation themes of,
8–11, 65–103, 259. *See also* Lyell,
Charles; Spiritualism; Transmutation;
Wallace, Alfred Russel

Spencer, Herbert, 24, 25, 71, 73, 223, 258, 285

Spiritualism, xv, xvii, 1, 33, 61, 62, 71, 269

Sterility, 99, 274; in social insects, 228–229

Stevens, Samuel, 21, 31, 38, 39, 143

Strickland, Hugh, 35, 285

Structure (morphology, anatomy, organization), 8, 33, 37, 54, 59, 69, 77–81, 90, 101, 102, 147, 186, 190, 193, 202, 215, 248, 268; and affinity, 170, 213; and design, 67–68, 71, 72, 112, 119, 131; and habit, 105, 121, 125, 142, 277, 292; law of generalized to specialized, 101–102; rudimentary, 30, 141, 146, 153, 170, 173. *See also* Affinity; Analogy; Homology

Struggle for existence, 11, 53, 194, 197, 201, 214, 228, 294; in *Principles of Geology*, 52, 176, 177, 197

Succession: geological, 9, 28, 29, 94, 101, 102–103, 105, 109, 176, 177, 230, 290; in Sarawak Law paper, 48, 55, 146, 153, 154, 158, 173; in Ternate essay, 213. *See also* Progressive development; Time

Sulawesi. *See* Celebes

Supersilience, 103

Survival of the fittest, 50, 258

Swainson, William John, 28

Synonyms/synonymy, plan to stop the proliferation of, 8

Systems of nature, compared to map, 154, 170

Teide, Mt., 118

Teneriffe, 74, 75, 118

Ternate essay, xiii, 5, 7, 10, 11, 12, 15, 49–56, 57, 58, 77, 96, 125, 134, 136, 137, 175–177, 219, 220, 221, 222, 223, 227, 230, 232, 233, 235, 240–241, 242, 245, 246, 248–249, 252, 258, 259, 260, 267, 273; facsimile and analysis, 194–213

Ternate, island of, xiii, 6, 11, 46, 47, 48, 49, 55, 56, 58, 60, 88, 89, 243, 244

Tertiary period, 34, 72

Theology, natural. *See* Natural theology; Religion

Thomson, James, 285

Tiedemann, Friedrich, 99–100

Time, geological, 11, 26, 39, 53, 54, 57, 93, 96, 106, 117, 149, 157, 161, 166, 169, 176, 190, 198, 210, 217

Timor, 56, 58, 60, 89

Traité de Paléontologie (Pictet), 29

Transmutation: and human origins, 34, 45–47, 90–91, 99, 128, 129; and Sarawak Law paper, 12, 27–31, 35, 144–173; argument of *Vestiges of the Natural History of Creation*, 7, 15–18, 19; Darwin's conversion to, 17, 236; domesticated varieties as evidence for, 95–96, 137–138, 139; evidence from geology for, 101–103, 109–110; evidence from islands for, 96–97, 140, 145; Herbert Spencer and, 24–25; Lyell's arguments against, 8, 10, 20, 29, 83, 92, 99–100, 175, 194, 218, 230; Notebooks (Darwin), 104, 107, 108, 111, 112, 114, 116, 117, 123, 128, 134, 135, 173, 246, 287–288, 290–294; popular resistance to, 17; similarity in Darwin and Wallace's paths to, 2–3, 5, 104–142, 229–231; themes of Wallace's Species Notebook, 8–11, 65–103, 259; Wallace's conversion to, 7, 15–18, 26, 150, 237; Wallace's planned book on, 3, 5, 8, 9, 93, 99, 259, 265, 277. *See also* Branching; Fossils; Gradualism; Islands; Sarawak Law; Ternate essay; Time; Trees

Trees: evolutionary, 35, 36, 54, 110, 111, 130, 153, 154, 162, 213, 246, 247; "gnarled oak" analogy, 28, 154; in *Vestiges*, 19; Tree of Life, 193, 226, 246, 265; unrooted, in Wallace's *Birds* paper, 35–36. *See also* Branching; Divergence

Trogonoptera, 29

Trogons, 29, 68, 157, 158

Tucotuco, 107, 112, 113

Ujung-Pandang. *See* Macassar, Makassar

Umbrella-bird, 22–23

Uniformity/uniformitarianism, 8, 20, 29, 30, 269

Uplift, geological, 39, 72, 101, 109, 149, 157, 290

Urville, Jules Dumont d', 107, 284, 287, 288

Variation/variability, 3, 16, 22, 32, 44, 53, 66, 68, 82, 105, 107, 125, 147, 176, 214, 225, 248, 255, 273, 274, 275, 277; and natural selection, 45, 190, 216, 218, 219–222, 224; and Principle of Divergence, 54, 250–252; Darwin's view of, 135, 186, 189, 190, 193, 194, 214–215, 219; in habit (vs. structure), 36–37, 68, 72, 98, 105, 115, 121, 125, 142, 202, 277, 291, 292; in nest architecture, 66, 82, 83, 85, 86, 87, 122, 123, 124, 125, 293; in instinct, 82, 84, 105, 123, 124, 186, 215, 292; Lyell and limits of, 9, 10, 67, 95–96, 135–139, 177, 194, 210, 294; Wallace's view of, 135, 194, 198, 209, 214–215, 220. *See also* Natural selection; Races, human

Variation of Plants and Animals Under Domestication, 268

Varieties: as incipient species, 141, 274, 294; classification of, by Blyth, 100, 140, 194; form via isolation, 74–75, 118; origin of new, 74–75, 95–96, 100, 108, 137–140, 277; reversion to parental form, 52, 95, 96, 135, 136, 205, 209, 213, 221. *See also* Domestication/ domestic varieties; Islands

Vestiges of the Natural History of Creation, x, 7, 15, 16, 17, 18, 19, 29, 31, 52, 54, 139, 150, 175, 237

Victoria, Queen, 91, 127

Victorian period, ix, xiii, xvi, 17, 256

Volcanic islands, 97, 117

Waigiou, New Guinea, 60, 89

Wallace, Alfred Russel: and social issues, xvi-xvii, 269; and spiritualism, xv, xvii, 1, 33, 61, 62, 71, 269; books by, 61, 269, 272; centrifugal governor analogy, 52–53, 213; citations in *On the Origin of Species,* 267, 297–299; conversion to transmutation, 7, 15–18, 26, 150, 237;

criticism of natural selection term, 50, 190, 224; critique of Lyell in Species Notebook, 8–12, 30, 66–67, 91–100; Darwin-Wallace medal, award of, 261; discovery of natural selection, ix, xii-xiii, 2, 11, 46, 49–50, 87–88, 125, 133, 147, 197; evolutionary themes of Species Notebook, 8–11, 65–103, 259; honors, 14, 261; letters with Darwin, 239–241; magnanimity, 257; path to evolution compared with Darwin's, 2–3, 5, 104–142, 229–231; planned evolution book, 3, 5, 8, 93, 259, 265, 277; posting of Ternate manuscript, 235, 242–245; praise for Darwin's *Origin,* 12, 60, 256–257, 265; pursuit of species origins, 3, 15–64; religious views, 62, 70–71, 73; reservations on natural selection and the human mind, xiv, xv-xvi, xvii, 62, 71, 86, 267; travels in South America, x-xi, 20–26; travels in Malay Archipelago, xi, 4, 26–60; view of natural selection compared with Darwin's, 144, 214–218, 219–227, 230. *See also* Consilience (of Inductions); Delicate arrangement; Natural selection; Sarawak Law; Spiritualism; Ternate essay; Transmutation; Wallace's Line

Wallaceism, 259, 273–274

Wallace's Line, ix, xii, 39, 58, 267

Wanumbai, Aru Islands, 40

Water ouzel, 121

Wedges metaphor, 185

Weismann, August, 278

Weissenborn, W., 139

Whewell, William, 12, 33, 65, 66, 103, 112, 142

White, Adam, 285

White, Gilbert, 285

Woburn Abbey, grass experiment at, 193

Wokan, Aru Islands, 40

Woodpeckers, 35, 68, 121, 193

Zizyphora: Z. dasyantha, 100; *Z. intermedia,* 100, 108

Zollinger, Heinrich, 113, 285